CONTROL SYSTEMS DESIGN OF BIO-ROBOTICS AND BIO-MECHATRONICS WITH ADVANCED APPLICATIONS

CONTROL SYSTEMS DESIGN OF BIO-ROBOTICS AND BIO-MECHATRONICS WITH ADVANCED APPLICATIONS

Edited by

AHMAD TAHER AZAR

Robotics and Internet-of-Things Lab (RIOTU), Prince Sultan University, Riyadh, Saudi Arabia; Faculty of Computers and Artificial Intelligence, Benha University, Benha, Egypt

ELSEVIER

ACADEMIC PRESS

An imprint of Elsevier

Academic Press is an imprint of Elsevier
125 London Wall, London EC2Y 5AS, United Kingdom
525 B Street, Suite 1650, San Diego, CA 92101, United States
50 Hampshire Street, 5th Floor, Cambridge, MA 02139, United States
The Boulevard, Langford Lane, Kidlington, Oxford OX5 1GB, United Kingdom

Notices
Knowledge and best practice in this field are constantly changing. As new research and experience
broaden our understanding, changes in research methods, professional practices, or medical
treatment may become necessary.

Practitioners and researchers must always rely on their own experience and knowledge in evaluating
and using any information, methods, compounds, or experiments described herein. In using such
information or methods they should be mindful of their own safety and the safety of others, including
parties for whom they have a professional responsibility.

To the fullest extent of the law, neither the Publisher nor the authors, contributors, or editors, assume
any liability for any injury and/or damage to persons or property as a matter of products liability,
negligence or otherwise, or from any use or operation of any methods, products, instructions, or ideas
contained in the material herein.

Library of Congress Cataloging-in-Publication Data
A catalog record for this book is available from the Library of Congress

British Library Cataloguing-in-Publication Data
A catalogue record for this book is available from the British Library

ISBN 978-0-12-817463-0

For information on all Academic Press publications
visit our website at https://www.elsevier.com/books-and-journals

Publisher: Mara Conner
Acquisition Editor: Sonnini R. Yura
Editorial Project Manager: Emma Hayes
Production Project Manager: Nirmala Arumugam
Cover Designer: Greg Harris

Typeset by SPi Global, India

Working together
to grow libraries in
developing countries

www.elsevier.com • www.bookaid.org

Contents

Contributors

BahaaAlDeen M. AboAlNaga
School of Engineering and Applied Sciences, Nile University, Giza, Egypt

Erhan Akdogan
Department of Mechatronics Engineering, Mechanical Engineering Faculty, Yıldız Technical University, Istanbul, Turkey

Mehmet Emin Aktan
Department of Mechatronics Engineering, Engineering Faculty, Bartın University, Bartın, Turkey

A. Altamirano
Departement de Recherche, Institution Nationale des Invalides CERAH—Ministere des Armees, Woippy, France

Ahmad Taher Azar
Robotics and Internet-of-Things Lab (RIOTU), Prince Sultan University, Riyadh, Saudi Arabia; Faculty of Computers and Artificial Intelligence, Benha University, Benha, Egypt

Francesco Belotti
Unit of Neurosurgery, Department of Surgical Specialties, Radiological Sciences, and Public Health, University of Brescia, Brescia, Italy

Iram Tariq Bhatti
School of Electrical Engineering and Computer Science, National University of Sciences and Technology, Islamabad, Pakistan

Andrea Bolzoni Villaret
Unit of Othorinolaryngology, Department of Surgical Specialties, Radiological Sciences, and Public Health, University of Brescia, Brescia, Italy

Barbara Buffoli
Section of Anatomy and Physiopathology, Department of Clinical and Experimental Sciences, University of Brescia, Brescia, Italy

Jonathan Casas
Department of Biomedical Engineering, Colombian School of Engineering Julio Garavito, Bogotá, Colombia

Riccardo Cassinis
Department of Information Engineering, University of Brescia, Brescia, Italy

Nathalia Cespedes
Department of Biomedical Engineering, Colombian School of Engineering Julio Garavito, Bogotá, Colombia

Lingbo Cheng
Department of Electrical and Computer Engineering, University of Alberta, Edmonton, AB, Canada

Carlos A. Cifuentes
Department of Biomedical Engineering, Colombian School of Engineering Julio Garavito, Bogotá, Colombia

Francesco Doglietto
Unit of Neurosurgery, Department of Surgical Specialties, Radiological Sciences, and Public Health, University of Brescia, Brescia, Italy

Bita Fallahi
Department of Electrical and Computer Engineering, University of Alberta, Edmonton, AB, Canada

Zahra Fathy
School of Engineering and Applied Sciences, Nile University, Giza, Egypt

Marco Ferrari
Unit of Othorinolaryngology, Department of Surgical Specialties, Radiological Sciences, and Public Health, University of Brescia, Brescia, Italy

Marco Maria Fontanella
Unit of Neurosurgery, Department of Surgical Specialties, Radiological Sciences, and Public Health, University of Brescia, Brescia, Italy

Osman Hasan
School of Electrical Engineering and Computer Science, National University of Sciences and Technology, Islamabad, Pakistan

Lena Hirtler
Center for Anatomy and Cell Biology, Medical University of Vienna, Vienna, Austria

Tad Hogg
Institute for Molecular Manufacturing, Palo Alto, CA, United States

Habiba Ibrahim
School of Engineering and Applied Sciences, Nile University, Giza, Egypt

Iyani N. Kalupahana
Department of Biomedical Engineering, National University of Singapore, Singapore; Department of Electronic and Telecommunication Engineering, University of Moratuwa, Moratuwa, Sri Lanka; NUS (Suzhou) Research Institute (NUSRI), Suzhou, China

Akhilesh Kumar
Department of Physics, Govt. Girls P. G. College, Lucknow, India

Amit Kumar
Department of ECE, THDC-Institute of Hydropower Engineering & Technology, Tehri, India

L. Leija
Centro de Investigación y de Estudios Avanzados del Instituto Politécnico Nacional, Mexico, Mexico

Changsheng Li
Department of Biomedical Engineering, National University of Singapore, Singapore

Xiaojian Li
School of Management, Hefei University of Technology, Hefei; The Department of Biomedical Engineering, City University of Hong Kong, Hong Kong, China

Ahmed Madian
School of Engineering and Applied Sciences, Nile University, Giza, Egypt

Alba Madoglio
Unit of Neurosurgery, Department of Surgical Specialties, Radiological Sciences, and Public Health, University of Brescia, Brescia, Italy

Davide Mattavelli
Unit of Othorinolaryngology, Department of Surgical Specialties, Radiological Sciences, and Public Health, University of Brescia, Brescia, Italy

Nada Ali Mohamed
School of Engineering and Applied Sciences, Nile University, Giza, Egypt

Marcela Múnera
Department of Biomedical Engineering, Colombian School of Engineering Julio Garavito, Bogotá, Colombia

R. Muñoz
Centro de Investigación y de Estudios Avanzados del Instituto Politécnico Nacional, Mexico, Mexico; Departement de Recherche, Institution Nationale des Invalides CERAH—Ministere des Armees, Woippy, France

Zbigniew Nawrat
Department of Biophysics, School of Medicine with the Division of Dentistry, Zabrze, Medical University of Silesia, Katowice; Zbigniew Religa Foundation of Cardiac Surgery Development & Director of Heart Prostheses Institute; International Society for Medical Robotics, Zabrze, Poland

Piero Nicolai
Unit of Othorinolaryngology, Department of Surgical Specialties, Radiological Sciences, and Public Health, University of Brescia, Brescia, Italy

Godwin Ponraj
Department of Biomedical Engineering, National University of Singapore, Singapore

Mohammad Javad Pourmand
School of Mechanical Engineering, Shiraz University, Shiraz, Iran

Vittorio Rampinelli
Unit of Othorinolaryngology, Department of Surgical Specialties, Radiological Sciences, and Public Health, University of Brescia, Brescia, Italy

Adnan Rashid
School of Electrical Engineering and Computer Science, National University of Sciences and Technology, Islamabad, Pakistan

Hongliang Ren
Department of Biomedical Engineering, National University of Singapore, Singapore; NUS (Suzhou) Research Institute (NUSRI), Suzhou, China

Luigi Fabrizio Rodella
Section of Anatomy and Physiopathology, Department of Clinical and Experimental Sciences, University of Brescia, Brescia, Italy

J. Antonio Ruvalcaba
Centro de Investigación y de Estudios Avanzados del Instituto Politécnico Nacional, Mexico, Mexico

Alberto Schreiber
Unit of Othorinolaryngology, Department of Surgical Specialties, Radiological Sciences, and
Public Health, University of Brescia, Brescia, Italy

Mojtaba Sharifi
Department of Mechanical Engineering, Sharif University of Technology, Tehran, Iran

Dong Sun
The Department of Biomedical Engineering, City University of Hong Kong, Hong Kong,
China

Mazen Ahmed Taha
School of Engineering and Applied Sciences, Nile University, Giza, Egypt

Fabio Tampalini
Department of Information Engineering, University of Brescia, Brescia, Italy

Mahdi Tavakoli
Department of Electrical and Computer Engineering, University of Alberta, Edmonton, AB,
Canada

Ramna Tripathi
Department of Physics, THDC-Institute of Hydropower Engineering & Technology, Tehri,
India

A. Vera
Centro de Investigación y de Estudios Avanzados del Instituto Politécnico Nacional, Mexico,
Mexico

Francesca Zappa
Unit of Neurosurgery, Department of Surgical Specialties, Radiological Sciences, and Public
Health, University of Brescia, Brescia, Italy

Guoniu Zhu
Department of Biomedical Engineering, National University of Singapore, Singapore

Foreword

The robot word was defined in the beginning of 1920 by the Czech writer Karel Čapek as an artificial man. In this sense, the robot imitates a human, with feeling, reacting to stimuli of the surroundings, moving and performing mechanical work, and pretending thinking software. Robot is an artificial man or part of it, a humanoid robot, artificial organs, or artificial intelligence. This means that the use of the name biorobots is only a reminder of the biosource of inspiration and the biopurpose option of its application.

A good example for bioinspired or bioapplied technology is using the neural network in computing methods (e.g., in decision-making systems) and—on other hand—using the wheel (not existing in natural organism motion systems) for robotics application. We can create robot drone inspired by the physiology of insects or birds or create medical transport devices based on wheels. We can use material technologies based on natural or artificial (e.g., shape memory alloys) components.

By expanding the scope of inspiration (from human center), the robotics also gain knowledge about anatomical (mechanics), physiological, and brain (control) solutions from animal observation, or from the analysis of plant structure and understanding natural phenomena. For Leonardo da Vinci (a genius who wrote a lot, but did not publish any books, he painted a lot, but only completed a few paintings, designed military machines, but used them only in the theatre), creating robot solutions was a consequence of the cognitive passion, it was an attempt to answer the following questions: how a person moves, how a bird flies (studying and analyzing anatomy, physiology, behavior).

At present, as a human being, we have moved our activity from the ground to a computer. From the natural world to the virtual one. Our activity is digital, so imitating human objects—**bots**—is another edition of robots. Bot specialist uses knowledge about the behavior of people to model and simulate digital characters. These characters are not only the heroes of the new field of art (computer games, which are an interactive version of the story) but, soon, above all, intelligent, pretending people, service provider interfaces. For example (chatbots, kowbots, avatars, agents, etc.), it is informant, hotel reception staff, our bank assistants, or a general practitioner.

After the great progress in medicine of telecommunications (distance transmission of information), it is now time for the development of teleaction (distance transmission of action, work). That means you need to use robots. A medical robot can be exactly there and when it is necessary, helpful, saving human lives or health.

Medical robots work as a diagnostic robots (digital image diagnostics), surgical robots (reducing the invasiveness of surgery), welfare and social robots (increasing the quality of life of elderly), rehabilitation robots (controlled movement), rescue robots (in various environments), artificial organs, biorobots (used for cognitive purposes—neurophysiology, brain study, or social self-organization), educational medical robots (interactive patient simulators).

The **art of robotics** is based on an intelligent combination of mechanical work and information management obtained by sensors. But among the solutions that can be used in medical applications, we have also fully mechanical devices—such as the designed by J.E.N Jaspers (Delft University) surgical telemanipulator (Jaspers et al., 2007), digital devices (dr Bot), as well as attempt to create a biological-technical **hybrid**.

Kevin Warwick, known as first human **cyborg** (after implanting an electronic interface), claims (Warwick, 2013) that "where a brain is involved it must … be seen … as part of an overall system—adapting to the system's needs." Warwick is a pioneer in studying the connection between the biological body and the robot. At Reading University, prof. Warwick's team has done a research with the use of naturally cultivated neurons as brains controlling small mechanical systems. "In terms of … learning and memory investigations are at an early stage. However, the robot can be seen to improve its performance over time in terms of its wall avoidance ability…" Warwick is considering robots with biological brains. This "approach allows for 'complete body engineering' in which brain size, body size, power, communications and other abilities are optimized for the requirements in hand." He concludes, "Maybe this technique will ultimately open up a future route for human development whereby humans can cast off the shackles and limitations imposed by the restrictions of having to live in a biological body."

The integration of engineering and biology is a fact. This raises the question as to the borderline between a biological organism and a technical device. Where is the beginning of consciousness and intelligence? In my opinion, my answer is: beginning of consciousness is when the question "why" appears between the analysis of information from the environment and start the actions (Nawrat, 2011). Of course, for standard-based medical activities, the "why" question (between fact-based assessment, e.g., diagnosis and action treatment) posed by a robot can be considered a system weakness. But that's what people are.

I believe ethics to be the art of making the right choices. For ages, philosophers have been analyzing issues connected with the man–versus–man and man–versus–the world relations in order to help us comprehend the reality and find the correct conduct. Ethics and morality mutely assume a human-to-human contact. Now we have to solve a number of ethical problems related to the fact that robots are among us, robots are moving, robots cowork with us (from robocars to robodoctors). The three laws of robotics proposed by I. Asimov are not enough. Perhaps a modification of the Declaration of Human Rights following the dissemination of robots will be necessary.

With the harmonious development of technical and biological sciences and their reasonable implementation, we can influence the evolution of species and the quality of life on Earth. I'm not sure that women are from Venus, men are from Mars, but I'm sure that robots are from Earth.

Let's do it! Robots have to be created to multiply our freedom.

Zbigniew Nawrat

Assistant Professor, Department of Biophysics, School of Medicine With the Division of Dentistry, Zabrze, Poland
Professor, Religa Foundation of Cardiac Surgery Development, Zabrze, Poland
Director of Heart Prostheses Institute, Zabrze, Poland
President, International Society for Medical Robotics, Zabrze, Poland

References

Jaspers, J., Diks, J., Wisselink, W., de Mol, B.A., Grimbergen, C.A., 2007. The Minimally Invasive Manipulator (MIM); an instrument improving the performance in standardized tasks for endoscopic surgery. In: Nawrat, Z. (Ed.), Adv. Biol. Technol. M-Studio, Zabrze, pp. 151–152.
Nawrat, Z., 2011. The ethics of artificial organs. In: Nawrat, Z. (Ed.), Implant Expert. M-Studio, Zabrze, pp. 127–135.
Warwick, K., 2013. The diminishing human–machine interface. Med. Robot. Rep. 2, pp. 4–11.

Preface

Biomedical robotics and biomechatronics research operations can be traced back to the 1970s and 1980s. Biomedical robotics and biomechatronics research cover a variety of fast-growing interdisciplinary areas, including bioinspired robots for industrial, military, medical, and rehabilitation applications. Biomedical robotics and biomechatronic devices are becoming a hot spot for research into technology and science. They have a methodological background in the fields of biomedical engineering and robotics, and are now applied to different engineers, basic and applied science, such as biology, neuroscience, medicine and humanities, as well as sociology, ethics, and philosophy. Knowledge acquisition of the biological system operating mechanism is a major objective of the study of biomedical robot systems and biomechatronic systems. As a consequence, the biological systems are frequently analyzed from a "biomechatronic" perspective. Knowledge is used to create technological and technological advancements that can lead to the development and construction, through the imitation of insects, pets, humans, and multiple lifetimes, of bioinspired devices and systems.

In future generations of biomedical systems and apps, the combination of robotic technology and in-depth biomedical sciences is also promising. A strategy based on biomedical robotics and biomechatronics is of excellent concern, with three key objectives: (1) enhance knowledge of underpinnings of sensing and acting in diverse animals, including our human beings; (2) build valid and helpful high-efficiency mechatronic and robotic systems and (3) develop efficient biological interactive systems, for example. Research into this direction clearly shows growing interest in humanoid technology; bioinspired and biomimetic robotics; human-robot cooperation and interaction; and biomechatronic endoscopy, intervention, aid, and recovery instruments. In addition to the growing importance of biological inspirational design in the advances in artificial structures, many applications in mechatronics and robotics in different areas present fresh difficulties, both in theory and in technology. Therefore, the technologies and models used in design and manufacture of biomechatronic equipment and biologically inspired robots are very important for further development.

We strive in this book (i) to highlight biomedical robotics and biomechatronical theoretical and practical problems; (ii) to bring together alternatives under distinct circumstances with particular attention to validation of these instruments in biorobotic environments using practical tests; and (iii) to launch important case studies.

About the book

The new Elsevier book, *Control Systems Design of Biorobotics and Biomechatronic with advanced applications*, consists of 13 contributed chapters by subject experts who are specialized in the various topics addressed in this book. The special chapters have been brought out in this book after a rigorous review process in the broad areas of biorobotics and biomechatronics. Special importance was given to chapters offering practical solutions and novel methods for the recent research problems in the mathematical modeling and control applications of robotic systems. This book aims at showcasing the most exciting and recent advances in the application of biorobotics and biomechatronics in various fields and brings together a broad spectrum of topics covering various definitions, developments, control, and deployment of biomechatronics/robot systems.

Objectives of the book

Through this book, we wish to deliver essential and advanced bioengineering information in applications of control and robotics technologies in life science. In the next few years, there will surely be much more exciting developments in this area. The first objective of this book is to focus on the engineering and scientific principles underlying the extraordinary performance of biomedical robotics and biomechatronics. The second objective is the application of the principles to design the corresponding algorithms that purposively operate in dynamic scenarios.

Organization of the book

This well-structured book consists of 13 full chapters.

Book features

- The book chapters deal with the recent research problems in the areas of biomedical robotics and biomechatronics.
- The book chapters present various applications for biomedical robotic systems.

- The book chapters contain a good literature survey with a long list of references.
- The book chapters are well written with a good exposition of the research problem, methodology, block diagrams, and mathematical techniques.
- The book chapters are lucidly illustrated with simulations.
- The book chapters discuss details of engineering applications and future research areas.

Audience

The book is primarily meant for researchers from academia and industry, who are working in the research areas—control engineering, biomedical engineering, electrical engineering, and computer Engineering. The book can also be used at the graduate or advanced undergraduate level as a textbook or major reference for courses such as Biorobotics, Biomechatronics, Selected topics in biomedical engineering, and many others.

Acknowledgments

As editor, I hope that the chapters in this well-structured book will stimulate further research in Biorobotics and Biomechatronic applications.

I hope that this book, covering so many different topics, will be very useful for all readers.

I would like to thank all the reviewers for their diligence in reviewing the chapters.

Special thanks go to Elsevier, especially the book Editorial Project Manager Emma Hayes and Production Project Manager Nirmala Arumugam.

No words can express my gratitude to the Acquisitions Editor, Sonnini Ruiz Yura, for her great effort and support during the publication process.

Special acknowledgment to Prince Sultan University and Robotics and Internet-of-Things Lab (RIOTU), Riyadh, Saudi Arabia for giving me the opportunity to finalize this book.

Ahmad Taher Azar
Robotics and Internet-of-Things Lab (RIOTU),
Prince Sultan University, Riyadh, Saudi Arabia
Faculty of Computers and Artificial Intelligence, Benha University, Benha, Egypt
http://www.bu.edu.eg/staff/ahmadazar14
https://sites.google.com/site/drahmadtaherazar/

CHAPTER ONE

Human-robot interaction for rehabilitation scenarios

Jonathan Casas, Nathalia Cespedes, Marcela Múnera, Carlos A. Cifuentes
Department of Biomedical Engineering, Colombian School of Engineering Julio Garavito, Bogotá, Colombia

1 Introduction

Robots are being introduced in an increasing variety of domains. In such areas they are used as a tool for social assistance to help people in their homes, to be a guide in public spaces, as a teacher in classrooms, or as a coach in rehabilitative settings. In this direction, researchers worldwide are studying the social factors related to the human-robot interaction (HRI) in human environments and great attention is being focused on the cognitive human-robot interaction (cHRI) (Santis, 2007). In this chapter, we focus on robots as platforms, companions, and coaches for helping people to exercise and increase their physical abilities after suffering diseases regarding the cardiovascular and/or the neurological systems.

As a brief introduction, the cardiovascular system allows the transport and delivery of nutrients, oxygen, hormones, and blood cells. In addition, it is a self-sealing circuit, that brings tools for repair and healing in case of damage. On the other hand, the nervous system coordinates actions by transmitting and receiving signals from several parts of the body. It consists of two main parts: the central nervous system (CNS) and the peripheral nervous system (PNS). The brain and spinal cord are part of the CNS and the PNS consists mainly of nerves, which connect the CNS to every part of the body. Those systems are susceptible to be affected by several diseases and disorders (see Fig. 1). Some of them will be introduced as follows.

Cardiovascular diseases (CVDs) are known as disorders of the heart and blood vessels that include coronary heart disease, cerebrovascular disease, rheumatic heart disease, and other conditions (World Health Organization, 2018a). These diseases are referred to conditions that involve narrowed or blocked vessels and can lead to heart attack, stroke, and heart failure (American Heart Association, 2017). Two groups of CVDs

Control Systems Design of Bio-Robotics and Bio-mechatronics with advanced applications
https://doi.org/10.1016/B978-0-12-817463-0.00001-0

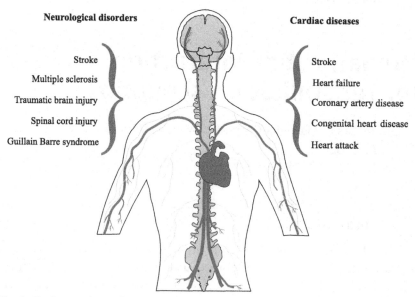

Fig. 1 Cardiovascular and neurological systems of the human body. Common cardiovascular and neurological disorders are illustrated.

are considered: (1) CVDs caused due to atherosclerosis, such as ischemic heart disease or coronary heart disease (heart attack), cerebrovascular disease (stroke), and diseases of the aorta and arteries that include hypertension and peripheral vascular disease; and (2) CVDs caused due to a different condition, including congenital or rheumatic heart disease, as well as cardiomyopathies and cardiac arrhythmias (World Health Organization, 2015). Among these groups, 70% of CVDs are caused by atherosclerosis (World Health Organization, 2015, 2018a). CVDs take the lives of 17.9 million people every year, an estimated 31% of all deaths worldwide (World Health Organization, 2018b). This situation affects quality of life, as demonstrated by 20% of patients who suffered a CVD event presenting prevalence of depression (Taylor et al., 2006).

Moreover, neurological disorders are diseases of the central and peripheral nervous systems, including the brain, spinal cord, nerves, neurological joints, and muscles, among others (WHO et al., 2006). Neurological disorders comprise an extensive group of heterogeneous (more than 600) pathologies, for example, spinal cord injuries, dementia, stroke, brain tumors, and multiple sclerosis (World Health Organization, 2016). As consequences of neurological disorders, upper and lower limbs can be affected, causing limitations within gait patterns and self-performance of the patients in social, economical, and physiological contexts (WHO et al., 2006). According to the World Health Organization, neurological disorders contributed to 92 million disabilities in 2005, and it is projected that this will increase to 103 million in 2030 (approximately 12%) (WHO et al., 2006). Within this group of diseases, stroke causes more than 6 million deaths each year (World Health Organization, 2016).

Stroke survivors typically show significantly reduced gait speed, shortened step length, and loss of balance in their gait patterns, and often experience falls (Potter et al., 1995). With the proven fact that repetitive and persistent stimulation could restore and reorganize defective motor functions caused by neurological disorders, there is a strong need for new therapeutic interventions (Johansson, 2011).

This chapter aims to present the latest advancements in the area of social robots for rehabilitation scenarios that involve pathologies with high prevalence worldwide, such as CVDs and neurological disorders. The main outlines showed that the use of a social robot in rehabilitation scenarios has a positive impact reflected in the improvement of physiological parameters (e.g., heart rate [HR] and spinal posture patterns) and patients' motivation to follow the treatment procedures. On the other hand, due to the monitoring provided by the cHRI and the robot interaction, therapists were more focused on other rehabilitation tasks.

This chapter is organized in five thematic sections, addressing relevant aspects regarding social robots for rehabilitation scenarios and the important interaction aspects involved in this process.

Section 2 addresses the literature review concerning social robotic agents (SARs), paying special attention to the interfaces that have been implemented or can be useful for applications in rehabilitation and health care.

Section 3 begins with the definition of human–robot interfaces and the proposed robot-based therapy model. Afterward, the current state of rehabilitation is described. Finally, cHRIs are presented in the context of cardiac and neurological rehabilitation (NR).

Section 4 addresses two experimental studies presented in this chapter: a cardiac rehabilitation (CR) longitudinal study and a study for NR based on repeated measurements.

Section 5 discusses the model developed for social HRI in the context of rehabilitation scenarios and analyzes outcomes obtained during the robot-therapy interventions.

Finally, Section 6 presents the conclusions and some recommendations for future works in this challenging field of rehabilitation robotics.

2 Related work

Socially assistive robotics (SAR) is the field dedicated to the development of robots that provide feedback, instructions, encouragement, and emotional support through social interaction to increase patients' motivation and performance within the therapies. SAR has been initially defined as the combination of assistive robotics (AR) and social interactive robotics (SIR). In the first place, SAR and AR are meant to provide assistance to human users. However, in SAR, this assistance is specifically achieved by means of social interaction with the user. From this perspective, SAR and SIR have the same goal, as they are focused on developing social interaction strategies that enable them to exhibit a closer and more effective interaction with the human user. Unlike SIR, the scope of SAR is limited to achieve major progress in the areas such as rehabilitation, health care, etc. (Feil–Seifer and Mataric, 2005).

2.1 Social robotic agents

The main role of SARs, or social robots, is to act as companions or assistants in a specific task. In rehabilitation and healthcare environments, social robots are regarded as training assistants, coaches, or motivator agents that help improve patients' performance or increase engagement during the therapy. With this in mind, social robots are required to contain a series of features that allow them to interact in an effective way, providing adaptability and flexibility to human environments. As these agents are designed to interact socially with humans, they must exhibit human–like behaviors and their appearance and functionality must be structured in a way that humans can interpret and be familiarized with Fong et al. (2003).

In this context, a considerable aspect that enables an effective social interaction is the *physical embodiment*, which allows the robot to perceive and experience the physical world. Hence, it will be able to interact with humans and engage with their activities in a more natural and intuitive way (Wainer et al., 2006b). The *embodiment* is a term considered to refer to the fact that intelligence cannot be limited to exist in the form of an abstract algorithm, but requires a physical instantiation or body (Pfeifer and Scheier, 1999; Dautenhahn et al., 2002). Different studies have demonstrated the effect and benefits that embodiment attributes to the robotic platforms over other types of social agents, such as virtual agents and screen-based avatars. It has been demonstrated that social robots receive more attention and the interaction with the users can be more engaging (Belpaeme et al., 2013; Powers et al., 2007; Wainer et al., 2006a).

Although all social robots are embodied (have a physical body that allow them to interact with the world), the degree of interaction may vary depending on the capabilities of the robot. Hence, a robot with more motor and sensor skills will present more capabilities to interact with the environment as it can establish more relationships with the world. Currently, there is a wide spectrum in the design features that social robots have. In this chapter, we consider the classification of social robots in two main categories: (1) *Real/Abstract*, which indicates the degree of similarity that the platform has with the nature (i.e., how similar the robot is to a living being), unlike the abstract design; and (2) *Animal/Human*, which describes their similarity to a human being or an animal creature. Fig. 2 illustrates some robots that are conventionally used. As can be observed, these platforms vary in their shape and appearance.

Although all these platforms can be regarded as SARs, their functionalities and field of applications can diverge, as each robot can be suitable for a specific task and a specific degree of interaction.

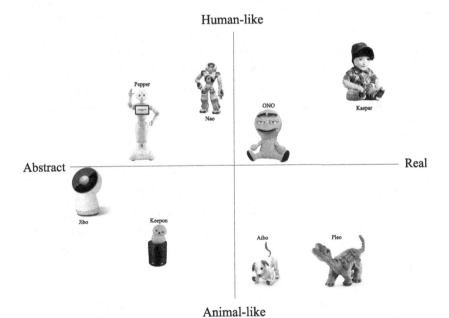

Fig. 2 Socially assistive robots classification. In this chapter we consider two main categories: *Real/Abstract*, referring to their similarity to living beings, and *Human/Animal*, referring to their similarity to humans or animals.

As every day more robotic platforms are designed, the application spectrum of SAR is expanding in a similar way, covering multiple areas in healthcare and rehabilitation scenarios. The next section presents a detailed overview of the applications and the relevant findings associated with this research.

2.2 Applications in rehabilitation and healthcare

SAR was initially explored in cardiovascular therapies with the development of CLARA, a hands-off physical therapy assistant whose aim was to reduce the effects of nursing shortages, provide motivation, and aid patients through the rehabilitation exercises as spirometry therapies. In this study, the researchers found high expectations over the robot's usefulness and an average overall satisfaction of the population of about 80% (Kang et al., 2005). Furthermore, SAR has been used in several applications focusing on elderly care (Bemelmans et al., 2012), dementia, mental health treatments (Rabbitt et al., 2015; Martín et al., 2013), and physical and poststroke rehabilitation (Matarić et al., 2007). Within elderly care areas, robots such as PARO were used in therapeutic scenarios, in order to achieve social exchanges and encourage patients during exercises (Marti et al., 2006). The study opens interesting perspectives about the use of a robot as a nonpharmacological therapeutic aid. It has been found that PARO was able to support the complexity of a clinical scenario in a flexible way allowing patients' engagement and sociorelational exchanges. In addition, effects such as the improvement of communication, cognitive skills (Tsardoulias et al., 2017), and reduction of anxiety (Louie et al., 2014) in elderly populations have been observed, showing in general positive attitudes toward social robots.

Another domain that has been broadly approached by SAR technology is poststroke rehabilitation. Autonomous robots (Matarić et al., 2007; Mead et al., 2010) and embodied agents (Jung et al., 2011) have been explored to monitor and supervise poststroke survivors during gait training and upper-limb exercises. The studies showed a positive impact within the users on their willingness to perform prescribed rehabilitation, changes in the motor functioning, and improvements in the average number of trials accomplished per minute.

Finally, physical training and coaching are areas of interest in social robotics as robots can be used as companions to guide different kind of exercises and improve the adherence to this programs using cognitive approaches. As an example, NAO robots were implemented into conventional physiotherapy practices in order to guide several body movements

(López Recio et al., 2013) and in upper-limb exercises for patients with physical impairments, such as cerebral palsy and obstetric brachial plexus palsy (Pulido et al., 2016). The results have demonstrated an accurate monitoring of the therapies, and fluent interaction with the robot. In addition, patients like to follow the exercises provided by the NAO and engage with the rehabilitation trying to perform the tasks (Pulido et al., 2016). In 2008, a long-term study showed the effects of HRI in coaching with the aim of reducing the rates of overweight and obesity. In this case, the robot asked patients their diet goals in terms of burning calories during exercise and data related to the food consumed during the day. The results showed that the participants assisted by the social robot were more interested in knowing their calorie consumption and exercise performed than those who used other methods (Kidd and Breazeal, 2008).

Adherence is an important factor to achieve exercise adoption, and different studies have shown positive results regarding this factor. Gadde et al. (2011) evaluated in the early stages an interactive personal robot trainer (RoboPhilo) to monitor and increase exercise adherence in older adults. The system was proved with 10 participants, showing initially a positive response and a favorable interaction. A complementary application where robots are being used to motivate and increase adherence in long-term therapies and medical self-care is diabetes mellitus treatments, where robots play the role of personal assistants to adults (Looije et al., 2006) and children (Baroni et al., 2014). This has shown potential results within motivational aspects and treatment engagement.

Summarizing, several authors have described SAR systems in terms of aiding patients in different areas showing great potential and results. This research focuses on deploying a social robot into cardiovascular and NR scenarios to provide monitoring and motivation during therapeutic treatment.

3 Human-robot interfaces for rehabilitation scenarios

Human beings interact with the environment through cognitive processes, sequences of tasks that include reasoning, planning, and finally the execution of a previously identified problem or goal. From this process, the robots may use information regarding human expressions and/or physiological phenomena to adapt, learn, and optimize their functions, or even to transmit back a response resulting from a cognitive process performed within the robot. This concept is named the cHRI (Pons et al., 2008).

cHRI systems often present bidirectional communication channels. On the one hand, robot's sensors measure the physiological parameters, human actions, and expressions. On the other hand, the actuators transmit the robot's cognitive information (social interaction) to the user. In other words, the user observes the state of the system through feedback sent immediately after the user command is executed. This configuration performs a closed-loop HRI in order to develop a natural cooperation during the rehabilitation task.

Humans perceive the environment in which they live through their senses: vision, hearing, touch, smell, and taste. They act on the environment using their actuators, for example, muscles, to control body segments, hands, face, and voice. Human-to-human interaction is based on sensory perception of actuator actions. A natural communication among humans also involves multiple and concurrent modes of communication (Sharma et al., 1998).

The goal of effective interaction between a user and their robot assistant makes it essential to provide a number of broadly utilizable and potentially redundant communication channels. This way, any HRI system that aspires to have the same naturalness should be multimodal. Different sensors can, in that case, be related to different communication modalities (Sharma et al., 1998). The integration of classic human-computer interfaces (HCi) like graphical input-output devices, with newer types of interfaces, such as speech or visual interfaces, tactile sensors, laser range finder (LRF) sensors, inertial measurement units (IMU), and physiological sensors, facilitates this task.

3.1 Proposed robot-based therapy model

The robot-therapy model proposed in this chapter is illustrated in Fig. 3. This schema considers two main components: (1) *Motivation*, which aims to provide intrinsic and extrinsic motivation, and (2) *Therapy control*, focused on the monitoring of the therapy performance, the management of warning events, and reduction of risk factors that can lead to emergencies. Each therapy has risk factors associated with the tasks that are performed. The model seeks to manage and reduce these risk factors while monitoring the development of physiological and spatiotemporal parameters that are relevant to each scenario. This goal is achieved by means of custom multimodal sensor interfaces that provide all relevant information to be processed by the social robot. Hence, the robot is able to generate feedback to the user appropriate to the context and the therapy conditions.

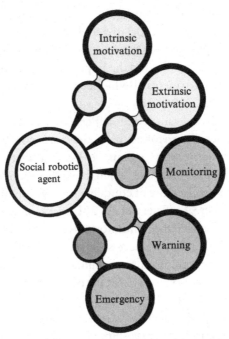

Fig. 3 Proposed robot-based therapy model. This model considers two main components: motivation and therapy control as the main features that the social robotic agent exhibits during the therapy.

The remainder of this section presents in detail both scenarios, cardiac and NR, where a social robotic platform was deployed in the framework of the model previously described.

3.2 Rehabilitation scenarios

Considering the robot-based therapy model described earlier, two interventions were carried out. The first intervention with a social robot was held in the CR service. Subsequently, this work has been extended to the NR scenario. This section describes the corresponding clinical context, in which the system was deployed, followed by the proposed custom cHRI designed for each application.

3.2.1 Current state of cardiac rehabilitation

CR programs are designed to prevent CVDs or to treat a patient after a cardiovascular event. CR covers different areas, such as nutrition, physical exercise, and health education. Even though these programs have proved successful in reducing and preventing the occurrence of a posterior cardio-vascular event, the adherence associated with the program does not reach a desirable level. Multiple studies have demonstrated the low adherence rates, which are not higher than 50%, and the implications that this situation has in the health condition of patients (Bethell et al., 2011; Sarrafzadegan et al., 2007; Worcester et al., 2004).

Different studies have evaluated the adherence in rehabilitation programs. From these studies, it has been found that patients are more likely to attend the programs when physiotherapists encourage them during the sessions and feel satisfied with the therapy. Likewise, when there is a perceived interest by the therapist and patients feel a sense of complete supervision, their performance and results tend to increase (Essery et al., 2017; Jackson et al., 2005). In this context, it is clear what role the continuous monitoring and encouragement of the medical staff plays in the success, in terms of adherence and performance, of patients attending to the rehabilitation therapies. Therefore, the work presented in this section focuses on the development of an assistive tool, based on SAR, that supports the work carried out by clinicians and aims to offer a more personalized service to the patients, through continuous monitoring, motivation, and companionship within the CR therapies.

Before introducing the proposed SAR system, it is worth mentioning the structure that a conventional CR program has. The structure and components differ depending on the country and institution. However, they traditionally consist of three phases: *inpatient (phase I), outpatient (phase II)*, and *community maintenance (phases III and IV)*. The *outpatient* phases, namely phases II and III, take place in a specialized center or institution and are carefully performed under the supervision of healthcare providers with monitoring based on exercise tolerance test results (Kim et al., 2011). Our work has been focused primarily on phase II of the CR programs. The features and structure of this phase are described in following sections.

Phase II

This phase is the first outpatient phase and begins immediately after the patient leaves the hospital. It consists of a combination of physical exercise on a treadmill and an education program oriented to prevention of risk factors, as well as adoption of healthy habits (e.g., controlling blood pressure, cholesterol, weight and stress management). This phase has an average duration of 3 months and is designed to provide a safe monitored environment for exercise. The monitoring consists of measuring the patient's blood pressure, HR, and eventually heart and lungs sounds. Additionally, it is important to monitor the perceived exertion level (i.e., fatigue or effort during the exercise). This measurement is carried out with the Borg Scale (BS), which is a qualitative measurement that estimates the perceived exertion of the patient (6 for low intensity and 20 for very high intensity). As a result from phase II, the patient should be able to self-monitor their physiological parameters and exertion levels. This aspect will return the confidence to the patient to continue a normal life, being aware of their health condition and the healthy lifestyle that is required to prevent a second cardiac event (Scherr et al., 2013).

Taking into account the CR context, as well as the structure of phase II, where the intervention takes place, a cHRI is proposed. This interface has been designed taking the adherence, motivational aspects, and therapy structure into consideration.

3.2.2 Cognitive human-robot interface proposal for CR

The design of the cHRI for CR has been carried out by means of different stages. In the first place, the system was developed and evaluated under laboratory conditions (Lara et al., 2017). As a second stage, the system was deployed at the clinic (Fundación Cardioinfantil-Instituto de Cardiologãa) and tested with a real patient during normal therapy. Results of this intervention are presented in a subsequent work (Casas et al., 2018). Finally, the system was deployed for a larger number of patients during the complete phase II of the CR program. This section describes the structure and components that conform the interface.

System modules

There are three main modules that comprise the system. A *sensor manager* is designed to handle the acquisition and recording of all the sensory data generated by the system. The HCi is incorporated to provide the possibility to interact with the information of the system, and finally the SAR responsible for acquiring the sensory information and providing appropriate feedback to the user during the therapy. The system is illustrated in Fig. 4, where the robot receives as input all the data generated by the sensors and the HCi, and provides feedback through social interaction according to the information processed. These modules are described later.

Sensor manager

The sensor manager is designed to measure two types of variables: physiological and spatiotemporal. Physiological variables are acquired by means of a

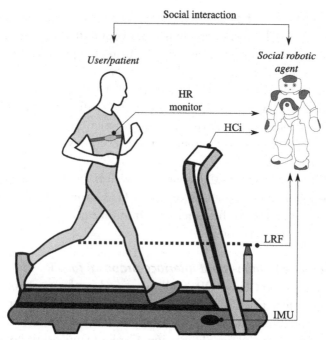

Fig. 4 Cognitive human-robot interface for cardiac rehabilitation. In this scenario, patients perform physical activity on a treadmill, while the sensor interface records their physiological and spatiotemporal parameters to be processed and analyzed by the socially assistive robot. The results of this analysis are provided to patients as feedback by means of social interactions.

HR monitor Zephyr HxM BT (Medtronic, Minneapolis, USA) attached to the chest of the patient (see Fig. 4). This sensor provides information about the HR variability and its evolution. The spatiotemporal variables are acquired with a LRF URG-04LX-UG01 (Hokuyo, Japan) that is placed in front of the patient to measure their legs. From this measurement, it is possible to obtain the speed, cadence (number of steps per second), and step length. Finally, an IMU sensor MPU9150 (Invensense, San Jose, CA, USA) is placed on the treadmill surface to measure the inclination of the band. This variable is associated with the exercise intensity.

Human-computer interface

This interface has been designed with two purposes. In the first instance, the interface allows the patient to observe the acquired sensory data during the therapy. Likewise, this system serves as a means of communication between the robot and the user. Thus, qualitative measurements such as the exertion rate (BS) can be delivered through the interface when the robot has requested these. Similarly, the patient has the capability to report an emergency or any anomaly during the session.

Social robotic agent

As shown in Fig. 4, the robotic platform that was incorporated in the system is the social humanoid robot NAO (SoftBank, Tokyo, Japan). According to the robot-therapy model previously defined in Section 3.1, the robot exhibits three main behaviors (i.e., motivation, warning, and emergency). In order to design behaviors that were familiar to the patients, our study considered an observation phase, where key therapist–patient interactions were identified. During this phase, it was possible to define the social features that the robot should exhibit, to ensure an effective communication. The robot's behaviors have been designed to be autonomous (e.g., no human intervention is required to trigger any of the behaviors).

The *Motivation* behavior is based on verbal expressions that are generated by the robot in different moments during the therapy. With these expressions, it is expected to provide encouragement and engage the patient with the exercise. The *Warning* behavior is designed to handle a potential occurrence of any of the risk factors associated with the therapy. In this case, two risk factors are carefully monitored: sudden increase of HR and risk of falling due to dizziness and bad posture. These risk factors are evaluated by means of the analysis of the HR sensory data and the image processing to identify appropriate postures. Finally, the *Emergency* behavior is adopted when one of the considered risk factors occurs or when the patient reports an unusual condition. In this scenario, the robot requests the immediate intervention of the medical staff to control the situation.

Evidence of a real intervention in the clinic can be observed in Fig. 5, where a real patient performs physical activity on the treadmill with the social robot acting as companion and assistant. As shown in the figure, the patient has access to the HCi that is placed on the front panel of the machine. Thus the patient can interact with the robotic system while performing the rehabilitation session as usual. Similarly to the CR application, the NR context, as well as the proposed cHRI for this scenario, is presented in following sections.

3.3 Current state of neurological rehabilitation

The World Health Organization defines physical rehabilitation (PR) as an active process to achieve a full recovery or, if full recovery is not possible,

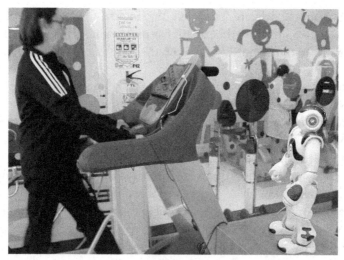

Fig. 5 Robot therapy for cardiac rehabilitation at Fundación Cardioinfantil-Instituto de Cardiología (FCI-IC).

to reach optimal physical, mental, and social potential to integrate people appropriately into society (WHO et al., 2006).

As a comprehensive strategy, PR has two approaches: (1) improve physical aspects of patients as cardiovascular functioning, aerobic capabilities, muscles functioning, and gait patterns during physical activity, and (2) improve cognitive aspects related to the cognition processes that include language, perception, motivation, and attention among others. These approaches are important to assess the patient's long-term performance and achieve a full recovery and a successful rehabilitative process.

Within PR methods, conventional and robot-assisted therapies can be found. The conventional therapeutic scenario is guided by a therapist and repetitive exercises are used to improved patient performance (Fisher et al., 2011): the results of conventional therapy depend on the expertise of the medical group and the intensity of the sessions (Hussain et al., 2011). Body weight support (BWS) treadmill training is one of multiple techniques used in neurorehabilitation. BWS combines recent findings that promote functional locomotor recovery and practices where the patient has to perform complete gait cycles at an early stage of gait rehabilitation (Werner et al., 2002; Winter, 1989). During the training, patients are mechanically supported in a harness to substitute the deficit of equilibrium reflexes and support a proportion of their body weight so that subjects can carry their remaining weight adequately (Winter, 1989). This method has shown improvements in spatiotemporal and kinematic gait parameters (Visintin et al., 1998).

On the other hand, robot-assisted therapy for gait rehabilitation combines BWS systems and a exoskeleton to train and recover gait patterns in patients with neurological limitations. Currently, a gold standard device in this type of therapy is Lokomat. This is a robotic device that adjusts to the lower limbs of the user to retrain the gait by means of neuroplasticity stimulation, intensive and effective exercises which enable optimal recovery (Hocoma, 2019). Different studies have shown the benefits of using Lokomat rather than conventional treatment. Husemann et al. (2007) showed that the Lokomat training increases muscle tissue and reduces fat mass in stroke patients. In addition, parameters such as cardiovascular functioning and oxygen uptake during the exercise (Krewer et al., 2007), muscular tone (Mayr et al., 2007), balance (Bang and Shin, 2016), motor control (Banz et al., 2008), and gait speed (Hwang et al., 2017) improved with the use of Lokomat.

In this context, some studies have presented concerns regarding Lokomat therapy, such as (1) lack of adherence to the programs, as around 42% of the patients deserted PR programs (Jack et al., 2010), and (2) the multitasking processes performed by the therapist during a session (Douglas et al., 2017; Munera et al., 2017). In healthcare areas, Appelbaum et al. (2008) demonstrate that multitasking processes have disadvantages, such as increasing time to task completion, stress, memory lapses, errors (Pashler, 2000), and accidents. We performed a previous study, where we found that Lokomat therapy requires parallel assistance of the medical group all the time. A session includes tasks such as device configuration, patient preparation, giving feedback to patients, assessment of gait performance, teaching, etc.

3.3.1 Cognitive human-robot interface proposal for NR

The design of cHRI for NR follows different stages. In the first place, initial observations at the clinic (Clínica Universidad de La Sabana; Fig. 7) were made in order to establish physiological and cognitive measurements according the needs and the interest of the medical and engineering team. In the second stage, a short-term study tested real patients during Lokomat sessions.

System modules
Similarly to cardiac cHRI, in NR there are three main modules that describe the system. A *sensor manager* module is used to acquire and record all the sensory data generated by the system. The HCi visualizes the data obtained

during the session, and finally the *SAR's* role is to interpret the sensory data and give feedback to the patient to motivate and improve patient performance within the therapy (Fig. 6). These modules are described here.

Sensor manager

This module records two variables: HR, essential to know the level of fatigue and the physical activity response (Achten and Jeukendrup, 2003); and spinal posture in order to monitor proper dynamic posture—maintaining this posture helps to promote back health, allows muscles to work properly, and decreases muscle fatigue (Bradley, 2004; Weaver and Ferg, 2010). For the HR measurement, the same sensor mentioned in Section 3.2.2 was used. Finally, two BNO055 IMUs (Adafruit, New York, USA) were used to measure cervical and thoracic inclination angles at pitch,

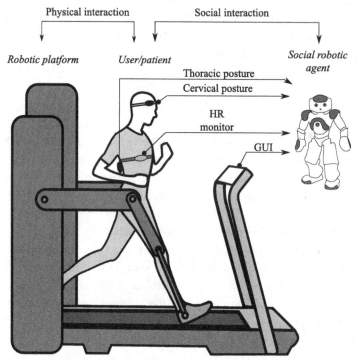

Fig. 6 Cognitive human-robot interface for neurological rehabilitation. The patient performs the assigned task with the support of a robotic platform, while the socially assistive robot receives the information generated by the sensor interface and the human-computer interface to interact subsequently with the patient through social interventions.

roll, and yaw rotations. The configuration of the IMUs' position determines the body planes. In this case, the sagittal plane corresponds to the pitch axis, transversal plane to roll axis, and coronal plane to yaw axis. One sensor was located on the user's forehead and the other one between the T6 and T7 spinal segments adjusted with velcro straps (Kim et al., 2013).

Human-computer interface

The HCi is used to visualize the data of the sensors, control the therapy acquisition times, and show the BS. According to the observations, the BS is represented with a range of 0–10, where 0 means a low level-fatigue perception and 10 a high-level fatigue perception.

Social robotic agent

The SAR approach was designed following the next robot roles: (1) motivate the patient to perform rehabilitation tasks as aerobic and anaerobic exercise, (2) monitor the patient's performance, and (3) provide feedback according to the parameters extracted from the sensory interface. The behaviors of robots are several routines depending on the patient's performance and the data acquired by the sensor manger module. As general actions, the robot has programmed greetings at the start and the end of each session where it presents itself and explains its tasks within the rehabilitation process. On the other hand, the feedback given by NAO is based on five routines. For *cervical posture feedback* and *thoracic posture feedback*, the robot performs verbal gestures and nonverbal actions to correct the patient (e.g., the robot tilts its head down and imitates the patient; then the robot shows the patient how to reach a healthy cervical posture). In the case of the *HR alert*, the robot uses verbal phrases to give advice to the therapist when this parameter exceeds a permissible threshold according to the medical prescription. Finally, the *BS request* and *motivational feedback* are given verbally to the patient by the robot in order to know the fatigue perception and to promote a proper posture. This motivational feedback is given randomly during the session.

Evidence of the intervention can be observed in Fig. 7, where a patient performs Lokomat-based therapy with an NAO robot. As shown in the figure, the patient has a sensory system to measure posture and an HCi that allows the therapist to visualize the data based on the previous cHRI presented for cardiac and NR.

The next section describes the experimental studies that were carried out in real scenarios, aiming to evaluate and validate the effect of SARs in these clinical contexts.

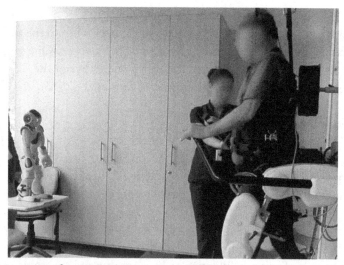

Fig. 7 Integration of an NAO robot in conventional therapy for neurological rehabilitation using Lokomat at Clínica Universidad de la Sabana.

4 Experimental studies

Following the same structure of the previous section of this chapter, this section presents, in the first place, the experimental study carried out for CR at FCI-IC. Subsequently, the experimental study deployed in the NR scenario is described.

4.1 Cardiac rehabilitation longitudinal study

In order to address the hypothesis stated in the human-robot interfaces proposed before, a longitudinal study is conducted. The system was tested during phase II of the CR program and each participant was for 18 weeks attending the sessions twice a week. In this study, the performance of the system was compared to a control group where the measurements were taken but the patient did not have any feedback. The purpose of this condition is to measure the performance of the patient during the therapy, without interfering with or altering the normal conditions of the session.

4.1.1 Experimental procedure

A standard session procedure divided in five stages was defined for the experiments (i.e., *Init*, *Warm-Up*, *Treadmill Exercise*, *Cool-Down*, and *End*). During these stages some physiological measurements are taken by the system or the medical staff. The interaction with the robot is mainly done in the treadmill exercise. During this stage, most of the interventions are present since the system provides *motivation*, requests the *BS*, and monitors all the *events* associated with warnings and emergencies.

The performance of the patients during the session was measured with quantitative variables related to the *nonattendance rate, physiological variables,* and *interaction variables*. The nonattendance rate indicates the number of absences of the participants during the study. The physiological variables are the *resting HR*, which is the HR level measured without any physical activity, and the *recovery HR*, measured a minute after the end of the activity during the cool down. The interaction variables quantify the interaction between the patient and the system through two measurements: the *response time (RT)*, which is the time (in seconds) between the robot's request for the BS and the patient's response, and the *posture corrections*, which express the amount of posture corrections that the robot requests of the patient. The patient's perception of the system was also important for the assessment and was obtained using a semistructured interview carried out at the end of the study. The questions are focused on gathering personal opinions and impressions of the system.

4.1.2 Results of the study

The study took place in a real clinical context throughout the complete phase II of the rehabilitation program. During this period, it was possible to evidence the impact that social assistive robots could generate in cardiac patients. According to the measurements considered during the study, there are some important aspects that are worthy of mention.

In the first place, the performance of the therapy was successfully measured and evaluated by means of the sensor interface previously described. As illustrated in Fig. 8, a sample of data recorded during one therapy is presented. Plot (a) shows the HR of one patient during the session. Plots (b), (c), and (d) represent the spatiotemporal parameters (speed [m/s], step length [m], and cadence [steps/s]). The inclination of the treadmill is plotted in diagram (e), and finally the BS values requested during the session are plotted in diagram (f). Additionally, two events are highlighted (A and B). Event A indicates when the therapists increment the exercise intensity (the speed and slope of the treadmill are modified). It can be observed that before this event the patient presented a low level in almost all their parameters, as well as the self-perception of the intensity estimated with the BS. However, once

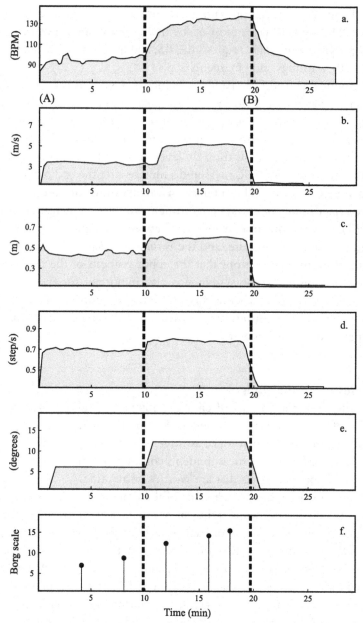

Fig. 8 Patient's data registered during a CR session. (a) Heart rate, (b) speed, (c) step length, (d) cadence, (e) inclination, and (f) BS.

the exercise parameters increase, the data recorded similarly reflect the increment as the HR and the spatiotemporal change with an increasing rate. Event B occurs when the exercising time has concluded and patients must step out the treadmill and perform the cool-down routine. As displayed in Fig. 8, after this event only the HR is monitored, since this variable is used to measure the patient's recovery. Thus, during the first minute of cool down, the HR difference is measured to estimate the rate at which the cardiovascular system returns to the resting point.

On the basis of the information that the sensor interface is able to record during the therapy, throughout the study it was possible to perform on-time assistance and monitoring of the therapy. Hence, the SAR provided continuously the appropriate feedback according to the context and the physiological conditions. In this way, it was also possible to reduce potential occurrence of risk factors, as the robot had access to the immediate state of patients to provide corrections and assistance.

Considering the multiple interactions that patients had with the robot during the therapies, it was possible to draw some metrics that could provide information about the progress in terms of the patient-robot interaction. Therefore, two variables were considered (i.e., *adaptation* and *correction*). The *adaptation* seeks to evaluate the patient's response to requests provided by the robot. In this case, *adaptation* was designed to be measured for the BS request. Patients must deliver their BS value through the HCi when the robot requests it. Hence, *adaptation* was measured as the time in seconds that patients spent delivering this value after the request. Results of this measure are illustrated in Fig. 9, where a decreasing rate of the RT is observed. In other words, patients take less time to deliver the value requested by the robot throughout the sessions, which is related to the adaptation that they have to the system and the interaction with the robot.

On the other hand, the *correction* was considered when the patient's posture was not ideal. This issue can generate dizziness, being likely to fall from the treadmill. In this way, during the study it was possible to quantify the amount of interventions related with the posture correction that took place along the sessions. In this period, it was also possible to perform over 50 corrections that clinicians could not monitor, due to the considerable volume of the therapy groups. Thus, the SAR system has proven to be an appropriate assistance tool that can support clinicians in their activities within the therapy. Likewise, the safety of patients during exercise can increase thanks to the continuous monitoring.

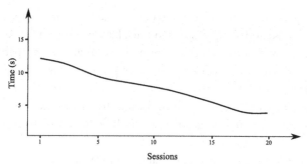

Fig. 9 Adaptation.

4.2 Neurological rehabilitation repeated measurements study

For the assessment of the NR interface proposed, a repeated measurements study was designed. In this study, the performance of the system was compared to the same patients with and without the robot. Without the robot the measurements were taken, but the patient had feedback only from the therapist. The purpose of this condition is to measure the performance of the patient during the therapy time (30 min), without interfering with or altering the normal conditions of the session.

4.3 Experimental procedure

The patients were asked to participate in a short-term study consisting of two sessions: one with the robot and one only with the measurement system. Without the robot the feedback is given only by the therapist, aiming to be similar to a session without any external intervention. The participants were users of Lokomat during their PR process and none of the patients had alterations in cognitive function that do not allow the understanding of the instructions and the NAO's behavior.

To assess the system, four parameters were taken into account. The *bad posture time* measures the time when the posture degrees overcome the good posture threshold. The *HR* and *BS* give an indication of the cardiovascular function. Finally, the perception of the patients and therapist was measured using a survey.

4.4 Results of the study

Regarding the human-robot interface for NR, four patients participated in the experimental study, and four parameters (i.e., HR, cervical posture, thoracic posture, and BS) were measured. According to the experimental design study mentioned in Section 4.2, the functionality of the interface was observed during 30 min of therapy. As an initial observation, the interface worked correctly in the sessions; the data was acquired online and saved in the database.

Regarding the data acquisition, the results of this section present the general behavior of the data recorded during a control session and the effects of the robot over this parameters. Fig. 10 shows the evolution of cervical and thoracic posture and BS of one patient during both scenarios. As it can be seen, the inclinations of the spinal cord were more noticeable at pitch axis (sagittal body plane) for both areas (thoracic and cervical sections) than yaw and roll axes; this behavior occurs due to the mechanical features of Lokomat that limit the movement of the patient's torso. Lokomat has a BWS system that allows a partial control of the body; rotations at the coronal (yaw axis) and transversal (roll axis) body planes are slightest.

On the other hand, to measure bad posture we establish a threshold that represents the maximum permissible degrees (10–15 degrees over 0 degree) of inclination in both areas, taking into account that a proper posture is reached when the patient has minimal inclinations during the gait training with Lokomat. Comparing between cervical and thoracic posture, we notice that cervical posture exceeds the threshold more times than the thoracic posture. This occurs because the head does not have any support, unlike the thoracic area. Concerning the HR and the BS, the results showed a constant behavior, as Lokomat sessions are not highly intensive, the patient's HR does not increase to a high value, and the perception of the fatigue corresponds to a neutral level in both cases (control and robot-assisted sessions).

During the control sessions we can observe that the tasks of the therapist were multiple and the corrections of the posture were not followed with accuracy due to this situation. As a consequence of these multitasking processes, the time where the patient kept a bad posture increased (Fig. 10A and D). The main aims of the robot within NR were to reduce therapist tasks and assist the patient by monitoring the parameters discussed before. The effects of using SAR in NR in this short-term study can be seen in the results obtained during the robot session. In this case, the time of bad posture in both areas (cervical and thoracic) decrease (Fig. 10B and E) compared to the control session (Fig. 10A and D). For the robot condition, the patient accomplished a proper posture more times than a control session.

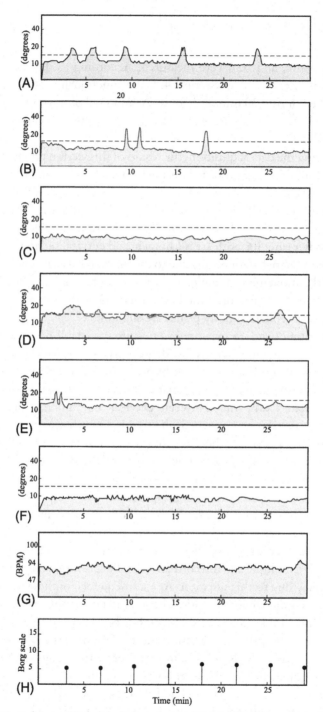

Fig. 10 Patient data registered during a Lokomat therapy. Thoracic posture at (A) pitch axis at control session, (B) pitch axis at robot-assisted session, and (C) yaw/roll behavior representation at control session. Cervical posture at (D) pitch axis at control session, (E) pitch axis at robot-assisted session, (F) yaw/roll axes at robot-assisted session, (G) heart rate mean during control and robot-assisted session, and (H) BS mean results during control and robot-assisted session.

5 Discussion

CVDs are considered the main cause of death worldwide, taking the lives of 17.9 million people each year. Similarly, neurological diseases are the major cause of disability in the world, with an increasing rate projected to reach over 100 million people with disability by 2030. Taking into account the context of this problematic, this chapter presented recent advancements in the field of SARs, where an SAR was incorporated in cardiac and NR therapies to act as a companion and assistant during the sessions. With the realization of this work, it was expected to increase patients' commitment to the therapy, aiming to improve their performance, and therefore ensure a greater recovery and quality of life.

Throughout the development of this chapter, the potential benefit that social robotic platforms exhibit in rehabilitation and physical training was discussed. Similarly, a proposed robot-based therapy model suitable for cardiac and NR was described and deployed in both scenarios. In the first place, a custom cHRI was developed for each application, followed by the experimental study that took place in real clinical contexts to assess the influence of the cHRI within the therapies. Regarding the results presented in the previous section, there are significant findings related to the therapy performance.

In both scenarios, one major feature that elicited positive outcomes was the continuous monitoring. In the cardiac therapy it was possible to detect and control potential risk factor events (e.g., increased HR or extremely high intensity of the exercise perceived through the BS). Likewise, for the NR, the continuous monitoring allowed the system to support the therapist's activities and reduce their workload. Hence, the therapist can focus on a more specific task while the robot can focus on monitoring general variables, such as the HR and posture parameters.

Based on the continuous monitoring, it was also possible to perform posture corrections in both scenarios. For the cardiac therapy, the robot is capable of performing appropriate posture corrections. In this context, the correction is carried out to reduce the risk of falling from the treadmill or experiencing dizziness that makes the exercise more difficult. As stated in the results of the study, the robot was able to perform more than 50 corrections. This result reflects the advantage of having such a system, since it can support the tasks of the clinical staff, and, more importantly, it can deliver safe therapy. On the other hand, for the NR, the posture correction was focused on complementing the task of the therapist. As the physical robotic platform Lokomat has partial control of the body (i.e., it controls the lower limbs unlike the upper limbs), the social robot seeks to monitor this part of the body and encourages the patient to adopt an appropriate posture while the therapist focuses on the correct execution of the exercises. As described in the results of the study, evidence was found that the patient adopts a correct posture more regularly in the presence of the robot. This result indicates the effectiveness of the robot-based therapy.

Finally, one aspect that is worth mentioning is the safety and motivation that this kind of system can deliver in this context. Throughout the studies that were carried out in both applications, we evidenced that the SAR can elicit a perception of safety by the patient during the therapy. Thus, this safety generates more confidence in the patient to perform the exercise and therefore their motivation increases. This effect holds a promising potential in terms of enhancing the physical performance exhibited by the patients.

6 Conclusions

In this chapter, the design and implementation of cHRI systems were presented. Two experimental studies were performed according to the needs of the cardiac and NR context. Each system has an individual sensory architecture that allows the use of a robot-based therapy model to improve the conventional rehabilitation and the patient's performance. Both systems allowed relevant patient data acquisition, processing, and an online feedback using SARs.

Important results showed that robot assistance during the therapy has a positive impact, which is reflected in the decreasing of risk factors such as high HR values or fatigue perception and the improvement of patients' performance indicators such as cervical and thoracic postures. As initial observations, in both cases, we found a positive attitude toward the robot role and patients exhibited a greater motivation to follow the rehabilitation treatment and improve their own performance according to the robot's requests and corrections.

Moreover, due to the monitoring capabilities of the cHRI system and the robot interaction, therapists were more focused on providing feedback to the patient about other parameters that are crucial in rehabilitation, or performed alternative tasks (e.g., assessed more patients and controlled the features of Lokomat), while the robot provided feedback and corrections.

This companionship is important, since the assessment and progress of the patients' performance can improve thanks to the accuracy of the monitoring and a thorough surveillance of the therapy.

Although the results presented in this chapter are promising, this research is still in an early stage. Thus, the findings of this work will serve as the basis to extend our studies to a larger scale. Therefore, as a further step we propose to test our cHRI systems in a long-term study, following a greater number of patients and sessions. This will allow a more robust statistical analysis, providing meaningful information associated with the adherence, motivation, and health evolution within the therapies. Moreover, regarding the HRI, it is expected to enhance the robot's behaviors, aiming to provide a more personalized interaction. This goal will be achieved by incorporating memory on the robot and additional robot features to achieve long-term interaction. Hence, the robot will be able to recognize each patient, their performance, and progress throughout the complete program.

References

Achten, J., Jeukendrup, A.E., 2003. Heart rate monitoring: applications and limitations. Sports Med. (Auckland, N.Z.) 33 (7), 517–538.

American Heart Association, 2017. What Is Cardiovascular Disease? American Heart Association, Dallas, TX.

Appelbaum, S.H., Marchionni, A., Fernandez, A., 2008. The multi-tasking paradox: perceptions, problems and strategies. Manag. Decis. 46 (9), 1313–1325. https://doi.org/10.1108/00251740810911966.

Bang, D.-H., Shin, W.-S., 2016. Effects of robot-assisted gait training on spatiotemporal gait parameters and balance in patients with chronic stroke: a randomized controlled pilot trial. NeuroRehabilitation 38 (4), 343–349. https://doi.org/10.3233/NRE-161325.

Banz, R., Bolliger, M., Colombo, G., Dietz, V., Lunenburger, L., 2008. Computerized visual feedback: an adjunct to robotic-assisted gait training. Phys. Ther. 88 (10), 1135–1145. https://doi.org/10.2522/ptj.20070203.

Baroni, I., Nalin, M., Baxter, P., Pozzi, C., Oleari, E., Sanna, A., Belpaeme, T., 2014. What a robotic companion could do for a diabetic child. In: 23rd IEEE International Symposium on Robot and Human Interactive Communication. https://doi.org/10.1109/ROMAN.2014.6926373.

Belpaeme, T., Baxter, P., Read, R., Wood, R., Cuayáhuitl, H., Kiefer, B., Racioppa, S., Kruijff-Korbayová, I., Athanasopoulos, G., Enescu, V., Looije, R., Neerincx, M., Demiris, Y., Ros-Espinoza, R., Beck, A., Cañamero, L., Hiolle, A., Lewis, M., Baroni, I., Nalin, M., Cosi, P., Paci, G., Tesser, F., Sommavilla, G., Humbert, R., 2013. Multimodal child-robot interaction: building social bonds. J. Hum. Robot Interact. 1 (2), 33–53. https://doi.org/10.5898/JHRI.1.2.Belpaeme.

Bemelmans, R., Gelderblom, G.J., Jonker, P., de Witte, L., 2012. Socially assistive robots in elderly care: a systematic review into effects and effectiveness. J. Am. Med. Dir. Assoc. 13, 114–120. e1.

Bethell, H.J., Turner, S.C., Evans, J.A., Rose, L., 2011. Cardiac rehabilitation in the United Kingdom. How complete is the provision? J. Cardiopulm. Rehabil. 21 (2), 111–115.

Bradley, W.G., 2004. Neurology in Clinical Practice. Butterworth-Heinemann, Oxford. ISBN: 9789997625885.

Casas, J., Irfan, B., Senft, E., Gutiérrez, L., Rincon-Roncancio, M., Munera, M., Belpaeme, T., Cifuentes, C.A., 2018. Social assistive robot for cardiac rehabilitation: a pilot study with patients with angioplasty. In: Companion of the 2018 ACM/IEEE International Conference on Human-Robot Interaction. ACM, pp. 79–80.

Dautenhahn, K., Ogden, B., Quick, T., 2002. From embodied to socially embedded agents—implications for interaction-aware robots. Cognit. Syst. Res. 3 (3), 397–428. https://doi.org/10.1016/S1389-0417(02)00050-5.

Douglas, H.E., Raban, M.Z., Walter, S.R., Westbrook, J.I., 2017. Improving our understanding of multi-tasking in healthcare: drawing together the cognitive psychology and healthcare literature. Appl. Ergon. 59, 45–55. https://doi.org/10.1016/j.apergo.2016.08.021.

Essery, R., Geraghty, A.W.A., Kirby, S., Yardley, L., 2017. Predictors of adherence to home-based physical therapies: a systematic review. Disabil. Rehabil. 39 (6), 519–534. https://doi.org/10.3109/09638288.2016.1153160.

Feil-Seifer, D., Mataric, M.J., 2005. Defining socially assistive robotics. In: 9th International Conference on Rehabilitation Robotics, 2005. ICORR 2005. IEEE, pp. 465–468.

Fisher, S., Lucas, L., Trasher, A., 2011. Robot-assisted gait training for patients with hemiparesis due to stroke. Top Stroke Rehabil. 18, 269–276. 3.

Fong, T., Nourbakhsh, I., Dautenhahn, K., 2003. A survey of socially interactive robots. Robot. Auton. Syst. 42, 143–166. https://doi.org/10.1016/S0921-8890(02)00372-X.

Gadde, P., Kharrazi, H., Patel, H., MacDorman, K.F., 2011. Toward monitoring and increasing exercise adherence in older adults by robotic intervention: a proof of concept study. J. Robot. 2011, 1–11. https://doi.org/10.1155/2011/438514.

Hocoma, 2019. Lokomat®—functional robotic gait therapy. Available from: https://www.hocoma.com/solutions/lokomat/. (Accessed 15 January 2019).

Husemann, B., Muller, F., Krewer, C., Heller, S., Koenig, E., 2007. Effects of locomotion training with assistance of a robot-driven gait orthosis in hemiparetic patients after stroke: a randomized controlled pilot study. Stroke 38 (2), 349–354. https://doi.org/10.1161/01.STR.0000254607.48765.cb.

Hussain, S., Xie, S.Q., Liu, G., 2011. Robot assisted treadmill training: mechanisms and training strategies. Med. Eng. Phys. 33 (5), 527–533. https://doi.org/10.1016/J.MEDENGPHY.2010.12.010.

Hwang, S., Kim, H.-R., Han, Z.-A., Lee, B.-S., Kim, S., Shin, H., Moon, J.-G., Yang, S.-P., Lim, M.-H., Cho, D.-Y., Kim, H., Lee, H.-J., 2017. Improved gait speed after robot-assisted gait training in patients with motor incomplete spinal cord injury: a preliminary study. Ann. Rehabil. Med. 41 (1), 34–41. https://doi.org/10.5535/arm.2017.41.1.34.

Jack, K., McLean, S.M., Moffett, J.K., Gardiner, E., 2010. Barriers to treatment adherence in physiotherapy outpatient clinics: a systematic review. Man Ther. 15 (3), 220–228. https://doi.org/10.1016/j.math.2009.12.004.

Jackson, L., Leclerc, J., Erskine, Y., Linden, W., 2005. Getting the most out of cardiac rehabilitation: a review of referral and adherence predictors. Heart 91 (1), 10–14. https://doi.org/10.1136/hrt.2004.045559.

Johansson, B.B., 2011. Current trends in stroke rehabilitation. A review with focus on brain plasticity. Acta Neurol. Scand. 123 (19), 147–159. https://doi.org/10.1111/j.1600-0404.2010.01417.x.

Jung, H.-T., Baird, J., Choe, Y.-K., Grupen, R.A., 2011. Upper-limb exercises for stroke patients through the direct engagement of an embodied agent. In: Proceedings of the 6th International Conference on Human-Robot Interaction—HRI'11, p. 157.

Kang, K.I., Freedman, S., Mataric, M.J., Cunningham, M.J., Lopez, B., 2005. A hands-off physical therapy assistance robot for cardiac patients. In: 9th International Conference on Rehabilitation Robotics, 2005. ICORR 2005, pp. 337–340.

Kidd, C.D., Breazeal, C., 2008. Robots at home: understanding long-term human-robot interaction. In: 2008 IEEE/RSJ International Conference on Intelligent Robots and Systems, IROS, pp. 3230–3235.

Kim, C., Youn, J.E., Choi, H.E., 2011. The effect of a self-exercise program in cardiac rehabilitation for patients with coronary artery disease. Ann. Rehabil. Med. 35 (3), 381. https://doi.org/10.5535/arm.2011.35.3.381.

Kim, H., Shin, S.-H., Kim, J.-K., Park, Y.-J., Oh, H.-S., Park, Y.-B., 2013. Cervical coupling motion characteristics in healthy people using a wireless inertial measurement unit. Evid. Based Complement. Alternat. Med. 2013, 1–8. https://doi.org/10.1155/2013/570428.

Krewer, C., Müller, F., Husemann, B., Heller, S., Quintern, J., Koenig, E., 2007. The influence of different Lokomat walking conditions on the energy expenditure of hemiparetic patients and healthy subjects. Gait Posture 26 (3), 372–377. https://doi.org/10.1016/j.gaitpost.2006.10.003.

Lara, J.S., Casas, J., Aguirre, A., Munera, M., Rincon-Roncancio, M., Irfan, B., Senft, E., Belpaeme, T., Cifuentes, C.A., 2017. Human-robot sensor interface for cardiac rehabilitation. In: 2017 International Conference on Rehabilitation Robotics (ICORR), pp. 1013–1018.

Looije, R., Cnossen, F., Neerincx, M.A., 2006. Incorporating guidelines for health assistance into a socially intelligent robot. In: ROMAN 2006—The 15th IEEE International Symposium on Robot and Human Interactive Communication, Hatfield, pp. 515–520. https://doi.org/10.1109/ROMAN.2006.314441.

López Recio, D., Márquez Segura, L., Márquez Segura, E., Waern, A., 2013. The NAO models for the elderly. In: 2013 8th ACM/IEEE International Conference on Human-Robot Interaction (HRI), Tokyo, pp. 187–188. https://doi.org/10.1109/HRI.2013.6483564.

Louie, W.Y., McColl, D., Nejat, G., 2014. Acceptance and attitudes toward a human-like socially assistive robot by older adults. Assist. Technol. 26 (3), 140–150.

Marti, P., Bacigalupo, M., Giusti, L., Mennecozzi, C., Shibata, T., 2006. Socially assistive robotics in the treatment of behavioural and psychological symptoms of dementia. In: First IEEE/RAS-EMBS International Conference on Biomedical Robotics and Biomechatronics, 2006. BioRob 2006. IEEE, pp. 483–488.

Martín, F., Agüero, C.E., Cañas, J.M., Valenti, M., Martínez-Martín, P., 2013. Robotherapy with dementia patients. Int. J. Adv. Robot. Syst. 10, 1–7.

Matarić, M.J., Eriksson, J., Feil-Seifer, D.J., Winstein, C.J., 2007. Socially assistive robotics for post-stroke rehabilitation. J. Neuroeng. Rehabil. 4 (1), 5. https://doi.org/10.1186/1743-0003-4-5.

Mayr, A., Kofler, M., Quirbach, E., Matzak, H., Fröhlich, K., Saltuari, L., 2007. Prospective, blinded, randomized crossover study of gait rehabilitation in stroke patients using the Lokomat gait orthosis. Curr. Neurol. Neurosci. Rep. 21 (4), 307–314. https://doi.org/10.1177/1545968307300697.

Mead, R., Wade, E., Johnson, P., St. Clair, A., Chen, S., Matarić, M.J., 2010. An architecture for rehabilitation task practice in socially assistive human-robot interaction. In: Proceedings—IEEE International Workshop on Robot and Human Interactive Communication, pp. 404–409.

Munera, M., Marroquin, A., Jimenez, L., Lara, J.S., Gomez, C., Rodriguez, S., Rodriguez, L.E., Cifuentes, C.A., 2017. Lokomat therapy in Colombia: current state and cognitive aspects. In: 2017 International Conference on Rehabilitation Robotics (ICORR), vol. 2017. IEEE, pp. 394–399.

Pashler, H., 2000. Task switching and multitask performance. In: Monsell, S., Driver, J. (Eds.), Attention and Performance XVIII: Control of Mental Processes. MIT Press, Cambridge, MA, pp. 277–307.

Pfeifer, R., Scheier, C., 1999. Understanding Intelligence. MIT Press, Cambridge, MA, ISBN: 0-262-16181-8.

Pons, J.L., Ceres, R., Calderón, L., 2008. Wearable robots and exoskeletons. Wearable Robots: Biomechatronic Exoskeletons, pp. 1–5.

Potter, J.M., Evans, A.L., Duncan, G., 1995. Gait speed and activities of daily living function in geriatric patients. Arch. Phys. Med. Rehabil. 76 (11), 997–999. https://doi.org/10.1016/S0003-9993(95)81036-6.

Powers, A., Kiesler, S., Fussell, S., Torrey, C., 2007. Comparing a computer agent with a humanoid robot. In: Proceedings of the ACM/IEEE International Conference on Human-Robot Interaction, HRI '07. ACM, New York, NY, pp. 145–152.

Pulido, J.C., González, J.C., Fernández, F., 2016. NAO therapist: autonomous assistance of physical rehabilitation therapies with a social humanoid robot. In: International Workshop on Assistive & Rehabilitation Technology (IWART 2016), pp. 15–16.

Rabbitt, S.M., Kazdin, A.E., Scassellati, B., 2015. Integrating socially assistive robotics into mental healthcare interventions: applications and recommendations for expanded use. Clin. Psychol. Rev. 35, 35–46. https://doi.org/10.1016/j.cpr.2014.07.001.

Santis, A.D., 2007. Modelling and Control for Human-Robot Interaction (Ph.D. thesis). Universita' Degli Studi Di Napoli Federico II Dottorato.

Sarrafzadegan, N., Rabiei, K., Shirani, S., Kabir, A., Mohammadifard, N., Roohafza, H., 2007. Drop-out predictors in cardiac rehabilitation programmes and the impact of sex differences among coronary heart disease patients in an Iranian sample: a cohort study. Clin. Rehabil. 21 (4), 362–372. https://doi.org/10.1177/0269215507072193.

Scherr, J., Wolfarth, B., Christle, J.W., Pressler, A., Wagenpfeil, S., Halle, M., 2013. Associations between Borg's rating of perceived exertion and physiological measures of exercise intensity. Eur. J. Appl. Physiol. 113 (1), 147–155. https://doi.org/10.1007/s00421-012-2421-x.

Sharma, R., Pavlovic, V.I., Huang, T.S., 1998. Toward multimodal human-computer interface. Proc. IEEE 86 (5), 853–869. https://doi.org/10.1109/5.664275.

Taylor, R.S., Unal, B., Critchley, J.A., Capewell, S., 2006. Mortality reductions in patients receiving exercise-based cardiac rehabilitation: how much can be attributed to cardiovascular risk factor improvements? Eur. J. Cardiovasc. Prev. Rehabil. 13 (3), 369–374. https://doi.org/10.1097/01.hjr.0000199492.00967.11.

Tsardoulias, E.G., Kintsakis, A.M., Panayiotou, K., Thallas, A.G., Reppou, S.E., Karagiannis, G.G., Iturburu, M., Arampatzis, S., Zielinski, C., Prunet, V., Psomopoulos, F.E., Symeonidis, A.L., Mitkas, P.A., 2017. Towards an integrated robotics architecture for social inclusion—the RAPP paradigm. Cogn. Syst. Res. 43, 157–173. https://doi.org/10.1016/j.cogsys.2016.08.004.

Visintin, M., Barbeau, H., Korner-Bitensky, N., Mayo, N.E., 1998. A new approach to retrain gait in stroke patients through body weight support and treadmill stimulation. Stroke 29 (6), 1122–1128. https://doi.org/10.1161/01.STR.29.6.1122.

Wainer, J., Feil-Seifer, D.J., Shell, D.A., Mataric, M.J., 2006. The role of physical embodiment in human-robot interaction. In: ROMAN 2006—The 15th IEEE International Symposium on Robot and Human Interactive Communication, pp. 117–122.

Wainer, J., Feil-Seifer, D.J., Shell, D.A., Matarić, M.J., 2006. The role of physical embodiment in human-robot interaction. In: Proceedings—IEEE International Workshop on Robot and Human Interactive Communication, pp. 117–122.

Weaver, L.J., Ferg, A.L., 2010. Therapeutic Measurement and Testing: The Basics of ROM, MMT, Posture, and Gait Analysis. Delmar Cengage Learning, ISBN: 9781418080808, p. 437.

Werner, C., Von Frankenberg, S., Treig, T., Konrad, M., Hesse, S., 2002. Treadmill training with partial body weight support and an electromechanical gait trainer for restoration of gait in subacute stroke patients: a randomized crossover study. Stroke 33 (12), 2895–2901.

WHO, Dua, T., Janca, A., Muscetta, A., 2006. Neurological disorders. Public health challenges. J. Nerv. Ment. Dis. 196, 7–25. https://doi.org/10.1097/NMD.0b013e31816372ab.

Winter, D.A., 1989. Biomechanics of normal and pathological gait: implications for understanding human locomotor control. J. Mot. Behav. 21 (4), 337–355.

Worcester, M.U., Murphy, B.M., Mee, V.K., Roberts, S.B., Goble, A.J., 2004. Cardiac rehabilitation programmes: predictors of non-attendance and drop-out. Eur. J. Cardiovasc. Prev. Rehabil. 11 (4), 328–335. https://doi.org/10.1097/01.hjr.0000137083.20844.54.

World Health Organization, 2015. Global Atlas on Cardiovascular Disease Prevention and Control. World Health Organization, Geneva.

World Health Organization, 2016. What Are Neurological Disorders? World Health Organization, Geneva.

World Health Organization, 2018. Cardiovascular diseases (CVDs). World Health Organization, Geneva.

World Health Organization, 2018. World Heart Day. World Health Organization, Geneva.

CHAPTER TWO

State observation and feedback control in robotic systems for therapy and surgery

Bita Fallahi, Lingbo Cheng and Mahdi Tavakoli
Department of Electrical and Computer Engineering, University of Alberta, Edmonton, AB, Canada

1 Introduction

Steerable needles are used in different minimally invasive procedures such as brachytherapy, biopsy, and neurosurgery. In these methods, hollow long flexible bevel-tipped needles are inserted into the human body for diagnosis, treatment, or sample removal. During these procedures, the targeted organs and the needles are monitored using different imaging methods such as ultrasound (US), fluoroscopy, and X-ray. Accurate needle positioning minimizes the undesirable side effects on the healthy and neighboring tissue and is a crucial factor in determining the efficiency of these methods. The desired needle path depends on the application. In biopsy, it is desired to reach a constant final deflection, whereas in brachytherapy the needle should follow a straight path. In case of having obstacles on the needle path (such as bones or nerves), the needle should follow a preplanned curved. In general, needle deflection, tissue deformation, and limitations in controlling the needle from outside the body are the challenges in accurate needle tip positioning, which are discussed in Section 2.

Section 3 is devoted to robot-assisted surgical systems for beating-heart surgery. Cardiovascular disease is one of the leading causes of death worldwide. Conventional extra-and intracardiac surgeries need the heart to be arrested by connecting the patient to a cardiopulmonary bypass (CPB) machine (Ruszkowski et al., 2015). However, arrested-heart surgery has adverse effects due to using CPB (Dacey et al., 2005; Newman et al., 2001; Paparella et al., 2002; Zeitlhofer et al., 1993; Bellinger et al., 1999). Different from arrested-heart surgery, beating-heart surgery could eliminate such negative effects of CPB by allowing the heart to beat normally (Angelini et al., 2002), and could also enable intraoperative evaluation of the heart tissue motion, which is critical to the assessment of reconstructive heart operations such as mitral valve surgery (Fix et al., 1993). The most prominent challenge to be addressed for beating-heart surgery is the rapid motions of the heart whose movement velocity and acceleration are approximately 210 mm/s and 3800 mm/s$_2$, respectively (Kettler et al., 2007). Manual tool position compensation according to the heart motion will not only lead to the human operator's fatigue and exhaustion but risks tool-tissue collision and tissue injury. The application of robot-assisted surgical systems and control methods for synchronizing the surgical robot's motion with the beating heart's motion are discussed in Section 3.

2 Needle insertion procedures

Prostate cancer is the second frequent cancer in men around the world, with an estimation of 1.1 million new cases to have occurred in 2012 (Torre et al., 2015). One leading treatment option for early-stage prostate cancer is US-guided brachytherapy, which is a type of radiotherapy. In this method, radioactive seeds are implanted around the prostate gland to deliver the radiations internally. The margins of the prostate and the target volume are found using preoperative axial images and are further used for preplanning the seed locations and the dosage distribution. The seed implantation is accompanied by intraoperative transrectal ultrasound (TRUS) images to provide visual information for needle guidance. These images, as well as the desired seed locations, are registered with respect to a 5-mm grid template. Several needles loaded with radioactive seeds are inserted through the template, and it is desired to insert the needle on a straight path to the final depth. The seeds are implanted on the needle track by retracting the needles and pushing out the seeds using a stylet. In brachytherapy, the preplanned location of the seeds is a determinant factor in defining radiation dosage and therefore is of great importance. Errors in the seed positioning reduce the method efficiency as instead of cancerous tissue, healthy tissue is imposed to the radiations.

Using a grid template, the desired path to the final depth is a straight line. However, this can only be true if there are no sensitive tissue such as nerves, blood vessels, or bones are on the needle path. In the case of pubic arch interference, which is common in patients with large prostate (Wallner et al., 1999), the needle path to the anterior prostate is obstructed. In such case, the needle should be steered on a desired curved path to go around the obstacle and reach the final depth. This scenario is in demand of finding feasible and collision-free trajectories from the insertion point to the final point and steering the needle on the desired path. Steering the needle on the desired path can be done using axial needle rotations to change the bevel orientation, insertion velocity, needle base lateral position, and needle base force/torque.

In manual insertions performed by experienced practitioners, the absolute seed positioning error is about ± 5 mm (Taschereau et al., 2000). To improve the seed positioning, intelligent assistant robots can be used to steer the needle toward the target and compensate for the errors caused by the needle and tissue deformation. In robot-assisted procedures, for safety reasons, it is desired to split the tasks between the clinician and the robotic system to keep the clinician in the loop. For example, the robotic system only controls the needle rotations, whereas the surgeon selects the initial insertion point and performs the insertion.

2.1 Related work

Depending on the application, different scenarios are possible in needle insertion procedures. For all situations, the first step is to plan the desired trajectory. The path planner should consider the insertion and target points, the location of the obstacles, and the constraints on the needle motion. Similar to path planners, motion planners provide not only a feasible path but also the control inputs required for following the path (Minhas et al., 2007; Schulman et al., 2014; Wang et al., 2014; Moreira et al., 2014a; Majewicz et al., 2014). However, in order to compensate for the errors caused by tissue nonhomogeneity and modeling uncertainties, replanning is necessary, which requires the motion planners to be computationally fast (Duan et al., 2014).

Another way to deal with the steering problems is to use feedback control strategies (Abolhassani et al., 2007; Haddadi et al., 2010; Abayazid et al., 2013; Rucker et al., 2013; Khadem et al., 2016; Sovizi et al., 2016; Fallahi et al., 2016b, 2017, 2018). This approach is based on the closed-loop feedback structure to calculate the control inputs. In motion planning, the steering commands are calculated for the current and future times, however, in feedback control methods, the commands are found only for the present time. By taking the desired path (found by any path planner), the controller is responsible for calculating the control commands to compensate the errors. The planner/controller should be provided with information about needle forces, needle shape, and tip pose (position and/or orientation) in real time. However, due to the nature of the procedures, in many situations, some of these variables cannot be measured directly. As the needle is inserted into the tissue, the needle is tracked using imaging modalities. The images are combined with image processing techniques to translate visual information into numerical values to be used by the computers. US imaging is a cost effective and widely used imaging modality, which can be used to track the needle and find the needle tip position. There have been different methods proposed in the literature for image registration and needle position and shape estimation using US images (Uherčík et al., 2009; Asadian et al., 2011; Zhao et al., 2012, 2013; Malekian et al., 2013; Kaya and Bebek, 2014; Waine et al., 2016; Rossa et al., 2016).

However, using the US images, it is not possible to measure the needle tip orientation as the low-resolution images, and the small diameter of the needle does not provide sufficient information for detecting the needle tip heading. Moreover, due to the limitations on the sensor dimensions and the sterilization issues, using needle-mounted pose or force sensors in clinical applications is not practical. This limitation is the motivation to employ observation methods and estimate the nonmeasurable variables using the mathematical models and the measured variables. In the sequel, we first provide an overview of needle steering modeling and then focus on different methodologies proposed for state observation, planning, and control in needle steering.

2.2 Modeling

The design and programming of needle steering systems require a suitable model that represents the system's behavior and describes the relation between the system inputs and outputs. There have been different modeling approaches proposed in the literature such as finite element methods (DiMaio and Salcudean, 2003, 2005; Goksel et al., 2006; Dehghan and Salcudean, 2009), flexible beam theory (Yan et al., 2006, 2009; Chentanez et al., 2009), energy-based methods (Misra et al., 2010), and the kinematic model (Webster et al., 2006). The kinematic model proposed by Webster et al. (2006) is a nonholonomic model. In this model, the needle is assumed to move on a curved path with constant curvature (circle). As the needle path curvature depends on the needle and tissue properties, for each needle/tissue combination, the needle path curvature should be found using curve-fitting methods. The main advantage of this model is its simplicity. It provides an intuitive and computationally efficient method for predicting needle deflection, which makes it suitable for planning and control purposes.

In this model, the motion of the needle resembles the motion of a unicycle and its kinematics are found as Webster et al. (2006)

$$\dot{g}_{ab}(t) = g_{ab}(t)\left(v\hat{V}_1 + u\hat{V}_2\right) \tag{1}$$

with

$$V_1 = \begin{bmatrix} e_3 \\ \kappa e_1 \end{bmatrix}, \quad V_2 = \begin{bmatrix} 0_{3\times 1} \\ e_3 \end{bmatrix} \tag{2}$$

where the operator $(\hat{\cdot})$ is defined as

$$\hat{} : \begin{bmatrix} \omega_1 \\ \omega_2 \\ \omega_3 \end{bmatrix} \mapsto \begin{bmatrix} 0 & -\omega_3 & \omega_2 \\ \omega_3 & 0 & -\omega_1 \\ -\omega_2 & \omega_1 & 0 \end{bmatrix} \tag{3}$$

In Eq. (1), $g_{ab} = \begin{bmatrix} R_{ab} & P_{ab} \\ 0^T & 1 \end{bmatrix}$ is the rigid transformation between frames $\{A\}$ and $\{B\}$, as shown in Fig. 1. The vector $P_{ab} = [x \ y \ z]^T$ and the matrix R_{ab} represent the position and the orientation of the moving frame $\{B\}$ with respect to the fixed frame $\{A\}$, respectively. v and u are the insertion velocity and the needle axial rotational velocity, respectively, and $e_i(i = 1, 2, 3)$ represent the standard basis vectors in $\mathbb{R}^3$. κ is the needle path curvature, which due to the tissue nonhomogeneity encounters uncertainty. This value, however, is considered to be bounded, and its bounds can be determined preoperatively and be written as $\underline{\kappa} < \kappa < \overline{\kappa}$.

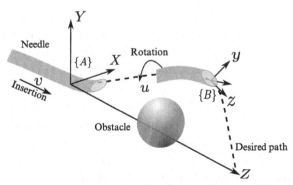

Fig. 1 The needle in 3D space, desired path and obstacle. Frame $\{A\}$ is the fixed frame and frame $\{B\}$ attached to the needle tip is the moving frame. v and u are the insertion and rotation velocity, respectively.

The coordinates free representation (1) can be expanded to

$$\dot{p} = R \begin{bmatrix} 0 \\ 0 \\ v \end{bmatrix} \tag{4a}$$

$$\dot{R} = R \begin{bmatrix} 0 & -u & 0 \\ u & 0 & -\kappa v \\ 0 & \kappa v & 0 \end{bmatrix} \tag{4b}$$

Defining the vector $\mathbf{q} = [x, \ y, \ z, \ \alpha, \ \beta, \ \gamma]^T$ as the generalized coordinates, which is well defined on

$$\mathcal{U} = \{\mathbf{q} \in \mathbb{R}^6 : \alpha, \gamma \in \mathbb{R}, \beta \in [-\pi/2, \pi/2]\} \tag{5}$$

Kallem and Cowan (2009) have presented the Z-Y-X representation of the needle kinematics (1) as

$$\dot{x} = v \sin \beta \tag{6a}$$
$$\dot{y} = -v \cos \beta \sin \alpha \tag{6b}$$
$$\dot{Z} = v \cos \alpha \cos \beta \tag{6c}$$
$$\dot{\alpha} = \kappa v \cos \gamma \sec \beta \tag{6d}$$
$$\dot{\beta} = \kappa v \sin \gamma \tag{6e}$$
$$\dot{\gamma} = -\kappa v \cos \gamma \tan \beta + u \tag{6f}$$

The values α, β, and γ are the yaw, pitch, and roll angles, respectively, and represent the orientation of the origin of the moving frame $\{B\}$ with respect to the fixed frame $\{A\}$. The relation between the rotation matrix R and the three angles is given by premultiplying the three basic rotations about the axes of the fixed frames (Taghirad, 2013). In reality, due to the presence of the tissue and the limited curvature of the needles, the angles α and β remain bounded, and more bending is related to larger values of these angles. The upper bound on these angles, which depend on the desired path traveled by the needle, can be determined in the planning level. Therefore, it is assumed that $|\alpha| < \alpha^*$ and $|\beta| < \beta^*$ with $\alpha^*, \beta^* \in [0, \pi/2]$.

In Eq. (6), by setting $\beta = 0$ degrees and $\gamma = 0$, 180 degrees, the two-dimensional (2D) needle motion can be found as

$$\begin{bmatrix} \dot{Z} \\ \dot{\gamma} \\ \dot{\alpha} \end{bmatrix} = \begin{bmatrix} \cos\alpha \\ \sin\alpha \\ \pm\kappa \end{bmatrix} v \tag{7}$$

The earlier equation represents the planar deflection of the needle in Y-Z plane. In this case, the $\pm$ sign determines the two possibilities for the bevel orientation as well as the concavity of the needle path curve in the plane as shown in Fig. 2. This model is later extended to account for paths with variable curvature by using an omnidirectional wheel (Fallahi et al., 2015).

2.3 Measurement and observation

In manual needle insertion procedures, the surgeon tracks the needle visually using imaging modalities. US imaging is a fast, widely used, and cost-effective method, which compared to other modalities such as magnetic resonance imaging and computed tomography, provides real-time tracking of the needle during the procedure. This visual information can be translated into numerical values using image processing techniques and be used in feedback computations. Different methods have been proposed in the literature for needle localization, that is, estimating the needle position in US images. Rossa et al. (2016) proposed a method for predicting the needle tip position using the deflection of a single point along the needle. Random sample consensus (RANSAC) is a robust method to fit polynomials on the curves in the three-dimensional (3D) space (Uherčík et al., 2009). This iterative method employs a set of observed data and deals with outliers to find the model parameters. Combining this method with Kalman filters reduces the search area and speeds up the algorithm (Zhao et al., 2012, 2013). Waine

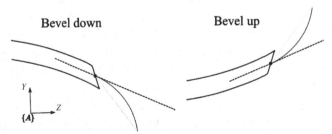

Fig. 2 Two possible orientations of the bevel in planar motion.

et al. (2016) have employed the RANSAC algorithm to find the needle tip position using 2D US images. Malekian et al. (2013) combine the RANSAC algorithm with a denoising method to increase the accuracy. Asadian et al. (2011) estimated the needle tip velocity from a noisy position using a high-gain observer.

In these methods, the needle tip position is acquired from the US images and the word "estimation" is used. However, since using needle-mounted sensors is not clinically feasible, the US images are the only source of position measurements, and therefore, the estimated values should be accepted as the true position "measurement."

Besides, according to Eq. (6), acquiring information about the needle tip heading is advantageous in controlling the needle tip position. However, it is not possible to measure the orientation parameters using US images, since due to the small diameter of the needle and the low resolution of the images, the bevel orientation is not detectable. Measuring the orientation requires utilizing sensors such as needle-mounted electromagnetic tracking sensor, which suffers from sterilization issues. The controller designed in Rucker et al. (2013) requires all the orientation parameters. This controller is implemented using a five-DOF magnetic tracking sensor and combined with a Kalman filter. However, since suitable sensors are not accessible, state observers can be employed to estimate the needle tip orientation. The state observers are computer-implemented systems, which run concurrently to the real system. The observer equations are formed using the system equations and additional corrective terms. If the observer is convergent, its states provide an estimate of the system's nonmeasurable state. Kallem and Cowan (2009) use the 3D kinematic equations (6) and design a linear observer/controller. Later, this observer is used in many other works. Reed et al. (2008) employed the linear observer to estimate the needle tip orientation and used the estimated variables in a low-level controller, which works along with a high-level 2D planner to steer the needle on the optimal path. In Motaharifar et al. (2015), the same transformation is used, and a nonlinear observer is designed to be used with an adaptive controller. In Kallem et al. (2011), this linear observer is designed for the reduced configuration space and is fed to the fiber space observer to estimate the full system states for a planar task. In Swensen et al. (2014), a model for torsional dynamics of the needle is presented and augmented with planar variables. The system is then linearized, and a Kalman filter is employed to estimate the system states and apply a state feedback control. Using Eq. (6) and the nonlinear transformation $s = \begin{bmatrix} x & v\sin\beta & -\kappa v^2 \cos\beta \sin\gamma \end{bmatrix}^T$, the system equations can be written as

$$\dot{s} = As + B\phi \tag{8a}$$

$$y = Cs \tag{8b}$$

with

$$A = \begin{bmatrix} 0 & 1 & 0 \\ 0 & 0 & 1 \\ 0 & 0 & 0 \end{bmatrix}, \quad B = \begin{bmatrix} 0 \\ 0 \\ 1 \end{bmatrix}, \quad C = \begin{bmatrix} 1 & 0 & 0 \end{bmatrix}, \quad \phi = v' \tag{9a}$$

where $v' = \kappa^2 \sin\beta - \kappa v^2 (\cos\beta \cos\gamma) u$ is defined as the new control variable and the observer is formulated as $\dot{\hat{s}} = A\hat{s} + B\phi + L(y - \hat{y})$. In this equation, $(\hat{\cdot})$ denotes the estimated values and L is the observer gain to be designed such that $A + LC$ is Hurwitz. Using the linear observer, an observer-based linear controller is designed in Kallem and Cowan (2009), which due to the singularities in the nonlinear system, can only be applied to stabilize the needle in one plane. In Fallahi et al. (2016a), the same transformation is used, however, instead of linearizing the whole system, the term ϕ is selected as

$$\phi = -(\kappa v)^2 s_2 \pm \kappa v^2 u \sqrt{1 - (s_2^2 + s_3^2)} \tag{10}$$

and its effect is considered by rewriting the observer equations as $\dot{\hat{s}} = A\hat{s} + \hat{\phi} + \Delta_\theta L(\hat{y} - y)$ with $\Delta_\theta = \text{diag}\{\theta, \theta^2, \theta^3\}$ and $\theta > 1$. The problem with the nonlinear observer in this form is that ϕ does not satisfy the Lipschitz continuity condition, and the convergence of the observer is guaranteed for certain assumptions to keep the states bounded. Due to the small number of researches performed for estimating the needle tip orientation, this subject remains open for further studies.

2.4 Planning

The nonholonomic constraints on the needle kinematics confine the needle moves to a set of reachable points and limit the trajectories that can be followed by the needle. Considering the location of the obstacles and the constraints on the needle, the path planner is responsible for finding a feasible, collision-free path from the insertion point to the target. Depending on the planning method used, other criteria such as parameter uncertainty, noise, and optimizations can be taken into account. A motion planner, however, not only plans the desired trajectory but also finds a sequence of commands for steering the needle on the planned trajectory. In other words, motion planning is a combination of path planning and an implementation method. In the ideal situation, these commands should steer the needle on the desired trajectory. However, in reality, due to the modeling and parametric uncertainties, noise, and tissue nonhomogeneity, the needle position deviates from the desired trajectory. In Schulman et al. (2014), the errors are compensated by insertion, partial retraction, rotation, and further insertion of the needle. However, retractions and reinsertions are not desirable as they increase the tissue trauma. Another solution for dealing with uncertainties is to perform replanning to compensate for the errors caused by tissue nonhomogeneity. Wang et al. (2014) propose the planner for dynamic environments using a mass–spring model for the deformable tissue. The dynamic planner is equipped with a vision system to track the radius of curvature, which is used in replanning and updating the path to adapt to the changes in the position and curvature. Such performance, however, requires the planner to be fast enough. Duan et al. (2014) show by experiments that the computations are done in 1.6 s which enables replanning for error compensation.

The desired path can be either a constant-curvature path or a variable-curvature path. A constant-curvature path is composed of a sequence of circular segments. Considering the axial needle rotations as the main control input, proper changes in the needle tip orientation at each segment steers the needle on the desired path. This can be done by the stop–turn strategy. In this strategy, the insertion and rotation commands are separated, meaning that for each segment, the needle is purely inserted to a certain depth and is stopped to perform the rotation, and the rotated frame is propagated by further insertion (Duindam et al., 2008). In this case, the output of the motion planner is the depth and the magnitude of rotation.

A variable-curvature path is a curved path with different curvatures along it. The strategy for steering the needle on such path is based on duty-cycle spinning. Minhas et al. (2007) showed that by duty-cycle spinning of the needle, paths with different curvatures can be achieved. In this method, the duty cycle, which determines the ratio between the simultaneous insertion and rotation time to the pure insertion time, is found as a function of the desired curvature. This method, however, requires the maximum value of the needle path curvature, which due to tissue inhomogeneity might be variable. Moreira et al. (2014a) have integrated online curvature estimation to 3D planners and duty-cycle spinning to overcome the curvature uncertainties. Majewicz et al. (2014) present two strategies for duty-cycle spinning to overcome the hardware limitations such as cable windup.

There have been different techniques used in the literature for motion planning of the needles such as optimization-based methods, rapidly exploring random tree (RRT) algorithms, and inverse kinematics. In optimization-based methods, efforts are made to optimize a predefined cost function subject to obstacle and needle constraints. One main criterion in clinical procedures is to minimize the trauma imposed on the tissue and to avoid collisions with any sensitive tissues. To this end, the planning method can be formulated as an optimization problem subject to equalities and inequalities constraints. Different optimization methods are used to find the optimal path in the sense of path length, the number of rotations, and distance to the obstacles (Bobrenkov et al., 2014; Wang et al., 2014; Schulman et al., 2014).

The RRT algorithms use the fast exploring methods to find all the feasible points. The RRT algorithm is a randomized path planning method proposed specifically for nonholonomic systems. This method is based on searching the state space, excluding the states that lie in the obstacle regions. Due to the nonholonomic constraints and the maximum curvature of the needle, the reachable region of the needle is a mushroom-like area (Vrooijink et al., 2014), so not all the configurations are reachable from the current configuration. To overcome this limitation, different modifications have been made on the RRT algorithm and used for needle path planning (Vrooijink et al., 2014; Bernardes et al., 2011, 2013, 2014; Shkolnik et al., 2009; Patil and Alterovitz, 2010).

The inverse kinematics of the robotics manipulators determine the relation between the Cartesian space variables and the joint space (Craig, 2005). Similarly, for a bevel-tipped needle the inverse kinematics can be defined as the relation between the needle tip pose, that is, position and orientation, and the translation and orientation at the needle base. This method is usually employed along with other methods such as numerical calculations, space discretization, geometric methods, and optimizations (Duindam et al., 2010; Glozman and Shoham, 2004).

2.5 Control

There are two main differences between the needle steering control and motion planning. In control, the steering is performed in a closed-loop

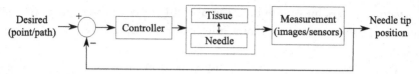

Fig. 3 Control loop for needle steering.

feedback structure. In this method, the controller calculates the control signal in real time based on the error, which is provided to the controller in the form of feedback, as shown in Fig. 3. However, unlike the motion planners that produce a sequence of control inputs for the current and future times, the controller only outputs the control action for the current time. Moreover, in planning methods, the errors caused by uncertainties and tissue non-homogeneity are compensated by replanning the desired path and the control sequence. In control, the desired input to the feedback loop is designed using any off-line path planner, whereas the controller is responsible for correcting the errors and steering the needle on the desired trajectory.

Due to the nonlinearity of the needle kinematics and dynamics, different control strategies such as model-based methods, probabilistic methods, and robust strategies have been applied to needle steering control. From a control perspective, in Haddadi et al. (2010), the controllability of the needle in soft tissue is studied using a dynamic model. Abolhassani et al. (2007) propose a method to perform the axial rotations when the needle deflection reaches some predefined threshold. In this method, the force data and the needle's model (flexible beam model) are employed to calculate the deflection and the rotation locations, and the goal is to keep the needle moving as straight as possible. Using a mechanics-based model of the needle, Khadem et al. (2016) developed a nonlinear model predictive controller for 2D needle steering, which is based on iterative optimization of the predictions. In Abayazid et al. (2013), fiber Bragg grating sensors are used to reconstruct the shape of the needle, which is used as feedback in the steering algorithm. The algorithm uses geometric methods to find the reachable regions and steers the needle such that the target lies in these regions. Considering the probabilistic approaches, Sovizi et al. (2016) propose a planner for 2D environment with obstacles using linear programming (LP). In this method, the uncertain system is approximated by a chain of discrete Markov process, considering the stochastic tissue motion. The optimal solution to this LP problem is found by minimizing the expectation of the cost function. Van Den Berg et al. model the needle motion and the sensor noise as a stochastic process and use linear quadratic Gaussian optimization to obtain a minimal probability of intersection with obstacles. Due to the nonlinearity of the equations, this method linearizes the model around the path and controls the deviation from the that. In this work, different paths are found, from which the optimal one that minimizes the probability of intersections with obstacles is selected. The path is then implemented by duty-cycle spinning. Approximating the probability density function of the needle as a Gaussian variable, Park et al. (2010) use a path planning-based method for steering through intermediate steps using the path of probability algorithm. At each step, the position of the needle is compared to the desired intermediate step and the next step is determined to maximize the probability that next points reach the target.

2.5.1 Sliding mode control in needle steering

Sliding mode control (SMC) is a technique, suitable for systems with disturbances and uncertainties. This method provides a robust approach for reaching the desired performance (Dodds et al., 2015). This is a discontinuous method, in which the control input switches between two limits. If the system under control is of degree n, the SMC redefines the problem as stabilizing a differential equation $S\left(h^{-1}\left(y, \dot{y}, \ldots, y^{(n-1)}\right), y_d\right)$ of order $n-1$, which is a relation between the system output and its derivatives. Here, h is the function relating the system states to the output y and y_d is the desired output value. The desired performance of the closed-loop dynamics is expressed as $S = 0$, which happens by proper design of the controller. The method proposed in Rucker et al. (2013) is a sliding-based method, where the measurements from a five-DOF magnetic tracking sensor are combined with a Kalman filter to estimate the full needle tip orientation. This information is then used to express the Cartesian position error in the needle tip frame. Based on the error, the sliding mode controller constantly rotates the needle and moves it on a helical path to reach the desired target. In the sequel, sliding-based methods are presented, which only require the needle tip position measurement.

For the system equations (4), the sliding functions along x- and y-axes, S_i, $(i = x, y)$, can be defined as

$$S_i = b\dot{e}_i + ce_i \quad (i = x, y) \tag{11}$$

in which $e_x = x - x_d$ and $e_y = y - y_d$. In this function, b and c are constant coefficients, which determine the convergence rate of the error and x_d and y_d

are the desired position along the x- and y-axes, respectively. If x_d and y_d are constant, their first and second time derivatives equal zero, which simplifies the time derivative of Eq. (11) as $\dot{S}_x = b\ddot{x} + c\dot{x}$ and $\dot{S}_y = b\ddot{y} + c\dot{y}$. Consider the Lyapunov function $V_i = \frac{1}{2}S_i^2 (i = x, y)$. The goal is to find the control input such that $\dot{V} = S_i\dot{S}_i < -\eta|S_i|$, which leads to S_i, or equivalently, the error along the desired axes approach zero.

2.5.2 2D switching control
Consider the planar control of the needle, using the Y-Z deflection equation (7) and the two possible inputs for the bevel orientation. It is required to have

$$\left\{ \begin{matrix} \dot{S}_y > \eta & \text{if } S_y < 0 \\ \dot{S}_y < -\eta & \text{if } S_y > 0 \end{matrix} \right\} \tag{12}$$

Using the assumption in Section 2.2, stating that $|\alpha| < \alpha^*$ with $0 < \alpha^* < \pi/4$, the sign of the first term in $\dot{S}_y$ only depends on the bevel orientation. If b and c are selected such that

$$\frac{b\kappa v}{c} > \tan\alpha^* \tag{13}$$

then by selecting $+\kappa$ when $S_y < 0$, and $-\kappa$ when $S_y > 0$, Eq. (12) is satisfied. This means that every time S_y changes sign, the needle should rotate to change the bevel orientation. As a result, when S_y is close to zero, chattering may happen, which increases tissue trauma. To prevent such behavior, the switching can be performed using the switching threshold S_s as shown in Table 1. A smaller switching threshold is beneficial in having smaller errors; however, it increases the number of rotations. Details about selecting the switching threshold can be found in Fallahi et al. (2016b).

Table 1 Selection of the operating mode based on S_y.

Region	System mode	$sign(\dot{S}_y)$
$S_y < -S_s$	Mode 1 ($\kappa > 0$)	+
$-S_s \leq S_y \leq S_s$	No mode change	+ or −
$S_y > S_s$	Mode 2 ($\kappa < 0$)	−

2.5.3 Three-dimensional sliding mode control

In order to extend the 2D method presented in the previous section to 3D environment, two sliding functions are considered and combined for compensating the error in the 3D space. The sliding functions S_x and S_y are defined using Eq. (11) for constant values of x_d and y_d with $c = 1$. Taking the time derivative of the sliding functions and substituting from Eq. (6) gives

$$\dot{S}_x = b_x \kappa v^2 \cos \beta \sin \gamma + v \sin \beta \tag{14}$$

$$\dot{S}_y = -b_y \kappa v^2 \cos \alpha \cos \gamma - b_y \kappa v^2 \sin \beta \sin \alpha \sin \gamma + v \cos \beta \sin \alpha \tag{15}$$

Considering the x-direction, there are two terms on the right-hand side of Eq. (14). If the absolute value of the first term is greater than the second term, that is, $|b_x \kappa v^2 \cos \beta \sin \gamma| > v \sin \beta$, the first term determines the sign of $\dot{S}_x$. According to Eq. (5), since $\cos \beta > 0$, using $\kappa > 0$ and $v > 0$, this condition can be written as

$$|\sin \gamma| > \frac{1}{b_x \kappa v} |\tan \beta| \tag{16}$$

Similarly, for the y-axis, if

$$|\cos \gamma| > |\sin \beta \tan \alpha \sin \gamma| + \frac{1}{b_y \kappa v} |\cos \beta \tan \alpha| \tag{17}$$

the term $-\text{sgn}(\cos \gamma)$ determines the sign of $\dot{S}_y$. Similar to the 2D case, according to $\text{sgn}(\sin \gamma)$ and $\text{sgn}(\cos \gamma)$, there are two possibilities for each direction, leading to the total number of four possibilities (four quadrants) for the 3D case, as shown in Fig. 4. For example, if $S_x > 0$ and $S_y < 0$ require $\dot{S}_x < 0$ and $\dot{S}_y > 0$, which is equivalent to $\sin \gamma < 0$ and $\cos \gamma < 0$, or having γ in the third quadrant. This can be written as

$$\gamma_d = \text{atan2}(-S_x, S_y) \tag{18}$$

In this equation, $\gamma_d \in [-\pi, \pi]$ is the desired roll angle and the function atan2 represents the tangent inverse function considering the sign of the two inputs to return the appropriate quadrant of the calculated angle. However, this is only true if Eqs. (16), (17) are satisfied. To ensure this, the parameters can be selected for the worst-case scenario. Using the assumption $|\alpha| < \alpha^*$ and $|\beta| < \beta^*$, Eqs. (16), (17) can be combined as

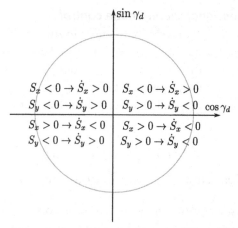

Fig. 4 The corresponding quadrant for γ_d based on sgn(S_x) and sgn(S_y).

$$\arcsin\left(\frac{1}{b_x \kappa v}\tan\beta^*\right) < |\gamma_d|$$

$$< \arccos\left(\sin\beta^*\tan\alpha^* + \frac{1}{b_y \kappa v}\cos\beta^*\tan\alpha^*\right)$$

(19)

in which α and β are substituted by α^* and β^*, respectively. This inequality gives the criteria for selecting the design parameters b_x and b_y. According to the definition of S_i $(i = x, y)$, smaller values of b_i are more desirable for having faster convergence; however, this will force α^* and β^* to be small, limiting the reachable workspace. From Eq. (19), it is clear that b_x, b_y, α^*, and β^* all affect the acceptable value of γ_d. The control input can be found by controlling the angle γ to the desired value γ_d in Eq. (18).

Note that since the needle system does not have any equilibrium points, the controller can only keep the sliding function very close to zero, because as long as the needle is moving the time derivative of the states changes. One way to deal with this property is to focus just on the deflection error and not on its time derivatives. This can be done by using a rotating sliding function, as shown in Fig. 5. The sliding surface (11) defines a line in the phase plane and can be rotated about the origin by changing the slope of the line $\frac{1}{b_i}(i = x, y)$. In Fig. 5, as the sliding function is rotated toward the vertical axis, the error decreases. The sliding function with variable slope is written as

$$\dot{S}_i = b_i \ddot{e}_i + \dot{e}_i(b_i + 1), \quad i = x, y$$

(20)

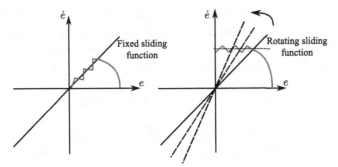

Fig. 5 Phase plane representation of the fixed and rotating sliding functions.

Selecting $b_i = -1$ gives

$$\dot{S}_x = b_x \kappa v \cos \beta \sin \gamma \tag{21}$$

$$\dot{S}_y = -b_y \kappa v \cos \alpha \cos \gamma + b_y \kappa v \sin \beta \sin \alpha \sin \gamma \tag{22}$$

Since b_i $(i = x, y)$ should be positive, the initial condition $b_i(0)$ is selected such that $b_i(0) \geq \frac{D}{v}$, where D is the final insertion depth and v is the insertion velocity. This structure relaxes the constraint (19) as

$$|\tan \gamma_d| < \frac{1}{\sin \beta^* \tan \alpha^*} \tag{23}$$

in which there is no dependence on the needle path curvature κ. Using the rotating sliding function, b_i gets smaller with time, which increases the weight of the position error in the sliding function, leading to smaller errors. The stability proof of this method and more details are provided in Fallahi et al. (2017).

2.5.4 PWM switching and sliding mode control

The application of the sliding mode technique for needle steering is not limited to the methods mentioned earlier. In the following, the application of SMC in an averaged model of the needle steering system is presented. The average-based structure models the 3D needle steering system as a four-mode switching system to transform the continuous input into a switching sequence. In this structure, the 3D system is divided into two 2-mode switching subsystems. Assuming the switching is performed in a pulse width modulation (PWM) structure, the time averaged of each subsystem across the PWM period represents the original subsystem. For each subsystem, a virtual input is defined, which can be designed individually and gives the duty cycle of switching for the subsystem. The duty cycles from both subsystems are then combined to steer the needle in the 3D space. Here, the sliding mode technique is used for designing the controllers for each subsystem, which shows another application of sliding mode technique in the context of needle steering.

The time-averaged model is given by Fallahi et al. (2018)

$$\ddot{x}_a = d_x b_x |\bar{u}_x| - (1 - d_x) b_x |\bar{u}_x| = b_x u_x \tag{24a}$$

$$\ddot{y}_a = d_y (f_y + b_y |\bar{u}_y|) + (1 - d_y)(f_y - b_y |\bar{u}_y|) = f_y + b_y u_y \tag{24b}$$

with

$$u_x = (2d_x - 1)|\bar{u}_x| \tag{25a}$$

$$u_y = (2d_y - 1)|\bar{u}_y| \tag{25b}$$

where $b_x = \kappa v^2 \cos\beta$, $f_y = \kappa v^2 \sin\alpha \sin\beta \sin\gamma_d$, $b_y = -\kappa v^2 \cos\alpha$, $\bar{u}_x = \sin\gamma_d$, and $\bar{u}_y = \cos\gamma_d$. The angle γ_d is the desired roll angle, which for each direction, can take only two fixed values (modes) separated by 180 degrees, such that $\gamma_d \in [0, \pi]$ for the x-axis and $\gamma_d \in [-\pi/2, \pi/2]$ for the y-axis. The previous equations are found with the assumption that the switching between these two modes is performed according to the normalized PWM period $D_i = [d_i \ 1 - d_i]$ with $\|D_i\| = 1$ ($i = x, y$). Here, u_x and u_y are the virtual inputs for the subsystems, which should be designed properly. Regardless of the method used in designing these control signals, these values can be converted into the duty cycles d_x and d_y to determine the duty cycle of switching between the two modes as

$$d_i = \frac{(u_i/|\bar{u}_i| + 1)}{2} \quad i = x, y \tag{26}$$

Since $\bar{u}_x$ and $\bar{u}_y$ are related to $\sin\gamma_d$ and $\cos\gamma_d$, respectively, it is possible to integrate the two 2-mode subsystems to build up a four-mode composite system. These four modes are selected based on the sign of $\sin\gamma_d$ and $\cos\gamma_d$, as shown in Fig. 6A. Through this selection, if $|\tan\gamma_d| = 1$, equal weights are given over the x- and y-directions. This weighting can be changed by selecting $|\tan\gamma_d| > 1$ or $|\tan\gamma_d| < 1$. The switching pattern is shown in Fig. 6B.

In this formulation, u_x and u_y can be designed using different control strategies and then transformed into the duty-cycle variables d_x and d_y to be used in the 3D switching framework. In the sequel, the sliding mode technique is used for designing the control signals u_x and u_y, which provides

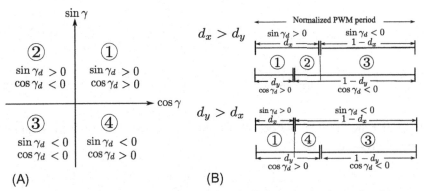

Fig. 6 The four-mode switching pattern. (A) The half plane modes for each subsystem and the resultant four modes: ① $\sin\gamma_d>0, \cos\gamma_d>0$, ② $\sin\gamma_d>0, \cos\gamma_d<0$, ③ $\sin\gamma_d<0, \cos\gamma_d<0$, ④ $\sin\gamma_d<0, \cos\gamma_d>0$. (B) The switching pattern when $d_x > d_y$ and $d_y > d_x$.

a suitable solution to deal with parameter uncertainties and disturbances. Moreover, due to the lack of proper measurements of the angles α and β their uncertainties should be considered in the equations. The terms dependent on these angles appear as bounded trigonometric functions, for which their bounds can be used to define the nominal and uncertain systems. Using the bounds on the needle curvature and the orientation-related terms, the bounds on b_x and b_y can be written as

$$\underline{\kappa} v^2 \cos\beta^* \le b_x \le \overline{\kappa} v^2 \tag{27a}$$

$$\underline{\kappa} v^2 \cos\alpha^* \le b_y \le \overline{\kappa} v^2 \tag{27b}$$

Assuming $\hat{\kappa}$, $\hat{b}_x$, and $\hat{b}_y$ as the nominal values of the parameters defined by the geometric mean of the previous bounds as $\hat{\kappa} = \sqrt{\underline{\kappa}\overline{\kappa}}$, $\hat{b}_x = \hat{\kappa} v^2 \sqrt{\cos\beta^*}$, and $\hat{b}_y = -\hat{\kappa} v^2 \sqrt{\cos\alpha^*}$, the nominal subsystems can be written as

$$\hat{\ddot{x}}_a = \hat{b}_x \hat{u}_x \tag{28}$$

with

$$\left(\frac{\underline{\kappa} \cos\beta^*}{\overline{\kappa}}\right)^{1/2} \le \frac{\hat{b}_x}{b_x} \le \left(\frac{\overline{\kappa} \cos\beta^*}{\underline{\kappa}}\right)^{1/2} \tag{29}$$

Similarly, the nominal subsystem in the y-direction can be written as

$$\hat{\ddot{y}}_a = \hat{f}_y + \hat{b}_y \hat{u}_y \tag{30}$$

with

$$|\hat{f}_y - f_y| \le 2\bar{\kappa}\nu^2 \tag{31a}$$

$$\left(\frac{\kappa \cos \alpha^*}{\bar{\kappa}}\right)^{1/2} \le \frac{\hat{b}_y}{b_y} \le \left(\frac{\bar{\kappa} \cos \alpha^*}{\underline{\kappa}}\right)^{1/2} \tag{31b}$$

Consider the sliding function (11) for the y-direction with $b = 1$. Taking its time derivative gives

$$\dot{S}_y = f_y + b_y u_y - \ddot{y}_d + c\dot{e}_y \tag{32}$$

Using Eq. (30) and solving for $\dot{S}_y = 0$ the control law u_y is found as

$$\hat{u}_y = \frac{1}{\hat{b}_y}\left(-\hat{f}_y + \ddot{y}_d - c\dot{e}_y\right) \tag{33}$$

Similarly for the x-direction

$$\hat{u}_x = \frac{1}{\hat{b}_x}(\ddot{x}_d - c\dot{e}_x) \tag{34}$$

However, these control signals only work for the nominal subsystems. Using the Lyapunov function $V_i = \frac{1}{2}S_i^2$ $(i = x, y)$, in Fallahi et al. (2018) it is shown that the following controllers ensure the convergence of the uncertain system:

$$u_x = \hat{u}_x - \frac{1}{\hat{b}_x}K_x \mathrm{sgn}(S_x) \tag{35a}$$

$$u_y = \hat{u}_y - \frac{1}{\hat{b}_y}K_y \mathrm{sgn}(S_y) \tag{35b}$$

with

$$K_x = \lambda_x \eta + (\lambda_x - 1)|\hat{b}_x \hat{u}_x| \tag{36a}$$

$$K_y = \lambda_y(F_y + \eta) + (\lambda_y - 1)|\hat{b}_y \hat{u}_y| \tag{36b}$$

where $\lambda_x = \left(\frac{\bar{\kappa} \cos \beta^*}{\underline{\kappa}}\right)^{1/2}$, $\lambda_y = \left(\frac{\bar{\kappa} \cos \alpha^*}{\underline{\kappa}}\right)^{1/2}$, $F_y = 2\bar{\kappa}\nu^2$, and $\eta > 0$.

3 Beating-heart surgery

Beating-heart surgery has significant advantages over conventional arrested-heart surgery. However, the fast motion of the beating heart introduces a challenge to the surgeon (human operator). To overcome this obstacle, a mechanical heart stabilizer (Bachta et al., 2009b) is usually used to keep the beating heart from moving. However, this device can only reduce the motion in a localized area on the exterior surface of the beating heart. To minimize the risks of tool-tissue collision and tissue injury, a robot-assisted system is necessary to automatically provide compensation for the fast beating heart's motion, so that to assist the human operator to perform operation accurately and precisely. Indeed, if the robotic system can move a surgical tool in synchrony with the target tissue while the heart beats, the oscillatory forces between the surgical tool and heart tissue may be small, and the human operator can then perform the surgical procedure as if the beating heart is stationary.

The robot-assisted system, however, introduces another issue: haptic feedback. As the human operator cannot make contact with the surgical tool, the tool-tissue interaction forces cannot be perceived by the human operator directly. To provide haptic feedback to the human operator, in some robot-assisted systems, a force sensor is attached to the end of the surgical robot to register forces. When the surgical robots' motions are synchronized with the hearts motions, the force sensor inertia will cause oscillatory forces, which should not be transmitted to the human operator. In other words, the haptic feedback should only contain the nonoscillatory portion of the environmental forces. To date, the state-of-the-art research on robot-assisted beating-heart surgical systems has been studied to compensate for the hearts motion and reflects nonoscillatory haptic forces to the human operator.

3.1 Related work

Depending on the intended surgical procedures, several robot-assisted surgical systems have been developed, which can be mainly categorized into two groups based on the interaction modes with the human operator (Tavakoli, 2008): handheld surgical robotic systems and teleoperated surgical robotic systems.

Handheld surgical systems require the human operator to hold the surgical system directly, which includes an actuator and a surgical tool attached at the end of the system so that the surgical tool can move with respect to the handle (Kettler et al., 2007; Yuen et al., 2009; Zahraee et al., 2010).

Different from the handheld surgical systems, a teleoperated surgical system involves a master robot that provides position and/or force commands and a slave robot that receives those commands and executes tasks on the heart tissue (Bowthorpe et al., 2014a, b). These systems have been shown to offer lots of advantages such as dexterity, fine and remote manipulation capability, and haptic feedback capability for the human operator. The DaVinci surgical robotic system (Guthart and Salisbury, 2000) by Intuitive Surgical Inc. (Sunnyvale, California, United States) is one of the most prominent commercial teleoperated surgical systems.

Teleoperated surgical systems can be divided into two categories depending on their features. In a unilateral teleoperation system, the human operator loses the sense of touch. In contrast, in a bilateral teleoperation system, the human operator can feel the interaction force between the slave robot and what it is touching, enabling the human operator to efficiently manipulate the master robot to provide appropriate commands.

Conventional surgical tools used in robot-assisted systems for cardiac surgeries are short and rigid. Surgical tools like scissors, forceps, and graspers are usually mounted on the end of the systems to perform surgical tasks. However, during intravascular interventions and minimally invasive surgeries, the dexterity of surgical robots can be enhanced by using flexible, thin, and lightweight surgical tools such as catheters while also reducing trauma, which is a benefit for postoperative recovery (Tavakoli et al., 2007). These flexible surgical tools can be combined with the above robot-assisted systems to perform intended surgical procedures (Kesner and Howe, 2011b; Khoshnam and Patel, 2017).

3.2 Measurements and feedbacks

To address the issue of beating-heart motion compensation, several types of sensors have been used to capture the position of the heart, so that the human operator perceives visual feedback of the surgical site through sensors, and the robotic surgical instruments track the beating heart's motion by utilizing the measured heart positions.

Nakamura et al. (2001) adopted one color camera to provide colorful visual feedback and one monochrome high-speed camera to measure the heart position. The human operator utilized the guidance of those two cameras to demonstrate automatic tracking of a point on the heart that was lit by laser. Ginhoux et al. (2004, 2005) and Gangloff et al. (2006) measured the 2D cardiac motions by using a 500-Hz camera to avoid aliasing. Richa et al. (2010, 2011) and Yang et al. (2015) extended 2D position tracking to 3D position tracking using a stereo camera system. In addition, the high-speed camera has been employed in other literatures about beating-heart surgery as well (Bachta et al., 2009a, 2011; Nakajima et al., 2014; Ruszkowski et al., 2016). These sensors provided real-time and accurate position information to compensate for the rapid movement of the beating heart. However, high-speed cameras can only visualize the outer surface of the heart and are not appropriate for surgeries performed inside the heart.

In Schweikard et al. (2000), a pair of X-ray cameras and an infrared tracking system were combined to obtain the positions of the internal markers attached to the heart tissue. Similarly, Mansouri et al. (2018) used an infrared tracker system to locate the 3D positions of the heart. These methods require passive markers attached on the point of interest of the heart tissue, which may be affected during tool-tissue interaction and further operations.

Another common sensor used for guiding intracardiac beating-heart repairs is US machine. Yuen et al. (2008, 2009) developed a 3D US-guided motion compensation system for beating-heart mitral valve repair. Kesner and Howe (2014) applied a robotic catheter system combining US guidance and force control to perform cardiac tissue ablation. Bowthorpe and Tavakoli (2016a, b, c) and Cheng and Tavakoli (2018b) developed a master-slave teleoperated system and combined US images with various controllers to compensate for the beating heart's motion. The acquisition and processing of US images cause a large time delay, which needs to be compensated for via control.

In addition to various image-based sensors, nonimage-based sensors such as force sensors and sonomicrometry crystals are proposed to solve the problem of motion compensation and/or haptic feedback.

Moreira et al. (2012, 2014b), Dominici and Cortesao (2014a, b), and Cortesao and Dominici (2017) utilized force sensors to compensate for the physiological motion by controlling the contact forces to track the desired ones. These methods were assumed that the surgical robot has somehow been initially controlled to come into contact with the heart tissue, and the control goals are maintaining contact between the tool and the tissue.

Tuna et al. (2013) and Bebek and Cavusoglu (2007) used sonomicrometry crystals to track the beating-heart motion in real time and generalized adaptive predictors to predict the hearts motion. By putting six and one sonomicrometry crystals under and on the surface of the heart, the electrocardiogram (ECG) biological signals of the heart surface can be measured based on the transmission and reception of US signals. This technique is feasible as the heart position can be captured through blood, although the calculation is complex and time consuming.

3.3 Control

Various position-, force-, and impedance-based control methods have been proposed for enabling tool-tissue motion compensation for beating-heart surgery and nonoscillatory haptic feedback in teleoperation systems.

3.3.1 Position-based control methods

The position-based controllers need the current beating heart's position and can be classified into predictive feed-forward controllers and predictive feedback controllers. Predictive feed-forward controllers use the hearts position as the set point to move the surgical tools. Predictive feedback controllers not only need the heart's current position but also take the tracking error into account.

Bebek and Cavusoglu (2007) proposed a control algorithm based on the previous quasiperiodic heart motions which are ECG signals detected through sonomicrometry crystals. Yuen et al. (2009) collected the heart positions from US images, and employed an extended Kalman filter (EKF) to compensate the time delay caused by image acquisition and processing. This method took advantage of the quasiperiodicity of the heart motion and modeled the heart motion as a time-varying Fourier series. Many of the predictive feed-forward controllers are used for handheld systems.

To further compensate for the position tracking errors, predictive feedback controllers are used. Bowthorpe et al. (2014b) developed a teleoperation system and proposed a feedback controller with a modified Smith predictor to ensure the distance between the surgical tool and the heart at desired values as commanded by the human operators hand position. Bowthorpe and Tavakoli (2016c) presented three different Smith-predictor-based feedback controllers to tackle issues such as time delays, different measurement rates, and unregistered sensor data.

3.3.2 Force-based control methods

Force control methods are benefit for applications that require contact such as heart biopsy with controlled depth. Considering the process of tool-tissue interaction in robot-assisted beating-heart surgery, precisely applying forces on the beating-heart tissue and enabling the surgical robot to comply with the beating heart's motion simultaneously is important. Therefore, several force control methods were proposed.

Moreira et al. (2012, 2014b) proposed a force control method using active observer based on a viscoelastic interaction model to compensate for the physiological motion. Dominici and Cortesao achieved motion compensation by designing a cascade model predictive control architecture with a Kalman active observer (Dominici and Cortesao, 2014a, b), and a double active observer architecture (Cortesao and Dominici, 2017). These systems use similar feedback controllers. In addition, Yuen et al. (2010) and Kesner and Howe (2011a, 2014) separately combined the US image guidance with a force controller incorporating a feed-forward term containing the estimated motion of the beating heart. These methods incorporated position control and force control to achieve beating-heart motion compensation.

Much of the earlier work focus on handheld systems instead of teleoperation systems, which are possible to enable haptic feedback to the human operator. Haptic feedback during a surgical operation is significant for the human operator to be able to accurately execute the surgical tasks especially in beating-heart surgeries involving tissue cutting and sewing, grasping, dissection, etc. (Wagner et al., 2007). During the operation of such surgical tasks, the tool-tissue interaction forces should be within a safety range to avoid potential tissue injury. To enable the human operator to perceive appropriate haptic feedback under contact, bilateral teleoperation systems were studied. As discussed earlier, the issue of oscillatory haptic feedback caused by force sensor inertia should be considered. For instance, Nakajima et al. (2014) performed haptic feedback using an acceleration-based bilateral control system. Mohareri et al. (2014) developed a force feedback control system for bimanual telerobotic surgery using the DaVinci surgical system (Intuitive Surgical Inc., Sunnyvale, California, United States).

3.3.3 Impedance-based control methods

Most successful applications of robot-assisted surgical systems to date have been performed based on the position or force control, in which the surgical robot is treated essentially as an isolated system. However, in robot-assisted beating-heart surgeries, control of the dynamic behavior between the surgical robot and the beating-heart tissue is also required. Given the beating heart contains inertial objects, the surgical robot and the beating heart can be expressed as an impedance and admittance, respectively (Hogan, 1984, 1985). Generally, the beating heart can be regarded as a source of disturbances "to the surgical robot, and the disturbance response" of the surgical robot can be modulated to control the dynamic behavior between the surgical robot and the beating-heart tissue by varying the parameters and/or structure of the impedance.

Zarrouk et al. (2010) proposed an adaptive control architecture based on the model reference adaptive control to solve the 3D physiological motion compensation in beating-heart surgery. A reference impedance model and an adaptive controller were designed for the surgical robot. In Cheng and Tavakoli (2018a, b), Sharifi et al. (2018), and Cheng et al. (2018), the model reference adaptive control was applied to a bilateral teleoperation system. The authors designed two reference impedance models for the master and slave robots to simultaneously make the slave robot compensate for the heart motion and ensure the human operator to perceive nonoscillatory haptic feedback. The main advantage of impedance-based control system is the desired performance can be achieved via appropriate parameter adjustment of the reference impedance models without any measurement or estimation of the beating heart's motion. In the following, two reference impedance model-based teleoperation systems are presented to describe the applications of observation and feedback control in robotic systems for beating-heart surgery.

Bilateral impedance control: The developed bilateral impedance-controlled teleoperation system is shown in Fig. 7. Here, f_h is the interaction force between the master robot and the human operator and f_e is the interaction

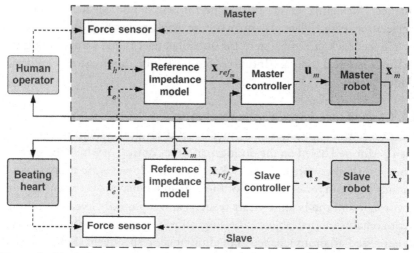

Fig. 7 The block diagram of the bilateral impedance-controlled teleoperation system with two reference impedance models for the master and slave robots.

force between the slave robot and the beating heart. They are measured directly through two force sensors. Also, x_{ref_m} and x_{ref_s} are the desired position for the master and slave robots and x_m and x_s are the actual position of the master and slave robots, respectively. The controllers receive the position errors between the desired positions generated by the reference models and the actual positions read from the robots and then output torque u_m and u_s to the robots.

The reference impedance models for the master and slave robots are designed and can be expressed as

$$m_m \ddot{x}_{ref_m} + c_m \dot{x}_{ref_m} + k_m x_{ref_m} = f_h - k_f f_e \qquad (37)$$

$$m_s \ddot{\tilde{x}}_{ref_s} + c_s \dot{\tilde{x}}_{ref_s} + k_s \tilde{x}_{ref_s} = -f_e \qquad (38)$$

where $\tilde{x}_{ref_s} = x_{ref_s} - k_p x_m$, and k_p is the position scaling factor. In the earlier, k_s, c_s and k_m, c_m, m_m are the virtual stiffness, damping, and mass parameters of the slave, and master impedance models, respectively. Also, k_f is the force scaling factor.

In order to accomplish the desired objectives, the parameters for the two reference impedance models should be adjusted appropriately. The damping ratios and the natural frequencies of the reference impedance models are introduced. Given one of the stiffness, damping, and mass parameters, the rest two parameters can be calculated through $\zeta_i = c_i / 2\sqrt{m_i k_i}$ and $\omega_{n_i} = \sqrt{k_i / m_i}$ ($i = m$ for the master, and $i = s$ for the slave). To ensure the impedance models have fast behaviors in response to the force inputs and small overshoots in response to the step force inputs, the damping ratios are set as 0.7.

The master impedance model (37) should be designed to provide feedback of the nonoscillatory part of the slave-heart interaction force for the human operator. For this purpose, the stiffness parameter (k_m) of the master impedance model should be chosen small, and the natural frequency of the master impedance model (ω_{n_m}) should be much lower than the frequency of the beating heart ω_h which has a range of 6.28–10.68 rad/s ($\omega_{n_m} \ll \omega_h$).

The slave impedance model (38) should be adjusted such that the slave robot complies with the physiological force during the interaction procedure. The stiffness of the slave impedance model (k_s) should be moderate as a too small value will lead to a super flexible slave robot and too large value will make the slave robot very rigid. The natural frequency of the slave impedance model (ω_{n_s}) should be much greater than the frequency of the heart ($\omega_{n_s} \gg \omega_h$). Detailed parameter adjustments can be found in

Table 2 Parameter adjustments of the master and slave impedance models

Characteristics	Master impedance adjustment	Slave impedance adjustment
Stiffness	$k_m = 5$ N/m	$k_s = 100$ N/m
Damping ratio	$\zeta_m = 0.7$	$\zeta_s = 0.7$
Natural frequency	$\omega_{n_m} = 0.5$ rad/s	$\omega_{n_s} = 50$ rad/s
Damping and mass	$m_m = 20$ kg, $c_m = 14$ Ns/m	$m_s = 0.04$ kg, $c_s = 2.8$ Ns/m
Scaling factor	$k_f = 1$	$k_p = 1$

Cheng et al. (2018). The adjusted parameters of the two impedance models are listed in Table 2.

By focusing on the direction of the major component of heart motion the proposed bilateral impedance-controlled teleoperation system (with motion compensation) was compared to the regular direct force reflection (DFR) teleoperation system (without motion compensation) (Liu and Tavakoli, 2011), which reflects the entire slave-heart contact force to the human operator and requires the operator to take care of the motion compensation manually. Fig. 8 shows the positions and forces during slave-heart interaction.

Without motion compensation, it is difficult to synchronize the slave robot with the beating heart. Comparatively, with motion compensation, the oscillatory force of the moving tissue was filtered, and only a stable baseline contact force was perceived by the human operator. Also, the position deviation between the slave robot and the heart during contact was very small. The human operator was easier to perform tasks, and the human-master interaction forces were steady and small.

Ultrasound image guidance and robot impedance control: In the earlier, as the stiffness of the slave impedance model (k_s) is chose to be moderate, the force applied on the heart tissue will not be very large, which limits the system's application. To this end, Cheng and Tavakoli (2018b) proposed that to combine the robot impedance control with US imaging algorithm to achieve the two objectives of the system (Fig. 9). The US imaging-based control algorithms are used to make motion compensation. The impedance model for the master robot is designed to provide the human operator with a feeling of operating on an idle heart.

In the slave site, an US imaging machine is used to obtain the position of the beating heart x_e. The summed position of the master robot x_m and the heart x_e transmitted to the slave robot as a reference signal $x_{ref_s}(= x_m + x_e)$. And then a slave controller is used to make the position of the slave robot x_s follow its reference trajectory x_{ref_s}. In the master site, the human-master interaction force f_h and the slave-heart interaction force f_e are transmitted directly to a reference impedance model, which as we discussed earlier

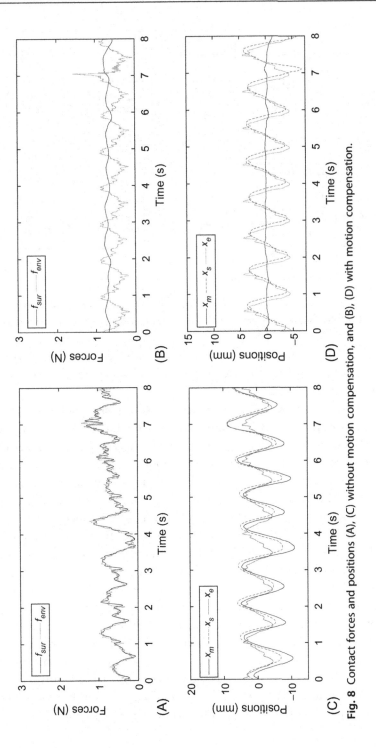

Fig. 8 Contact forces and positions (A), (C) without motion compensation, and (B), (D) with motion compensation.

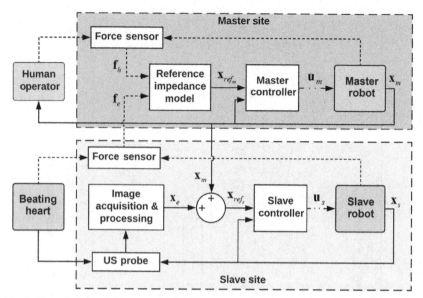

Fig. 9 The teleoperation system with US image guidance and robot impedance control.

can filter out the high-frequency portion of f_e and achieve f_h equals the filtered f_e. The impedance model generates a reference position x_{ref_m} for the master robot to follow.

For the sake of brevity, we will focus on the direction of the major component of heart motion. The motion compensation system is designed to make the slave robot to follow the combined trajectory of the master robot and the beating heart. The beating-heart position can be calculated based on the position of the slave robot and the measured robot-heart distance captured by US imaging along the surgical tools axis. The slow sampled robot-heart distance can be measured directly from each US image (the detailed algorithm is in Cheng and Tavakoli, 2018b). As the low sampling rate of the US image, the measured robot-heart distance needs to be upsampled to the system control sampling rate first by using cubic polynomial interpolation. Then, the delayed upsampled heart position can be obtained by delaying the position of the slave robot and adding it to the upsampled robot-heart distance. To further compensate for the time delay, the delayed quasiperiodic heart position is modeled as a time-varying Fourier series and predicted by an EKF. The reference impedance model for the master robot is designed the same as shown in the bilateral impedance control.

In experiments, the proposed method (Fig. 10C and F) was compared with the regular DFR teleoperation without and with automatic motion

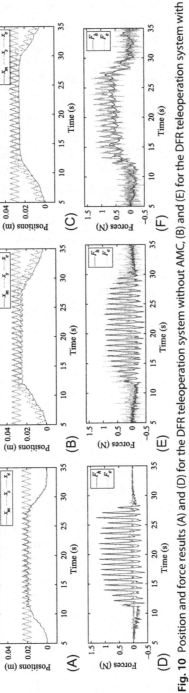

Fig. 10 Position and force results (A) and (D) for the DFR teleoperation system without AMC, (B) and (E) for the DFR teleoperation system with AMC, and (C) and (F) for the proposed teleoperation system.

compensation (AMC). The DFR teleoperation system without AMC (Fig. 10A and D) requires the human operator to perform motion compensation manually, while the DFR teleoperation system with AMC (Fig. 10B and E) compensates for the hearts motion automatically.

In Fig. 10A and D, the slave robot tracks both positions and forces of the master robot during the entire operation. However, the tracking of the beating-heart motion is poor as the human operator must manually compensate for the heart motion. Moreover, the oscillatory human-master interaction force suggests that the human operator receives unsteady haptic feedback, which makes it more challenging to synchronize the motion of the slave robot with the hearts motion. In Fig. 10B and E, the position tracking result is much better in this case than that in the first case. Nevertheless, the haptic feedback to the human operator is still oscillatory. An oscillatory motion with small amplitude remains in the master robot position due to the poor quality of haptic feedback. In Fig. 10C and F, the position tracking result is significantly better than that in Fig. 10A and D. Both the position and force of the master robot are much steadier as the oscillatory portions have been filtered using the proposed impedance model for the master robot. The human operator is able to operate on a beating heart without manual compensation, and simultaneously has a sense of operation on a seemingly idle heart.

4 Discussion

The discussion of this chapter is twofold: Section 2 addresses the needle insertion procedures and Section 3 discusses the challenges in beating heart surgery. The control strategies mentioned in Section 2, show the application of the sliding mode control in needle steering. The 2D controller in Section 2.5.2 is supplied with the needle deflection error obtained from ultrasound images and switches the system to the proper mode for reducing the targeting error. The constraints on switching parameters are derived using kinematic unicycle equations for the needle to ensure the stability of the system and convergence of the error. Similarly, the 3D controller in Section 2.5.3 finds the desired value of the needle base angle to steer the needle on the desired path in the 3D environment and to reduce the number of rotations and the tissue trauma. In this method, for each subsystem, a sliding surface is defined, and the two sliding surfaces are combined. In these methods the controller is not dependent on the exact knowledge of the system parameters, and only the maximum value of the needle path curvature is needed. These structures represent simple and non-model-based control strategies to improve needle tip positioning.

The sliding mode controller in Section 2.5.4 is designed for a special framework, where the 3D needle equations are divided into two subsystems, representing the in-and out-of-plane motions. Each subsystem is considered as a planar switching system with two modes determining the two possible bevel orientations, which are 180 degrees apart. Assuming the switching between the two modes is performed according to some duty cycle period in a PWM framework, the performance of each subsystem is approximated by the averaged subsystem in the PWM period. Each averaged subsystem has its virtual input, for which controllers are designed using sliding mode control technique.

These mentioned methods use the position data obtained from ultrasound images. However, since the path traveled by the needle depends on the needle tip orientation, having information about the needle tip orientation will be helpful in the controller design process. Due to the small diameter of the needle and the low resolution of the images, it is not possible to retrieve the orientation information from the ultrasound images. This problem is dealt with by using state observers as explained in Section 2.3, where linear and nonlinear observers are introduced for estimating the needle tip orientation. In this method, nonlinear transformations are applied on a 3D unicycle model. Since these equations do not satisfy the Lipschitz continuity condition, designing a convergent observer is not possible. Further studies are required in this field for designing nonlinear observers and their stable combination with controllers.

Regarding the beating heart surgery, in Section 3, two impedance-based teleoperation systems were presented to achieve heart motion compensation and nonoscillatory robot-heart interaction feedback simultaneously. The bilateral impedance-controlled teleoperation system designed two reference impedance models for the master and slave robots, respectively. By tuning the parameters of the impedance models, the slave robot was able to comply with the movement of the beating heart, and the oscillatory portion of the slave-heart interaction force was filtered out, which made the human operator only perceive the nonoscillatory contact force. In the experiments, compared to the conventional DFR teleoperation system, the proposed system was able to provide the human operator a feeling of operating on an arrested heart and make the anchor deployment task much easier to perform. It should be noted that for the parameters of the slave robot's impedance model, they were adjusted to be moderate because too small values would not apply enough forces to the heart and too large values would lead to the motion compensation be inaccurate. Therefore, this system is more suitable for surgeries that require less slave-heart contact forces such as mitral valve annuloplasty, blunt resection, ablation, etc.

To extend the applications for beating-heart surgery, the robot imped- ance control was combined with US image guidance, and the second system was proposed. This teleoperation system retained the reference impedance model for the master robot to attain nonoscillatory force feedback but rep- laced the reference impedance model for the slave robot with US image guidance for position control purposes. In this system, the slave robot pro- vided motion compensation for the heart motion by following the position of the heart which was measured through US images. The low frame rate and time delay caused by image acquisition and processing were addressed by cubic polynomial interpolation and EKF, respectively. As the reference trajectory for the slave robot is the sum trajectory of the master robot and beating heart, large slave-heart contact force can be exerted on the heart tis- sue by increasing the master robot's position commands. Consequently, the applications of this system can be extended to surgeries that need large slave- heart contact forces such as tissue cutting, penetration, and so on.

5 Conclusion

The subsections in Section 2 provided an overview of the related observer and controller methods proposed for robot-assisted, image-guided needle steering. The goal of these approaches is to improve the needle tip positioning and increase the efficiency of the clinical needle insertion pro- cedures. From the control perspective, the main challenges in designing a controller for needle/tissue system arise from the under-actuation property and the nonholonomic constraints imposed on the needle kinematics. Moreover, due to the small diameter of the needle and the low resolution of the ultrasound images, it is not possible to retrieve the orientation infor- mation from the ultrasound images, which can be dealt with by using state observers. The proposed control methods can be further expanded to have moving targets and obstacles, online estimation of the system parameters and observer/controller combination for trajectory tracking in the 3D environ- ment. Further developments and experiments are required to verify the application of the proposed structures in clinical settings.

In Section 3, another application of robotic assistive systems in beating- heart surgery is discussed. A robot-assisted beating-heart surgical system has the potential to improve the outcome of many surgical procedures performed on the heart. There are different ways to control such a system as each surgical procedure has different requirements. For instance, the procedure could be ablation which requires small exerted force or tissue cutting which requires large exerted force. This would affect the choice of the motion-capture module and, in turn, affects the choice of a controller. Based on the requirements for a specific surgical procedure, the robot-assisted beating-heart surgical system can be designed. The controllers for the system should automatically compensate for the beating heart's motion while providing nonoscillatory haptic feedback for the human operator.

References

Abayazid, M., Kemp, M., Misra, S., 2013. 3D flexible needle steering in soft-tissue phantoms using fiber Bragg grating sensors. In: IEEE International Conference on Robotics and Automation (ICRA), IEEE, pp. 5843–5849.

Abolhassani, N., Patel, R.V., Ayazi, F., 2007. Minimization of needle deflection in robot-assisted percutaneous therapy. Int. J. Med. Robot. Comput. Assist. Surg. 3 (2), 140–148.

Angelini, G.D., Taylor, F.C., Reeves, B.C., Ascione, R., 2002. Early and midterm outcome after off-pump and on-pump surgery in beating heart against cardioplegic arrest studies (BHACAS 1 and 2): a pooled analysis of two randomised controlled trials. Lancet 359 (9313), 1194–1199.

Asadian, A., Patel, R.V., Kermani, M.R., 2011. A distributed model for needle-tissue friction in percutaneous interventions. In: 2011 IEEE International Conference on Robotics and Automation (ICRA), IEEE, pp. 1896–1901.

Bachta, W., Renaud, P., Cuvillon, L., Laroche, E., Forgione, A., Gangloff, J., 2009. Motion prediction for computer-assisted beating heart surgery. IEEE Trans. Biomed. Eng. 56 (11), 2551–2563.

Bachta, W., Renaud, P., Laroche, E., Gangloff, J., 2009. Cardiolock2: parallel singularities for the design of an active heart stabilizer. ICRA, pp. 3839–3844.

Bachta, W., Renaud, P., Laroche, E., Forgione, A., Gangloff, J., 2011. Active stabilization for robotized beating heart surgery. IEEE Trans. Robot. 27 (4), 757–768.

Bebek, O., Cavusoglu, M.C., 2007. Intelligent control algorithms for robotic-assisted beating heart surgery. IEEE Trans. Robot. 23 (3), 468–480.

Bellinger, D.C., Wypij, D., Kuban, K.C.K., Rappaport, L.A., Hickey, P.R., Wernovsky, G., Jonas, R.A., Newburger, J.W., 1999. Developmental and neurological status of children at 4 years of age after heart surgery with hypothermic circulatory arrest or low-flow cardiopulmonary bypass. Circulation 100 (5), 526–532.

Bernardes, M.C., Adorno, B.V., Poignet, P., Zemiti, N., Borges, G.A., 2011. Adaptive path planning for steerable needles using duty-cycling. In: 2011 IEEE/RSJ International Conference on Intelligent Robots and Systems (IROS), IEEE, pp. 2545–2550.

Bernardes, M.C., Adorno, B.V., Poignet, P., Borges, G.A., 2013. Robot-assisted automatic insertion of steerable needles with closed-loop imaging feedback and intraoperative trajectory replanning. Mechatronics 23 (6), 630–645.

Bernardes, M.C., Adorno, B.V., Borges, G.A., Poignet, P., 2014. 3D robust online motion planning for steerable needles in dynamic workspaces using duty-cycled rotation. J. Control Autom. Electr. Syst. 25 (2), 216–227.

Bobrenkov, O.A., Lee, J., Park, W., 2014. A new geometry-based plan for inserting flexible needles to reach multiple targets. Robotica 32 (6), 985–1004.

Bowthorpe, M., Tavakoli, M., 2016. Generalized predictive control of a surgical robot for beating-heart surgery under delayed and slowly-sampled ultrasound image data. IEEE Robot. Autom. Lett. 1 (2), 892–899.

Bowthorpe, M., Tavakoli, M., 2016. Physiological organ motion prediction and compensation based on multirate, delayed, and unregistered measurements in robot-assisted surgery and therapy. IEEE/ASME Trans. Mechatron. 21 (2), 900–911.

Bowthorpe, M., Tavakoli, M., 2016. Ultrasound-based image guidance and motion compensating control for robot-assisted beating-heart surgery. J. Med. Robot. Res. 1 (1), 1640002.

Bowthorpe, M., Castonguay-Siu, V., Tavakoli, M., 2014. Development of a robotic system to enable beating-heart surgery. J. Robot. Soc. Jpn 32 (4), 339–346.

Bowthorpe, M., Tavakoli, M., Becher, H., Howe, R., 2014. Smith predictor-based robot control for ultrasound-guided teleoperated beating-heart surgery. IEEE J. Biomed. Health Inform. 18 (1), 157–166.

Cheng, L., Tavakoli, M., 2018. Switched-impedance control of surgical robots in teleoperated beating-heart surgery. J. Med. Robot. Res. 3, 1841003.

Cheng, L., Tavakoli, M., 2018. Ultrasound image guidance and robot impedance control for beating-heart surgery. Control. Eng. Pract. 81, 9–17.

Cheng, L., Sharifi, M., Tavakoli, M., 2018. Towards robot-assisted anchor deployment in beating-heart mitral valve surgery. Int. J. Med. Robot. Comput. Assist. Surg. 14 (3), e1900.

Chentanez, N., Alterovitz, R., Ritchie, D., Cho, L., Hauser, K.K., Goldberg, K., Shewchuk, J.R., O'Brien, J.F., 2009. Interactive simulation of surgical needle insertion and steering. ACM Trans. Graph. 28(3). https://doi.org/10.1145/1531326.1531394.

Cortesao, R., Dominici, M., 2017. Robot force control on a beating heart. IEEE/ASME Trans. Mechatron. 22 (4), 1736–1743.

Craig, J.J., 2005. Introduction to Robotics: Mechanics and Control. vol. 3 Pearson Prentice Hall, Upper Saddle River, NJ.

Dacey, L.J., Likosky, D.S., Leavitt, B.J., Lahey, S.J., Quinn, R.D., Hernandez Jr., F., Quinton, H.B., Desimone, J.P., Ross, C.S., O'Connor, G.T., et al., 2005. Perioperative stroke and long-term survival after coronary bypass graft surgery. Ann. Thorac. Surg. 79 (2), 532–536.

Dehghan, E., Salcudean, S.E., 2009. Needle insertion parameter optimization for brachytherapy. IEEE Trans. Robot. 25 (2), 303–315.

DiMaio, S.P., Salcudean, S.E., 2003. Needle insertion modeling and simulation. IEEE Trans. Robot. Autom. 19 (5), 864–875.

DiMaio, S.P., Salcudean, S.E., 2005. Interactive simulation of needle insertion models. IEEE Trans. Biomed. Eng. 52 (7), 1167–1179.

Dodds, S.J., et al., 2015. Feedback Control. Springer, London, p. 5.

Dominici, M., Cortesao, R., 2014. Cascade robot force control architecture for autonomous beating heart motion compensation with model predictive control and active observer. In: 2014 5th IEEE RAS & EMBS International Conference on Biomedical Robotics and Biomechatronics, IEEE, pp. 745–751.

Dominici, M., Cortesao, R., 2014. Model predictive control architectures with force feedback for robotic-assisted beating heart surgery. In: 2014 IEEE International Conference on Robotics and Automation (ICRA), IEEE, pp. 2276–2282.

Duan, Y., Patil, S., Schulman, J., Goldberg, K., Abbeel, P., 2014. Planning locally optimal, curvature-constrained trajectories in 3D using sequential convex optimization. In: 2014 IEEE International Conference on Robotics and Automation (ICRA), IEEE, pp. 5889–5895.

Duindam, V., Alterovitz, R., Sastry, S., Goldberg, K., 2008. Screw-based motion planning for bevel-tip flexible needles in 3D environments with obstacles. In: IEEE International Conference on Robotics and Automation, 2008. ICRA 2008, IEEE, pp. 2483–2488.

Duindam, V., Xu, J., Alterovitz, R., Sastry, S., Goldberg, K., 2010. Three-dimensional motion planning algorithms for steerable needles using inverse kinematics. Int. J. Robot. Res. 29 (7), 789–800.

Fallahi, B., Khadem, M., Rossa, C., Sloboda, R., Usmani, N., Tavakoli, M., 2015. Extended bicycle model for needle steering in soft tissue. 2015 IEEE/RSJ International Conference on Intelligent Robots and Systems (IROS), IEEE, pp. 4375–4380.

Fallahi, B., Rossa, C., Sloboda, R., Usmani, N., Tavakoli, M., 2016. Partial estimation of needle tip orientation in generalized coordinates in ultrasound image-guided needle insertion. In: IEEE International Conference on Advanced Intelligent Mechatronics (AIM), 2016, IEEE, pp. 1604–1609.

Fallahi, B., Rossa, C., Sloboda, R.S., Usmani, N., Tavakoli, M., 2016. Sliding-based switching control for image-guided needle steering in soft tissue. IEEE Robot. Autom. Lett. 1 (2), 860–867.

Fallahi, B., Rossa, C., Sloboda, R.S., Usmani, N., Tavakoli, M., 2017. Sliding-based image-guided 3D needle steering in soft tissue. Control. Eng. Pract. 63, 34–43.

Fallahi, B., Sloboda, R., Usmani, N., Tavakoli, M., 2018. Model averaging and input transformation for 3D needle steering. J. Med. Robot. Res. 3, 1841004.

Fix, J., Isada, L., Cosgrove, D., Miller, D.P., Savage, R., Blum, J., Stewart, W., 1993. Do patients with less than "echo-perfect" results from mitral valve repair by intraoperative echocardiography have a different outcome? Circulation 88 (5 Pt. 2), II39–II48.

Gangloff, J., Ginhoux, R., de Mathelin, M., Soler, L., Marescaux, J., 2006. Model predictive control for compensation of cyclic organ motions in teleoperated laparoscopic surgery. IEEE Trans. Control Syst. Technol. 14 (2), 235–246.

Ginhoux, R., Gangloff, J.A., De Mathelin, M.F., Soler, L., Sanchez, M.M.A., Marescaux, J., 2004. Beating heart tracking in robotic surgery using 500 Hz visual servoing, model predictive control and an adaptive observer. In: 2004 IEEE International Conference on Robotics and Automation, 2004. Proceedings. ICRA'04, vol. 1. IEEE, pp. 274–279.

Ginhoux, R., Gangloff, J., de Mathelin, M., Soler, L., Sanchez, M.M.A., Marescaux, J., et al., 2005. Active filtering of physiological motion in robotized surgery using predictive control. IEEE Trans. Robot. 21 (1), 67–79.

Glozman, D., Shoham, M., 2004. Flexible needle steering and optimal trajectory planning for percutaneous therapies. International Conference on Medical Image Computing and Computer-Assisted Intervention, Springer, pp. 137–144.

Goksel, O., Salcudean, S.E., Dimaio, S.P., 2006. 3D simulation of needle-tissue interaction with application to prostate brachytherapy. Comput. Aided Surg. 11 (6), 279–288.

Guthart, G.S., Salisbury, J.K., 2000. The intuitive^{TM} telesurgery system: overview and application. IEEE International Conference on Robotics and Automation, 2000. Proceedings. ICRA'00, vol. 1. IEEE, pp. 618–621.

Haddadi, A., Goksel, O., Salcudean, S.E., Hashtrudi-Zaad, K., 2010. On the controllability of dynamic model-based needle insertion in soft tissue. In: 2010 Annual International Conference of the IEEE Engineering in Medicine and Biology Society (EMBC), IEEE, pp. 2287–2291.

Hogan, N., 1984. Impedance control: an approach to manipulation. American Control Conference, 1984, IEEE, pp. 304–313.

Hogan, N., 1985. Impedance control: an approach to manipulation: part II—implementation. J. Dyn. Syst. Meas. Control 107 (1), 8–16.

Kallem, V., Cowan, N.J., 2009. Image guidance of flexible tip-steerable needles. IEEE Trans. Robot. 25 (1), 191–196.

Kallem, V., Chang, D.E., Cowan, N.J., 2011. Observer design for needle steering using task-induced symmetry and reduction. IFAC Proc. Vol. 44 (1), 8028–8033.

Kaya, M., Bebek, O., 2014. Needle localization using Gabor filtering in 2D ultrasound images. In: 2014 IEEE International Conference on Robotics and Automation (ICRA), IEEE, pp. 4881–4886.

Kesner, S.B., Howe, R.D., 2011. Force control of flexible catheter robots for beating heart surgery. IEEE International Conference on Robotics and Automation. ICRA, NIH Public Access, p. 1589.

Kesner, S.B., Howe, R.D., 2011. Position control of motion compensation cardiac catheters. IEEE Trans. Robot. 27 (6), 1045–1055.

Kesner, S.B., Howe, R.D., 2014. Robotic catheter cardiac ablation combining ultrasound guidance and force control. Int. J. Robot. Res. 33 (4), 631–644.

Kettler, D.T., Plowes, R.D., Novotny, P.M., Vasilyev, N.V., Pedro, J., Howe, R.D., 2007. An active motion compensation instrument for beating heart mitral valve surgery. In: IEEE/RSJ International Conference on Intelligent Robots and Systems, 2007. IROS 2007, IEEE, pp. 1290–1295.

Khadem, M., Rossa, C., Sloboda, R.S., Usmani, N., Tavakoli, M., 2016. Ultrasound-guided model predictive control of needle steering in biological tissue. J. Med. Robot. Res. 1 (1), 1640007.

Khoshnam, M., Patel, R.V., 2017. Robotics-assisted control of steerable ablation catheters based on the analysis of tendon-sheath transmission mechanisms. IEEE/ASME Trans. Mechatron. 22 (3), 1473–1484.

Liu, X., Tavakoli, M., 2011. Adaptive inverse dynamics four-channel control of uncertain nonlinear teleoperation systems. Adv. Robot. 25 (13–14), 1729–1750.

Majewicz, A., Siegel, J.J., Stanley, A.A., Okamura, A.M., 2014. Design and evaluation of duty-cycling steering algorithms for robotically-driven steerable needles. In: IEEE International Conference on Robotics and Automation (ICRA), IEEE, pp. 5883–5888.

Malekian, L., Talebi, H.A., Towhidkhah, F., 2013. Needle detection in 3D ultrasound images using anisotropic diffusion and robust fitting. International Symposium on Artificial Intelligence and Signal Processing, Springer, pp. 111–120.

Mansouri, S., Farahmand, F., Vossoughi, G., Ghavidel, A.A., 2018. A hybrid algorithm for prediction of varying heart rate motion in computer-assisted beating heart surgery. J. Med. Syst. 42 (10), 200.

Minhas, D.S., Engh, J.A., Fenske, M.M., Riviere, C.N., 2007. Modeling of needle steering via duty-cycled spinning. In: 29th Annual International Conference of the IEEE Engineering in Medicine and Biology Society, 2007. EMBS 2007, IEEE, pp. 2756–2759.

Misra, S., Reed, K.B., Schafer, B.W., Ramesh, K.T., Okamura, A.M., 2010. Mechanics of flexible needles robotically steered through soft tissue. Int. J. Robot. Res. 29 (13), 1640–1660.

Mohareri, O., Schneider, C., Salcudean, S., 2014. Bimanual telerobotic surgery with asymmetric force feedback: a DaVinci® surgical system implementation. In: 2014 IEEE/RSJ International Conference on Intelligent Robots and Systems (IROS 2014), IEEE, pp. 4272–4277.

Moreira, P., Liu, C., Zemiti, N., Poignet, P., 2012. Beating heart motion compensation using active observers and disturbance estimation. IFAC Proc. Vol. 45 (22), 741–746.

Moreira, P., Patil, S., Alterovitz, R., Misra, S., 2014. Needle steering in biological tissue using ultrasound-based online curvature estimation. In: IEEE International Conference on Robotics and Automation (ICRA), 2014, IEEE, pp. 4368–4373.

Moreira, P., Zemiti, N., Liu, C., Poignet, P., 2014. Viscoelastic model based force control for soft tissue interaction and its application in physiological motion compensation. Comput. Methods Prog. Biomed. 116 (2), 52–67.

Motaharifar, M., Talebi, H.A., Abdollahi, F., Afshar, A., 2015. Nonlinear adaptive output-feedback controller design for guidance of flexible needles. IEEE/ASME Trans. Mechatron. 20 (4), 1912–1919.

Nakajima, Y., Nozaki, T., Ohnishi, K., 2014. Heartbeat synchronization with haptic feedback for telesurgical robot. IEEE Trans. Ind. Electron. 61 (7), 3753–3764.

Nakamura, Y., Kishi, K., Kawakami, H., et al., 2001. Heartbeat synchronization for robotic cardiac surgery. ICRA, vol. 2, pp. 2014–2019.

Newman, M.F., Kirchner, J.L., Phillips-Bute, B., Gaver, V., Grocott, H., Jones, R.H., Mark, D.B., Reves, J.G., Blumenthal, J.A., 2001. Longitudinal assessment of neurocognitive function after coronary-artery bypass surgery. N. Engl. J. Med. 344 (6), 395–402.

Paparella, D., Yau, T.M., Young, E., 2002. Cardiopulmonary bypass induced inflammation: pathophysiology and treatment. An update. Eur. J. Cardiothorac. Surg. 21 (2), 232–244.

Park, W., Wang, Y., Chirikjian, G.S., 2010. The path-of-probability algorithm for steering and feedback control of flexible needles. Int. J. Robot. Res. 29 (7), 813–830.

Patil, S., Alterovitz, R., 2010. Interactive motion planning for steerable needles in 3D environments with obstacles. In: 3rd IEEE RAS and EMBS International Conference on Biomedical Robotics and Biomechatronics (BioRob), 2010, IEEE, pp. 893–899.

Reed, K.B., Kallem, V., Alterovitz, R., Goldbergxz, K., Okamura, A.M., Cowan, N.J., 2008. Integrated planning and image-guided control for planar needle steering. In: 2nd IEEE RAS & EMBS International Conference on Biomedical Robotics and Biomechatronics, 2008. BioRob 2008, IEEE, pp. 819–824.

Richa, R., Poignet, P., Liu, C., 2010. Three-dimensional motion tracking for beating heart surgery using a thin-plate spline deformable model. Int. J. Robot. Res. 29 (2–3), 218–230.

Richa, R., Bó, A.P.L., Poignet, P., 2011. Towards robust 3D visual tracking for motion compensation in beating heart surgery. Med. Image Anal. 15 (3), 302–315.

Rossa, C., Sloboda, R., Usmani, N., Tavakoli, M., 2016. Estimating needle tip deflection in biological tissue from a single transverse ultrasound image: application to brachytherapy. Int. J. Comput. Assist. Radiol. Surg. 11 (7), 1347–1359.

Rucker, D.C., Das, J., Gilbert, H.B., Swaney, P.J., Miga, M.I., Sarkar, N., Webster, R.J., 2013. Sliding mode control of steerable needles. IEEE Trans. Robot. 29 (5), 1289–1299.

Ruszkowski, A., Mohareri, O., Lichtenstein, S., Cook, R., Salcudean, S., 2015. On the feasibility of heart motion compensation on the DaVinci® surgical robot for coronary artery bypass surgery: implementation and user studies. 2015 IEEE International Conference on Robotics and Automation (ICRA), IEEE, pp. 4432–4439.

Ruszkowski, A., Schneider, C., Mohareri, O., Salcudean, S., 2016. Bimanual teleoperation with heart motion compensation on the DaVinci® research kit: implementation and preliminary experiments. In: IEEE International Conference on Robotics and Automation (ICRA), 2016, IEEE, pp. 4101–4108.

Schulman, J., Duan, Y., Ho, J., Lee, A., Awwal, I., Bradlow, H., Pan, J., Patil, S., Goldberg, K., Abbeel, P., 2014. Motion planning with sequential convex optimization and convex collision checking. Int. J. Robot. Res. 33 (9), 1251–1270.

Schweikard, A., Glosser, G., Bodduluri, M., Murphy, M.J., Adler, J.R., 2000. Robotic motion compensation for respiratory movement during radiosurgery. Comput. Aided Surg. 5 (4), 263–277.

Sharifi, M., Salarieh, H., Behzadipour, S., Tavakoli, M., 2018. Beating-heart robotic surgery using bilateral impedance control: theory and experiments. Biomed. Signal Process. Control 45, 256–266.

Shkolnik, A., Walter, M., Tedrake, R., 2009. Reachability-guided sampling for planning under differential constraints. In: IEEE International Conference on Robotics and Automation, 2009. ICRA'09, IEEE, pp. 2859–2865.

Sovizi, J., Kumar, S., Krovi, V., 2016. Approximating Markov chain approach to optimal feedback control of a flexible needle. J. Dyn. Syst. Meas. Control 138 (11), 111006.

Swensen, J.P., Lin, M., Okamura, A.M., Cowan, N.J., 2014. Torsional dynamics of steerable needles: modeling and fluoroscopic guidance. IEEE Trans. Biomed. Eng. 61 (11), 2707–2717.

Taghirad, H.D., 2013. Parallel Robots: Mechanics and Control. CRC Press, Boca Raton, FL.

Taschereau, R., Pouliot, J., Roy, J., Tremblay, D., 2000. Seed misplacement and stabilizing needles in transperineal permanent prostate implants. Radiother. Oncol. 55 (1), 59–63.

Tavakoli, M., 2008. Haptics for Teleoperated Surgical Robotic Systems. vol. 1. World Scientific, Singapore.

Tavakoli, M., Aziminejad, A., Patel, R.V., Moallem, M., 2007. High-fidelity bilateral teleoperation systems and the effect of multimodal haptics. IEEE Trans. Syst. Man Cybern. B Cybern. 37 (6), 1512–1528.

Torre, L.A., Bray, F., Siegel, R.L., Ferlay, J., Lortet-Tieulent, J., Jemal, A., 2015. Global cancer statistics, 2012. CA Cancer J. Clin. 65 (2), 87–108.

Tuna, E.E., Franke, T.J., Bebek, O., Shiose, A., Fukamachi, K., Cavusoglu, M.C., 2013. Heart motion prediction based on adaptive estimation algorithms for robotic-assisted beating heart surgery. IEEE Trans. Robot. 29 (1), 261–276.

Uherčík, M., Liebgott, H., Kybic, J., Cachard, C., 2009. Needle localization methods in 3D ultrasound data. International Congress on Ultrasonics, pp. 11–17.

Vrooijink, G.J., Abayazid, M., Patil, S., Alterovitz, R., Misra, S., 2014. Needle path planning and steering in a three-dimensional non-static environment using two-dimensional ultrasound images. Int. J. Robot. Res. 33 (10), 1361–1374.

Wagner, C.R., Stylopoulos, N., Jackson, P.G., Howe, R.D., 2007. The benefit of force feedback in surgery: examination of blunt dissection. Presence Teleop. Virt. Environ. 16 (3), 252–262.

Waine, M., Rossa, C., Sloboda, R., Usmani, N., Tavakoli, M., 2016. Three-dimensional needle shape estimation in TRUS-guided prostate brachytherapy using 2-D ultrasound images. IEEE J. Biomed. Health Inform. 20 (6), 1621–1631.

Wallner, K., Ellis, W., Russell, K., Cavanagh, W., Blasko, J., 1999. Use of TRUS to predict pubic arch interference of prostate brachytherapy. Int. J. Radiat. Oncol. Biol. Phys. 43 (3), 583–585.

Wang, J., Li, X., Zheng, J., Sun, D., 2014. Dynamic path planning for inserting a steerable needle into a soft tissue. IEEE/ASME Trans. Mechatron. 19 (2), 549–558.

Webster, R.J., Kim, J.S., Cowan, N.J., Chirikjian, G.S., Okamura, A.M., 2006. Nonholonomic modeling of needle steering. Int. J. Robot. Res. 25 (5–6), 509–525.

Yan, K., Ng, W.S., Ling, K.V., Yu, Y., Podder, T., Liu, T.-I., Cheng, C.W.S., 2006. Needle steering modeling and analysis using unconstrained modal analysis. In: The First IEEE/RAS-EMBS International Conference on Biomedical Robotics and Biomechatronics, BioRob 2006, IEEE, pp. 87–92.

Yan, K.G., Podder, T., Yu, Y., Liu, T.-I., Cheng, C.W.S., Ng, W.S., 2009. Flexible needle-tissue interaction modeling with depth-varying mean parameter: preliminary study. IEEE Trans. Biomed. Eng. 56 (2), 255–262.

Yang, B., Liu, C., Poignet, P., Zheng, W., Liu, S., 2015. Motion prediction using dual Kalman filter for robust beating heart tracking. 37th Annual International Conference of the IEEE Engineering in Medicine and Biology Society (EMBC), IEEE, pp. 4875–4878.

Yuen, S., Kesner, S., Vasilyev, N., Del Nido, P., Howe, R., 2008. 3D ultrasound-guided motion compensation system for beating heart mitral valve repair. Medical Image Computing and Computer-Assisted Intervention—MICCAI 2008, 711–719.

Yuen, S.G., Kettler, D.T., Novotny, P.M., Plowes, R.D., Howe, R.D., 2009. Robotic motion compensation for beating heart intracardiac surgery. Int. J. Robot. Res. 28 (10), 1355–1372.

Yuen, S.G., Perrin, D.P., Vasilyev, N.V., Pedro, J., Howe, R.D., 2010. Force tracking with feed-forward motion estimation for beating heart surgery. IEEE Trans. Robot. 26 (5), 888–896.

Zahraee, A.H., Paik, J.K., Szewczyk, J., Morel, G., 2010. Toward the development of a hand-held surgical robot for laparoscopy. IEEE ASME Trans. Mechatron. 15 (6), 853.

Zarrouk, Z., Chemori, A., Poignet, P., 2010. Adaptive force feedback control for 3D compensation of physiological motion in beating heart surgery. In: IEEE/RSJ International Conference on Intelligent Robots and Systems (IROS), IEEE, pp. 1856–1861.

Zeitlhofer, J., Asenbaum, S., Spiss, C., Wimmer, A., Mayr, N., Wolnar, E., Deecke, L., 1993. Central nervous system function after cardiopulmonary bypass. Eur. Heart J. 14 (7), 885–890.

Zhao, Y., Liebgott, H., Cachard, C., 2012. Tracking micro tool in a dynamic 3D ultrasound situation using Kalman filter and RANSAC algorithm. In: 9th IEEE International Symposium on Biomedical Imaging (ISBI), IEEE, pp. 1076–1079.

Zhao, Y., Cachard, C., Liebgott, H., 2013. Automatic needle detection and tracking in 3D ultrasound using an ROI-based RANSAC and Kalman method. Ultrason. Imaging 35 (4), 283–306.

CHAPTER THREE

Robin Heart surgical robot: Description and future challenges

Zbigniew Nawrat[a,b,c]

[a]Department of Biophysics, School of Medicine with the Division of Dentistry, Zabrze, Medical University of Silesia, Katowice, Poland
[b]Professor Zbigniew Religa Foundation of Cardiac Surgery Development, Zabrze, Poland
[c]International Society for Medical Robotics, Zabrze, Poland

Abbreviations

AI	artificial intelligence
CABG	coronary artery bypass grafting
CAD	computer aided design
CEM	Centre of Experimental Medicine
DOF	degrees of freedom
FRK	Fundacja Rozwoju Kardiochirurgii im. Prof. Zbigniewa Religi (Professor Zbigniew Religa Foundation of Cardiac Surgery Development)
HCR	hybrid coronary revascularization
PVA	PortVisionAble (Robin Heart robot)
RALS	robotic-assisted laparoscopic surgery
RCM	remote center of motion
TECAB	totally endoscopic coronary artery bypass surgery

1 Introduction
1.1 From telecommunication to teleaction

Modern medical techniques are now successfully implemented in order to raise standards, improve access to services, and introduce staff optimization in the care, treatment, and rehabilitation of patients. Everyone realizes that the challenges of demographic change, increased needs, and social expectations cannot be solved with the current health care system. There is a shortage of qualified medical personnel, and human and financial capital.

Medical services (often in emergency mode) should be provided at the right time and in a place often away from competent medical centers. Both organizational and technical innovations are necessary. The development of medical robotics is an opportunity to solve some of the problems; this is similar to the early stage of the development of industrial civilization, when the automation of production processes solved the problem of the lack of hands to work and adequate production efficiency. After the progress of tele-communication, i.e., sending information at a distance now is the time for the development of teleaction, i.e., remote transmission of the action. Robots are necessary for this.

Robot is one of the few words of Slavic origin to enter the global language of modern science and technology. Robotics is a technical discipline devoted in principle to mechanisms—robots performing selected human activities. The robot, unlike the automaton, is a smart combination of perception (sensors) with action (mechanical work). The robot consists of a mechanical manipulator, a control, and a programming system. A fully autonomous robot acquires and processes information and takes specific physical action.

Medical robotics, as a technical discipline, deals with the synthesis of certain functions of the doctor or nurse by means of using some mechanisms, sensors, actuators, and computers. It includes the manipulators and robots dedicated to support surgery, therapy, prosthetics, and rehabilitation. Medical robots improve quality and create an opportunity to introduce new standards. Automated diagnostic devices (e.g., tomography) or radio-therapeutic devices are already a well-spread standard. Other applications of robots, e.g., surgical and rehabilitation robots, require control through direct supervision or remote control. Will it be possible to automate all diagnostic or surgical procedures? To a large extent, it depends on the development of sensor techniques and artificial intelligence.

One of the pioneers of innovative surgery and robotics, Dr. Richard Satava from the University of Washington and the Defense Advanced Research Projects Agency (DARPA), predicts that in the next 40–50 years, operations will be completely automated. The role of the surgeon will evolve, including the management of a full information system built around the surgical environment. "The future of technology and medicine is not in the blood and guts at all, but in bits and bytes" (Sant'Anna et al., 2004). "A robot is not a machine—it is an information system." "One of the principle advantages of the robotic surgical system is the ability to integrate the many aspects of the surgical care of the patient into a single place ... and at a single time" (Satava, 2011).

The first description of a surgical robot can be found in Alexander's paper (Alexander, 1974). The first tests were based on adapted industrial robots, used for brain biopsy and orthopedic surgery. The challenge of constructing a robot for heart surgery (and other soft tissues) was undertaken by two competing American companies from California: Intuitive Surgical (da Vinci) and Computer Motion (Zeus, AESOP). Robotic-assisted laparoscopic surgery (RALS) has evolved since its inception in 1985. Since the introduction (US Food and Drug Administration approval in 2000 for adult and pediatric surgeries) of the da Vinci system (Intuitive Surgical, Sunnyvale, California), most cases are now dedicated toward urology and urologic oncology procedures. Disadvantages of RALS include relatively longer operative times, absence of tactile feedback, and instrument collisions when traversing broader operative fields (Williams et al., 2014).

Minimally invasive surgical endoscopic procedures require new mechatronic and robotic tools. Surgical robots are increasingly used. A million operations per year have been achieved using the market leader da Vinci (Intuitive Surgical, USA) (www.intuitive.com). The robot, constructed 30 years ago with a view to cardiac surgery, is today mainly used for urological and gynecological operations. The heart, as the object of operation, is still a technical challenge.

The Robin Heart robot, developed in Poland in Zabrze, thanks to the innovations being developed, will bring robotic technology closer to surgeons' expectations in this field.

The benefits of using surgical robots in preference to conventional open surgery include:

a. decreased hospitalization time;

b. lowered complication rates;

c. reduced postoperative pain;

d. improved cosmetic results; and

e. faster return to normal daily activities.

Robotic systems are not yet available for the full spectrum of cardiac disease surgery. Gao indicates many contraindications also for other thoracic surgery. In respect to the ever-advancing surgical robotic systems, with a price tag of several million dollars, their costs seem prohibitive, let alone the costs of maintenance and upgrading. … "Absence of haptic sensation and consequent loss of tactile feedback impair the manipulation of tissue as well as suturing material, which may be the most technical obstacle for surgeons to perform delicate suturing because of their inability to judge qualitatively. Lack of more compatible instruments for retraction, exposure and visioning increases the reliance on tableside assistance to perform certain part of the surgery such as knotting, retracting, replacing instruments, etc." … Much remains to be done before full potential of robotic cardiac surgery can be realized (Gao, 2014).

Currently, complex mitral valve repair or replacement (Chitwood et al., 1997; Gao et al., 2012; Mohr et al., 2001; Mehmanesh et al., 2002), atrial septal defect repair (Argenziano et al., 2003; Torracca et al., 2001), coronary artery bypass grafting (CABG) on arrested heart (de Canniere et al., 2007; Falk et al., 2000; Loulmet et al., 1999), or on beating heart can be performed with the use of robotic assistance. New possibilities for heart treatment have been created; as Srivastava explains: "Hybrid coronary revascularization (HCR) is a treatment strategy for the revascularization of multi-vessel coronary artery disease that utilizes minimally-invasive CABG techniques in conjunction with percutaneous coronary intervention, integrating the advantages of both." "The Heart Team approach merges the insight and skill-sets of both the cardiac surgeon and the interventional cardiologist to allow for optimized revascularization based on the coronary anatomy and clinical characteristics of the patient. The limitations of this approach include the availability of hybrid operating suites and the advanced technologies and skill sets required for the performance of HCR. Additionally, specific features of the approach such as optimal anti-coagulation strategies and sequence of revascularization remain to be elucidated" (Srivastava et al., 2014).

New heart treatment options require new ideas and appropriate tools. Robin Heart is an example of a robot that is created to solve the problems described. The heart remains a challenge for the constructors of surgical robots.

The chapter is devoted to the description of research and construction works on the development of Polish Robin Heart robots, which the author and the team have been running since 2000. The scope of FRK experience is a good example of the multidisciplinary teamwork to create a modern surgical tool, testing new ideas on the borderline of medicine and techniques, mechanics, and software. and to overcome the natural limitations of man concerning movement, precision, access to tissues requiring surgical intervention and operation space, and distance between the doctor and the patient, as well as the "distance" between knowledge and practice.

1.2 Surgical robots in Poland—Important dates and facts

In Poland, the interest in the development of robotics was associated with doctors—mainly Silesian surgeons—such as Religa, Bochenek, and Witkiewicz:

2000 Prof. Zbigniew Religa and Zbigniew Nawrat start the Robin Heart project.

2001 Prof. Andrzej Bochenek in Katowice implements the AESOP assisting robot (Computer Motion, US).

2002 Prof. Andrzej Bochenek performs first 10 cardiac surgery operations in Poland by ZEUS robot (borrowed from Computer Motion, US).

2010 The first da Vinci robot appears in Poland. During the first half of 2011. with the assistance of a robot surgeon from the Provincial Specialist Hospital in Wroclaw under the direction of Prof. Wojciech Witkiewicz, 36 operations were performed in the field of gynecology, urology, and vascular surgery. To date, more than 300 have been performed.

Currently, there are six da Vinci robots in Poland, mainly in private hospitals.

Analysis [average data from countries with high HDI (human development index), after rejected extreme data from Poland and the United States] shows acceptance of the index 1 robot for 1.15 million inhabitants is considered justified at the current stage of development of da Vinci robot applications. This would mean that there should be 34 robots in Poland, including four robots used in cardiac surgery (Kroczek et al., 2017). According to the plan, the first Polish Robin Heart robots will appear on the Polish market soon; maybe they'll fill this gap.

2 Surgical robots

The telemanipulator control system works as a master-slave manipulator system, reflecting the movements of the operator (surgeon) on the movement of the arm and tools by developing appropriate control signals for its drives. According to the assumptions, the control provides the required accuracy, it allows the scaling of the set value to increase the positioning accuracy, and also eliminates the operator's hand tremor.

The control system of the robot's drive units allows the robot's effector to perform specific tasks. The manipulator drive units are powered from power amplifiers in the manner imposed by the controller. The control system (the analyzed signals are n-dimension vectors) creates a closed control system, evaluating the current compliance of the actual implementation of the tasks with the planned ones. If the priority is to reach the position (e.g., when the robot does not come into contact with the environment)—the positional control is chosen. Its use requires good quality image information of the operation field for the telemanipulator operator. If the priority is to accomplish the task (despite the occurrence of obstacles), force control is used. Its use requires real-time information about the value of the force action of the tool in the space of the operation. There are currently no suitable sensors for surgical robots—force sensors suitably small, precise, and resistant to critical working conditions of tools in the human body.

In medical robots, it is very important to define the hierarchy of management and to enable patient and staff protection in every situation. In the event of a robot failure, power outage, or other critical situations, doctors withdraw the robot (constructors must choose a technical solution that allows this) and continue operation with classical methods. Creating a control program requires the use of data on the phenomena of mechanical interaction with the tissue (empirical), simple and inverse kinematics of the mechanical system, and characteristics of the drive system determining, for example, accuracy (resolution) for each degree of freedom.

The surgeon uses both conditioning and coordination motor skills during work. The comfortable position and way of working in the robot control console allows the overcoming of ergonomics problems of the surgeon's work. Performing precise motion is associated with proper planning and online control. Correcting the position and functions of the tools requires a good image quality and all current additional information from the operating field, including force feedback.

The information is processed during telemanipulator control. The operator of a surgical robot making decisions and performing specific physical tasks with a tool uses:

(1) medical data and history of patients (optimization through personalization);

(2) diagnostic data, factual data based on facts, i.e., current information on geometry and on the state of tissues (optimization by defining the operating space);

(3) feedback of the tool's response to tissues through visual observation (image analysis) and/or force coupling (optimization of motion and action by the senses);

(4) a set of data characterizing the current status of vital functions of the patient's organs during surgery (optimization by interactive evaluation of the patient's condition in real time); and

(5) information obtained in the process of education and practice (optimization through knowledge and experience). Supported by a planning system based on computer simulations, they can be extended, updated using artificial intelligence (AI) and advisory programs, which are the sum of the surgeons' and potentially robots' experiences of similar cases.

The effect of the operation (accomplishment of the assumed goal) should be measurable and verified—then it can be used to develop the standard of the performed service, and in the future may lead to the automation and independence of robots.

3 Robin Heart

The work presented here is an attempt to summarize the research and development of the Polish surgery robots project carried out by a multidisciplinary team led by the author. The results of the project initiated by the author are the family of Robin Heart robots and universal mechatronic tools series Robin Heart Uni System for use during minimally invasive surgery on the heart and other soft tissues. The Polish project began in 2000. It is possible to find reports on the progress of work (cited partially below) available on the Internet (edited by the author of this chapter, manager of the Robin Heart project) the book series *Medical Robots, Advances in Biomedical Technology*, and the scientific journal *Medical Robotics Reports*. The most important summary is presented in a book published in Polish (Nawrat, 2011).

This chapter presents the design, prototyping, and testing of innovative surgical tools. An assessment of the achieved results and prospects for development of the Robin Heart project robot against the current state of scientific knowledge and global experience in the field of robotic surgical systems are presented.

The process of projecting a robot starts by determining the tool–tissue reaction (mechanical characteristic, the forces for specific operations, and dynamic analysis of the work of a tool) and the person-tool man-machine contact (kinematic analysis of the surgeon's motion). The surgeon's motion and tool trajectory in the natural environment are analyzed with the use of optical biometry techniques. The forces applied during the impact of tools on tissue during typical surgical activities are measured in simulation of natural conditions. The construction assumptions, as well as the functionality and ergonomics of the innovative tools, are verified experimentally. As a result, a user-friendly surgical console and an efficient surgical tool are constructed (Nawrat, 2011, 2012).

The following medical robots have been prepared so far (Fig. 1):

(1) Spherical model (2001);

(2) Robin Heart 0 (2002);

(3) Robin Heart 1 (2003);

Fig. 1 The Robin Heart family of robots (from left and top): Robin Heart 0, Robin Heart Vision, Robin Heart Spheric + Robin Heart 0, Robin Heart Junior, Robin Heart Pelikan, Robin Heart PVA, Robin Heart 1, Robin Heart 8, Robin Heart Tele + Console Robin Stiff-Flop, Robin Heart Tele tool platform, Robin Heart mc^2, Robin Heart mc^2 during in vivo test.

(4) Robin Heart 2 (2003);

(5) Robin Heart Vision (2008);

(6) Robin Heart Junior (2009);

(7) Robin Heart mc^2 and mechatronic tools Robin Heart Uni System (2009);

(8) Robin Heart PortVisionAble model 0,1,2,3 (2012–2015);

(9) Robin Heart Pelikan (2013);

(10) TeleRobin (2015); and

(11) Robin 8 (2016).

The Robin Heart 0 and Robin Heart 1 telemanipulators have an independent base and are controlled via industrial computer and specialist software. The Robin Heart 2 is fixed to the operating table and has two arms, on which one can fix various surgical instruments. The control system uses its own software as well as signal and specialist microprocessors. The Robin Heart PortVisionAble (PVA) will become the surgeon's partner in the operating room next year. It will replace a human assistant who usually holds the endoscope to enable the observation of the operative field of laparoscopic instruments. The portable Robin Heart PVA is easy to use and install.

Robin Heart mc^2 is the biggest robot—it can work for three surgeons at the operating table. The biggest invention here is a tool platform consisting of two surgical instruments and an endoscope mounted on the robot's middle arm. This idea solves the collision problem of the robot arms. This modular robot can be set for various operations in a selected way. The Robin Heart Tele robot solution can be developed as an extension of the tool platform concept. This robot was created for the implementation of a telemanipulation test project at various distances and under different physical conditions. It offers a completely new design solution, kinematics, and a larger workspace.

The Robin Heart family of robots is mainly spherical. Each movement maintains a fixed point of rotation for the surgical instrument (remote center of motion—RCM). The RCM mechanism is based on a parallelogram structure. They should be set in a proper manner to the patient's body, so that the construction pivot point is located where the tool passes through the patient's body surface.

Robin Heart 8 is an exception here. It was developed as a robotic tester of robots and surgical instruments, simulating any contact with the tissue (tissue physics) and any assumed limitations of the working space. However, currently, due to its kinematic properties—full freedom of choice of the trajectory—it is considered as a robot potentially usable in ophthalmology.

The Polish surgery robot is an original design. Thanks to its modular structure, it can be adjusted for surgery of different types. The Robin Heart Shell console is equipped with an advisory system that makes it possible to obtain all patients' diagnostic information during the operation, as well as elements of the operation planning on the screen. The virtual operating theater introduced in our laboratory allows surgeons to practice some elements of an operation and check the best placement of the ports in order to avoid internal collisions. Exhaustive 3D visualization means this system can be helpful in planning an operation on a given patient. New, semi-automatic tools are in the process of emerging—Robin Heart Uni System.

Using virtual reality technology, an interactive model of a surgery room equipped with a Robin Heart system was created with the use of EON Professional software. This computer modeling method allows advance procedure training and will be used as a low-cost training station for surgeons. The model allows understanding better the process of less invasive surgery treatment and robot behavior. The link between this type of modeling and a computer aided design (CAD) technique is using accurate CAD robot models in VR software together with a precise reflection of workspace geometry (Nawrat and Koźlak, 2007). This approach gives a surgeon an easy and intuitive way to understand technical information and uses it to optimize and plan medical process. The presented model of an operating room in a virtual reality environment has been performed in FRK and successfully used since 2006.

The process of developing the robot starts from the determination of the tools-tissue reaction (mechanical characteristic, the forces for specific operations, and dynamic analysis of tool work) and person—a tool and then man-machine contact (kinematic analysis of surgeon motion). Robin Heart Shell console includes not only ergonomic handles with microjoysticks but also the advisory system and the possibility of a full visual and voice communication with the operating theater.

The Robin Heart manipulator has a very good and relatively large working space, in which the surgeon can select a small subspace with very good isotropic kinematics, as presented by Podsędkowski (2003), and properties for manipulating of objects with good position accuracy. The system has been verified both functionally and technically. Standard technical evaluation allowed estimating the value of positioning resolution equal to 0.1 (mm). The milestones of the project were animal experiments, carried out in January 2009 (Robin Heart model 1, 2, Vision) and May 2010 (Robin Heart mc^2). The operations were performed on pigs at the Centre of Experimental

Medicine (CEM), the Silesian Medical University in Katowice. The Robin Heart system experiment carried out on pigs allowed us to verify many aspects of a very complex project and was the source of hints for future development. A preoperation planning stage included surgeon training on physical and virtual anatomy models with the use of real pig tissues (Nawrat, 2011, 2012).

In the course of the animal experiment, the surgical task achieved in the abdominal space was cholecystectomy, and in the chest and heart, the repair of heart valves (with extracorporeal circulation) using the Robin Heart models 1, 2, and Vision. In the last phase of the experiment, the efficiency of the Robin Heart Uni System mechatronic tools was checked—they were mounted on the robot's arm (and controlled from the console) or held in the surgeon's hand (manual control). Elements of the operations in the abdomen and heart cavities were successfully performed (Fig. 2). Robotically supported totally endoscopic coronary artery bypass surgery (TECAB) was not performed due to collision of robots during the mammary harvesting procedure. It was decided to build a new robot. After 9 months, in 2010, the first experiment was conducted, by means of a specially designed robot—Robin Heart mc^2. The purpose of the experiments was achieved: the surgical team accepted new construction functionality (Nawrat, 2011, 2012).

Robin Heart mc^2 creates completely new job opportunities for a surgeon—both in the local area and globally. It can be assembled as an arm of the platform (a small robot with two endoscopic tools and endoscope for observation) or as telemanipulator working for three people—the main surgeon, assisting surgeon, and an assistant holding the endoscope (controlled from console by one operator). It is a really new solution for robotically assisted surgery.

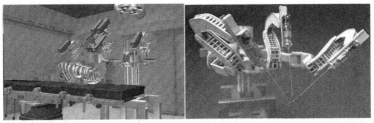

Fig. 2 Research. Robin Heart mc^2 in version for four tools plus endoscope and classic two tools plus endoscope. Robin Heart mc^2—modular robot hardware allows optimal set up into the operation. The robot can work in a set of four tools plus endoscope, or as a standalone platform with two tools plus endoscope working in the local work area.

As conclusions from the experiment, users (surgeons) have expressed positive opinions on the ergonomics and possibilities of controlling the robotic arms by means of the Robin Heart Shell console. The opportunities of operating by means of the Robin Heart Uni System mechatronic tool are very promising, as it may be mounted on the robot's arm or controlled manually (Nawrat, 2011).

The first model of teleoperation was performed in December 2010 between two cities—Zabrze (FRK) and Katowice (CEM)—successfully. Robin Heart system experiments allowed verifying robots before the phase of preparing serial production and clinical initiating (Nawrat, 2011, 2012).

Our new robot, Robin Heart PortVisionAble (PVA), enables some parts of surgery to be performed in the "solo" mode, i.e., independently by a single surgeon, while the surgeon advisor or student participates at a distance. Robin PVA, a lightweight, portable, and inexpensive robot can replace one of the assistants for mini-invasive surgery and enable the distance-participation of either experts/advisor or student during surgery. Three prototype robots ready for clinical trials have been prepared and tested. Developed research and technological documentation was used to start the series production and clinical implementation of this robot (Lis et al., 2015). In 2019, this robot was sold to the Polish company Meden-Inmed, which is preparing its industrial implementation.

3.1 Robin Heart innovation

The Robin Heart system includes a planning system, training system, experts' program, and advisory system, as well as telemanipulators and automatic surgical tools. In the Polish Robin Heart surgical robot, many original solutions were introduced (PL 208 988, PL 208 430, PL 208 432, PL 208 433, PL 208 693, PL 217 349, PL. 220015, PL. 220329, PL. 409735, PL 422691 EP3097839, US 9610135, EP 2990005B1, US 9393688, EP 3 150 184B1, EP3146930, EP 15179357.7, EP 2949967A1, US2015/034560, e-00090358-D (RCD), W-123845, W-123846, W-124541): telescopic sliding motion to move the tool (2002), mechatronic tool "for the hand (2006) and the robot", and Robin Heart mc^2 (2010) is the first surgical robot that can work for three persons (two surgeons and assistant responsible for endoscope orientation) (Nawrat, 2011). Robin Heart PortVisionAble (2012–2016) and Pelikan (2013) are examples of robots for endoscope assistance. The Tele Robin Heart (2015) with a new multitool platform (modular type) is an example of development and progress in construction of new robots.

Heart surgery is still a challenge for robotics. Thanks to introduced innovations, such as modularity of structures, effective force feedback, or mechatronic tools, one can expect an increase in interest in surgical robots in areas where currently used robots do not meet the expectations of recipients, such as cardiac surgery. Safety and precision depend on the introduction of a new quality of sensitivity and comfort of robot control.

3.2 Construction—Modularity of structures

In the Robin Heart mc^2 model, a technical solution in the form of a tool platform (two tools and an endoscope) placed on one arm was introduced, and then improved in the TeleRobin model (Figs. 2–4). "The aim of the invention is to provide a universal tool arms assembly of a surgical robot equipped with surgical instruments and fixing elements for a single support arm of any surgical robot, which should first of all eliminate low functionality of surgical robot's single support arm usually equipped only one surgical instrument, and furthermore eliminate problems resulting from limited work area and the possibility of collision occurring when surgical robots with multi-instrument arm assemblies are used. Another aim of the invention is to provide an extended range of work of each of the tool arms … and reduce the space and weight of surgical robot" (Nawrat et al., 2018a).

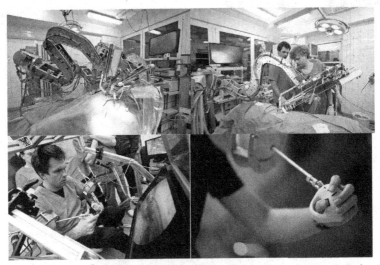

Fig. 3 Robin Heart mc^2—modular robot hardware allows optimal set up into the operation. The robot can work in a set of four tools plus endoscope, or as a standalone platform with two tools plus endoscope working in the local work area. Animal tests using new Robin Heart robot mc^2 (mammary artery harvesting and bypass surgery) were carried out.

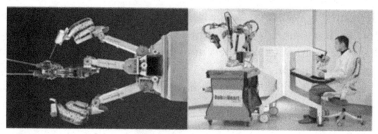

Fig. 4 Robin Heart mc² and TeleRobin—the new version of tools platform on one robotic arm.

3.3 New tools

A. For manual and robotic manipulations

The opportunities of operating by means of the Robin Heart Uni System, a universal mechatronic tool for both; robotic arm and hand (it may be mounted on the robot's arm and controlled by console or controlled manually), are very promising (Fig. 5).

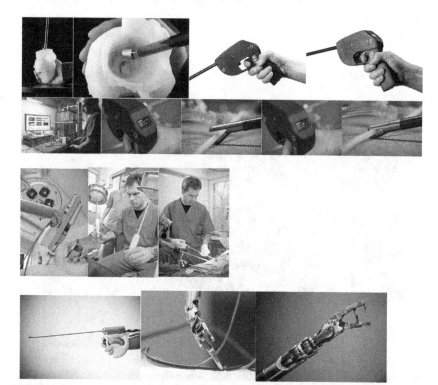

Fig. 5 At the top is shown a mechatronic tool with monitored reaction force with tissues. Research. Robin Heart Uni System. Animal test and new version of holder (2009). The universal system—a tool mounted on the robot arm can be used manually. Robin Heart Uni System v.1 and tools for classical and automatic sew.

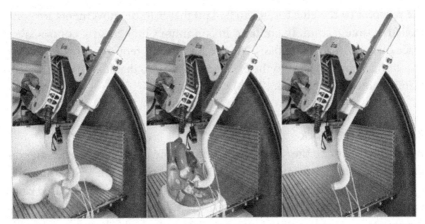

Fig. 6 A model of innovative robot tools—Robin Heart with variable geometry (STIFF-FLOP project EU).

B. For control of shape and physical parameters (flexibility, stiffness)

Surgery needs new tools with controlled geometry and stiffness that can work effectively among soft tissues, limiting the destruction of healthy tissues during penetration and reaching the place of surgical intervention (Fig. 6) This concept of the robot was developed as part of the EU STIFF-FLOP project, of which the FRK was a participant (Arezzo et al., 2017).

3.4 Control system

The Robin Heart is a telemanipulator. Surgical telemanipulators work as a master-slave manipulator system. The basic task of the control module, the system working in this configuration, is to map the movements—the surgeon's hands (position/velocity adjuster and possibly other physical quantities)—on the movement of the actuating arm by developing appropriate control signals for its drives. An additional feature of the system, which is highly sought by potential system users (surgeons), is the incorporation of a feedback system for transmitting the sensation of sensing the force/touch of the tool to the operator. To obtain the highest accuracy of movement it is necessary to have information about the state of tension of individual muscle groups, the speed of movements, and the angular position in the joints. It means that a good interface should offer a balanced resistance.

To control robot motion, an appropriate information structure is required, understood as the transmission of a series of ordered data in time, containing a fixed amount of motion information and conditions for its implementation. Based on the analysis of the speed of information flow from the receptor to the effector, it can be said that ballistic movements are controlled by ante factum. In contrast, in continuous movements, control takes place on an in-real-world basis. Muscle stimulation requires constant adjustment (differential values) to the existing situation so that it is possible to perform a motor task (Nawrat, 2011).

3.4.1 Force feedback

In order to improve the safety of the endoscopic intervention, the surgeon must get on-line information about the physical parameters during the operation. The principal parameter required is the force as measured during interaction endoscopic tools with the tissue. The sensor built inside the gripper can measure the strength with which the laparoscope holds the surgical tool. The sensor placed on the tip can provide information about the hardness and surface roughness of the tissue the endoscopic tools touches.

The present task activity is focused on integrating a tip head 3D tactile sensor and an array of force sensors, on the grasper head and inside the grasper, respectively. The first sensor facilitates steering and identification of the tissue the laparoscope touches, whereas the inside sensors provide appropriate information about the force with which the grasper holds a needle, surgical sutures, or tissue. The feasibility of application in a laparoscope arm has been demonstrated in the frame of the "INCITE" project (2012–2017) by integrating two piezoresistive force sensors in a magnified grasper (Fig. 7). Three different functions are targeted: (1) microjoystick actuator to be integrated in the hilt of the endoscopic tool to provide easy robotic movement control during operation; (2) force sensor inside the endoscopic jaw to provide feedback to the surgeon by measuring the grasping strength; and (3) 3D force/tactile sensor, which facilitates palpation for tissue diagnostics during operation. A model of the robot controller using a prototype 3D sensor force has been successfully tested during the study of functional robot. The studies of prototype sensors have demonstrated their usefulness in the robot force feedback system to assess the state of tissue and to assess the clamping force of the grasper surgical system (Nawrat et al., 2016).

The endoscopic tool was fixed on the arm of the Robin Heart surgery robot. The signals of both sensors were utilized to control the movement of the arm and the jaw of the gripper. In ongoing follow-up research, the group of FRK will elaborate adequate methodologies for human-robot interactions.

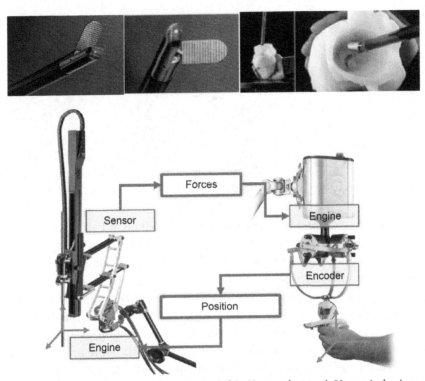

Fig. 7 Innovative sensors and their use in Robin Heart robot tool. "A surgical gripper with embedded force and tactile sensors and readout electronic circuit was developed. Piezoresistive, vectorial forces sensors were designed by … according to the proposed force ranges and manufactured by 3D bulk micromachining technology. Two different MEMS sensors were electromechanically integrated into a metal endoscope tweezers together with the pre-processing electronics performing the analogues-digital data conversion and the communication towards the robot control system. According to the proposed medical application and functional requirements the sensors were embedded in biocompatible elastic polymer. The accomplished smart laparoscope was integrated with the Robin Heart surgery robot. The functionality of sensor system was validated by biomechanical tests" (Radó et al., 2018).

One of the next Polish inventions is the Robin Heart Uni System. with force feedback information monitor. The Robin Heart Team plans include implementing clinically innovative mechatronic tools with force feedback.

3.4.2 User interfaces

The Robin Heart Team (FRK) tested the control of surgical robots using gestures, voice, remote control, and a set of haptic models. A completely new, ergonomic solution to this problem—motion haptic system Robin

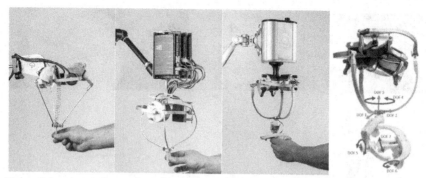

Fig. 8 Robin Heart Tele and RobinHand user interfaces (Mucha et al., 2015).

Hand (haptic system)—has been developed, presented in Fig. 8. The operators must feel as if they were performing remote control tasks directly in the right environment, and moving freely in the free space, feeling contact with the obstacles encountered.

The Robin Heart project introduces the idea of control, in which the operator intuitively determines the position of the working tip, and the operation itself (the developed concept of task automation) is activated by means of electronic buttons and microjoysticks (with the logic of control prepared for various tools). This concept allowed the natural introduction of a universal tool system (Mucha et al., 2015, 2018a,b).

3.4.3 Robotic safety system

According to Kirkpatrick's summary, robotic surgery is generally safe with low overall complication rates, but adding a robot to the surgical equation inserts another potential entry point for error into an already complex and risk-fraught arena. In general, surgical outcomes are ultimately a direct manifestation of the skill and experience of the surgeon, not the technology or approach used. Potential areas for improvement and reduction of error in robotic surgery include more standardized training and credentialing practices, improved reporting systems for robotic-associated adverse events, and enhanced patient education (Kirkpatrick and LaGrange, 2016).

The Robin Heart telemanipulator safety system includes a collection of mechanical solutions, an electronic drive system, and software. A significant part of the time and computing power of the telemanipulator system control system is intended for related tasks with support for system security modules. The basic elements of the safety system include: monitoring the operation of the drive units (detection of current overloads and detection of exceeding the extreme limits, possible shutdown of the entire system, and unblocking of drive units for manual handling of the arms) and duplicate detection sensors (accelerometers, gyroscopes or rotary-pulse absolute position transducers) of the actual shift at individual degrees of freedom connected by a fast serial peripheral interface bus.

3.4.4 Robotics and remote action

After the successes and development of many medical applications related to distance information technology, i.e., telemedicine, it is now possible to use technologies for remote transfer, i.e., medical telerobotics.

One of the most interesting fields of development, for many reasons, is tele-surgery. First of all, we have to achieve the possibility of professional surgical intervention when and where it is necessary. But—which is often forgotten—in many cases, also the safety of medical personnel during surgery can and must be increased by moving the medical team away from the patient. The biggest technical problem is the delay in image transmission, which may prevent proper control of the robot's operation. During the December conference "Medical Robots 2010" the first teleoperation experiment in Poland was carried out. The surgeon (Joanna Śliwka, a cardiac surgeon at the Silesian Center for Heart Diseases in Zabrze) from behind a console in the FRK in Zabrze operated a Robin Heart robot placed in the Center for Experimental Medicine of the Medical University of Silesia in Katowice-Ligota (see Fig. 9). In the same operating room a few months earlier, an experiment was carried out on a pig using the Robin Heart mc^2 robot (Nawrat, 2011).

As a part of cooperation with the FRK, Emitel company set up a distance information transmission system consisting of a wireless point–to-point radio link (13 km), data transmission system, audio–video signal from several sources along with conversion of analogue signal to digital. Measurement of the delay of the control signal between Katowice CEM and FRK in Zabrze robot (1 ms) and image (280 ms) indicates that gallbladder excision can be performed today. However, the delay in image transmission, on the basis of which the surgeon operates, is too large to perform heart surgery. The Zabrze team performed rail tests using an internet network, including robot control, at the Guido mine in 2016 (Fig. 9). Procedures for selected operations and security systems during teleoperation are being developed.

2010   2016

Fig. 9 RobinHeart Team (FRK) has pioneered research in Poland in the field of teleoperation. In the photo on the left, the Zabrz-Katowice teleoperation model from FRK to the Center for Experimental Medicine of the Medical University of Silesia in Kato-wice; on the right, a remote control experiment of the Robin Heart Pelikan robot at the Guido Mine in Zabrze.

The first teleoperation with the da Vinci robot prototype (developed at the SRI in Stanford) by Richard Satava and John Bowersox (Bowersox et al., 1996) was done in 1996. The first spectacular success of tele-surgery was the Lindbergh Operation in September 2001, in which the signals controlling the robot were transmitted to a distance of 7000 km, from New York to the hospital in Strasbourg (France). Jacques Marescaux, with his team IRCAD, using a Zeus robot, removed the gallbladder of a 68-year-old patient. It was the first RARTS (robot-assisted remote telepresence surgery) operation. The delay of the ATM OC-3 undersea fiber network with a 10-Mb/s bandwidth transferring signals did not exceed 155 ms (for security reasons, the delay may not exceed 200 ms) (Marescaux and Rubino, 2004).

M. Anvari introduced the first practical implementation of teleoperation for routine medical care in Canada (Anvari et al., 2005). Technological progress concerning both the improvement of robots (operating tools) and devices for transmission, which will reduce the delay time of signals, will allow achieving proper system performance and its dissemination (Anvari, 2007). The most important factor influencing the progress in the field of teleoperation is to reduce the delay in signal transmission related to the distance between the operator and the robot, doctor, and patient. The introduction of the 5G standard is one of these opportunities. The first proof of this thesis is reports about an animal experiment carried out in China. "The operation was performed from a distance of 30 miles using a 5G connection with a latency of just 100 milliseconds" (Humphries, 2019).

3.5 Surgery planning

The introduction of standardization for surgery requires planning and supervising the implementation of the procedure. Depending on the choice of the tool (with different degrees of freedom and geometry), we can operate it in a different space (workspace). Therefore, it is very important to optimize the process; for example, the selection of the right places for making openings in the patient's body shells—Fig. 10. The Robin Heart Team used virtual space technology and created a virtual operating room for planning and training of robot operations. In addition, computer simulations of biophysical effects are performed, which are used in the advisory program for the Robin Heart robot operator.

The model allows for a better understanding process of less invasive surgery treatment and robot behavior. The link between this type of modeling and computer aided design (CAD) techniques is using accurate CAD robot models in VR software together with a precise reflection of workspace geometry (Nawrat and Koźlak, 2007). This approach gives a surgeon an easy and intuitive way to understand technical information and use it to optimize and plan medical process. The presented model of an operating room in a VR environment has been performed in FRK and successfully used since 2006.

The surgeon's training system includes both a virtual computer operating room and hybrid systems connecting the physical console with the virtual system. The advisory system includes an operation planning system and computer simulations of the effects of the operation (for example; pressure and flow in blood vessels after a by-pass operation).

3.6 Surgery training

Work prepared by Małota et al. presents the latest FRK experience in the technical support of surgical training methods. There are a variety of simulators: synthetic models and box trainers, live animal models, cadaveric models, ex vivo animal tissue models, virtual reality (computer-based) models, hybrid simulators, procedure-specific trainers, and robotic simulators (Fig. 11). The most important factor for obtaining an appropriate surgical simulation is creating quasi-natural geometry of surgical scene and physical characteristics of the used materials. Constructing simulators for a completely new type of tool is a challenge for both engineers researching devices for new tools as well as trainers (surgeons) to discover the optimal use of new functional features of tools for various types of operations. The fidelity of a simulator is determined by the extent to which it provides

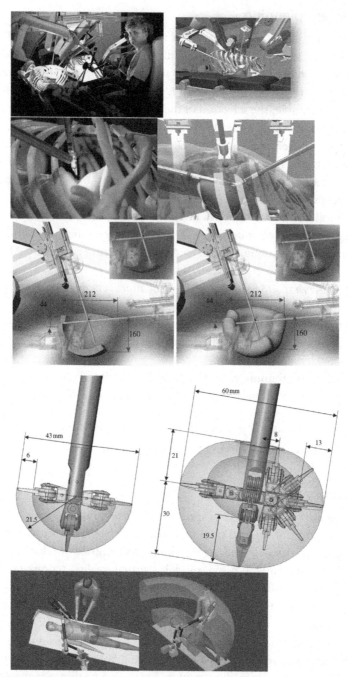

Fig. 10 Virtual operating room, physical model for surgical tests and studying the work space of the surgical Robin Heart 1 robot and Robin Heart PVA during endoscopic surgery (ergonomy study).

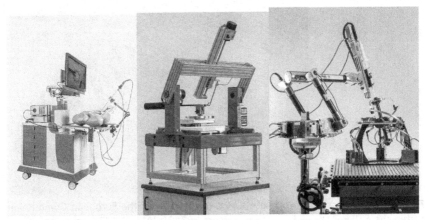

Fig. 11 The figures illustrate prepared in FRK training and testing systems: endoscopic tools training system, and two robotic testing system (mechatronic tissue simulation).

realism through characteristics such as visual cues, tactile features, feedback capabilities, and interaction with the trainee.

The modeling of biological tissues is very difficult because the biological soft tissues have nonlinear force-deformation properties and show viscous behavior. Additionally, dissected tissue changes its mechanical properties and all properties strongly depend on many factors, including temperature, pressure and patient health (pathology). The artificial surgical scene and described devices for testing tools and surgeons create the possibility of standardization for the educational and research process. The new solution is to test robots using robotic devices that simulate tool behavior in various clinical situations. All these experiences constitute a unique support for constructors and users of mechatronic and robotic tools created in the FRK (Małota et al., 2018).

An especially interesting solution is the test robot Robin 8 (Nawrat et al., 2018b), shown in Fig. 11 on the right. It allows testing of surgical instruments and surgical robots by simulating the loads and range of motion suitable for various tasks performed in the patient's body. A computer, to which an appropriate model of physical properties of tissue and ambient geometry are introduced, controls the system.

4 Future directions for the Robin Heart project
4.1 Flex tools and STIFF-FLOP

Surgery needs new tools with controlled geometry and stiffness that can work effectively among soft tissues, limiting the destruction of healthy

Fig. 12 The project of flexible tool supported in part by the European Commission within the STIFF-FLOP FP7 European project FP7/ICT-2011-7-287728 presentation at King's College London.

tissues during penetration and reaching the place of surgical intervention. This concept of the robot was developed as a part of the EU STIFF-FLOP project, of which the FRK was a participant (Fig. 12). The project ended with a robot demonstration made on cadaver (Arezzo et al., 2017).

In various publications (Mucha et al., 2015, 2018a,b), the FRK haptic system has been presented. "Designing a surgical robot operator's workstation requires an in-depth analysis of the specificity of a surgeon's work; experience, psychology (related to the decision making processes and factors that have an impact on the precision of work), ergonomics, anatomy, physiology, tissue biophysics and the robot itself (kinematic structure, robot control system characteristics). It has been assumed for Robin Heart robots that actuating the position of the tool tip will be performed intuitively and the robot's operator will acquire a complete set of information: visual and haptic feedback, which will enable accurate control of the new-type robot. The challenge of the Stiff Flop project (the pneumatically controlled soft robot) was in the need to control the tool tip mounted on an arm with a variable, controllable stiffness and geometry. A set of sensors on the surface of the octopus-inspired arm tool give access to information about an interaction with organs, which are transmitted to a special sleeve on the surgeon's arm. Guided by data from visual monitors, the operator moves and actuates certain tool actions. FRK team have developed a specially equipped console – an ergonomic surgeon's workstation" (Mucha et al., 2018b).

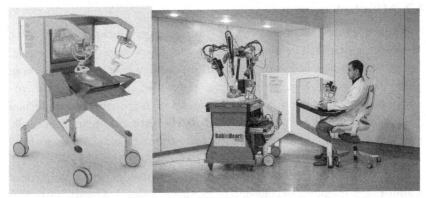

Fig. 13 Robin Stiff-Flop Console: CAD model and real console with Robin Heart Tele robot.

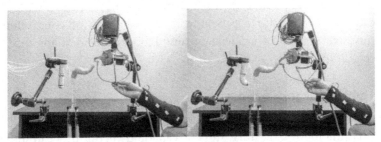

Fig. 14 The full haptic system: flexible manipulator plus vibration sleeve test of feedback.

The system was launched and tested with force feedback acting both on the haptic console and the vibrating sleeve. The integrated system (Figs. 13 and 14) used for the testing comprised of:

- two pneumatically operated robot arms equipped with both a lateral flexi force sensor as well as frontal and circular lateral pressure sensors; and
- a completely integrated system composed of the FRK Robin Hand haptic, a haptic vibration sleeve and a soft robot arm (made from Dacron—Polyester Fiber vessel prosthesis and Ecoflex—30 silicone).

Encoders in the haptic Robin Hand unit capture the movement performed by the operator's hand. Information from the encoder is processed by a microcontroller and is sent in the form of Cartesian coordinates to the microcontroller, which operates the pressure valves used to control the robot arm's movements.

4.2 AORobAS project

Potential applications of robotics in cardiac surgery include: aortic valve replacement (standard or percutaneous technology); tricuspid valve repair; descending thoracic aortic surgery; transmyocardial laser revascularization; ventricular septal defect, patent ductus arteriosus, coarctation; intramyocardial delivery platform for biological agents such as stem cells, molecular therapeutics, genetic vectors; and **AORobAS**—artificial organs robotically assisted surgery. There are plans to use the robot for implanting artificial organs, prostheses, and ventricular assist devices.

The future plans connected with development of the Polish robot Robin Heart include carrying out a robotically assisted minimally invasive surgery to implant artificial organs; VADs, TAHs, valve, vessel prostheses. Currently, clinically used blood pumps, valves, and pace makers require to be replaced and repaired during the patient's life. This is a temporary application. The best for a patient will be realization of conception of mini-invasive service of artificial organs. Robots are ideal for this task (Nawrat, 2001).

4.3 Hybrid surgery robots

The author proposed a novel system for hybrid operations (Fig. 15), in which a cardiologist and cardiac surgeon can jointly perform a heart surgery, in 2017. The system consists of two robots that offer vascular and surgical access to the heart, two control consoles, and an appropriate information system, creating a synergy of information exchange and interoperability. The idea will be implemented after obtaining financing. The developed system is a challenge for the ergonomics of control.

4.4 Robin Heart Synergy

The aim of the Robin Heart Synergy project is to improve the functionality, safety, ergonomic, and economic aspects of using a surgical robot. The most important task of the project will be to solve the problem of assembling the robot with the operating table by integrating the robot's position control with the current position of the operating table.

Robin Heart Synergy is a modular telemanipulator equipped with a universal system of mechatronic surgical tools, synchronized with the operating table. The main objective of the project is to study the possibilities of developing the concept of a surgical robot as a device cooperating both with elements of the operating room and with all members of the operating team.

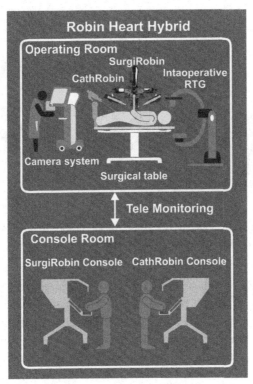

Fig. 15 Schematic representation of the idea of a hybrid robot (SurgRobin + CathRobin) and Robin Heart Cardio robot model (for invasive cardiology).

The synergy action will be achieved by introducing innovative solutions to technological, structural, and organizational processes of surgery. It will aim to achieve shorter operating time, greater comfort and safety, and the opportunity to develop teleoperation. Unfortunately, the project, announced in 2014, has not been financed to date.

4.5 Robin Heart Pelikan and lightweight robots technology

The Lis work presents a new, light Robin Heart robot. "To meet the expectations of veterinary physicians, a team of researchers from the Cardiac Surgery Development Foundation and the Silesian University of Technology decided to assess the applicability of the Robin Heart "Pelikan" robot in veterinary procedures performed on animals. For the purpose of the project, a special support frame was made to mount the robot on a standard operating table and to pre-position it conveniently for the patient and the operating

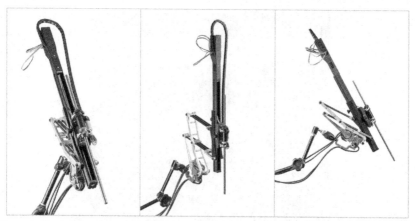

Fig. 16 The Robin Heart "Pelikan" is a vision tracking robot that has 4 degrees of freedom (DOF 4) enabling it to navigate the camera inside the abdominal animal cavity.

surgeon. The Robin Heart "Pelikan" is a vision-tracking robot that has 4 degrees of freedom (DOF 4) enabling it to navigate the camera inside the abdominal cavity. DOF 1 and 2 make it possible to change the position of the robot (tilting front-back and left-right), while DOF 3 and 4 are responsible for rotating the camera and for pulling it forward and backward" (Lis et al., 2017).

This robot has a mechanical constant fixed position at the camera's entrance port on the patient's skin (remote center of motion; RCM). Composite materials made it possible to minimize the weight of the robot (which is approx. 4 kg), permitting transportation in a small suitcase. Steering the robot is done with a manipulator mounted directly on a traditional laparoscopic tool. The surgeon uses the index finger to exert pressure on a force sensor. This solution makes it possible to manipulate the robot freely without putting away the tools held in the hand. Fig. 16 shows the robot's degrees of freedom, the mounting system (the supporting frame) adjustment, and the steering system options (direction controller).

5 Discussion "challenges and limitations"

The current state of development of artificial intelligence and sensory devices is insufficient, so currently used surgical robots are telemanipulators. Designing a robotic tool is a challenge for a multidisciplinary team in which ergonomics is as important to success as precision and operational safety.

5.1 Decision-making and artificial intelligence

High hopes are connected with the influence of artificial intelligence on the development of medicine. In my opinion, artificial intelligence is a robotics department. The source of the word robot is associated with the figure of an artificial man introduced by Karel Čapek in the science fiction play. R. U.R.—Rossumovi Univerzální Roboti (Rossum's Universal Robots), in 1920. As is well known, rational man (*Homo sapiens*) is the only contemporary species of the *Homo* genus. This means that reason and intelligence played a decisive role in human survival. If we want robots to replace human beings, they must be available in a quality (standard) similar to that of a human being.

Robotics is a technical discipline that deals with mechanisms, executive assemblies, and computers. Because humans have brains, artificial intelligence is an integral part of robotics. Simplifying: the robot is a smart combination of sensory and executive devices.

Intelligence is the ability to perceive, analyze, and adapt to changes in the environment. Intelligence is an ability to understand, learn and use your knowledge and skills in different situations (appropriate reacting and solving problems); according to L.L. Thurstone: verbal fluency, numerical abilities, spatial abilities, reasoning, memory, and speed of perception (Thurstone, 1938). Therefore, the key to introducing intelligence to robotics is to acquire information (sensors, speech, and image recognition) and its skillful use for action, that is, proper response to external stimuli, communication, and language, etc. Essentially, artificial intelligence was created to communicate machines and computers with human intelligence.

The space in which we live is a space of information (in a given point in time and space we can assign information, data, history), space:
– geometric (we can determine, or measure, positions x, y, z in the Cartesian system of each point); and
– physical (we can determine, or measure, the physical values of an electromagnetic field, etc.).
For the physician, the area of surgery is also a space:
– biological (we can determine, i.e., diagnose and describe, physiological and pathological traits—related to the development of the disease—the organism, organs, tissues, cells, etc.).
For the patient, doctor, and administrator, important role plays also space:
– economic (in which each activity is assigned a value of financial burden);
– legal/ethical (in which rights and threats are defined for each decision); and

– statistical (evaluation of chances and threats for the treatment strategy being pursued).

Returning to intelligence—this is our device to optimize the decisions we make in this multidimensional information space.

Man is information – about 1 Tb during the year. Sick man (for instance who suffers from diabetes or heart) collects even more, very important information.Data analysis, the so-called big data, shows spectacular successes (and failures) by analyzing the impact of drugs, diet, or behavior on the effectiveness of treatment. Where the scope of cause-and-effect analysis, application and verification of theory, and strict algorithms ends, a field for the use of multilayer computer neural networks, self-learning, and artificial intelligence applications appears.

Social work and service robots require cooperation with humans, so for them, artificial intelligence is a measure of the success of communication with a human being. But for surgical robots, a human appears in the form of a complex biological work space, tissues, proteins, genes, and physiology, reacting to a specific situation requiring the surgeon's knowledge or ability to assess what the effects will be of, for example, incisions in a specific manner or removal of a specific body part (e.g., in the case of oncological diseases).

A human is just as good as his or her senses and the ability to use information (reason, experience). The robot is only as good as its ability to use information. Since we still do not have possibilities to use artificial intelligence in all seven features (following L.L. Thurstone), therefore, today surgical robots are only telemanipulators controlled by surgeons (we expect that they have all intelligence features, which are obviously verified during medical career stages).

5.2 How the surgeon's decision is made

Let's try to analyze the decision making process by the operator from the point of view of artificial intelligence methods used in process automation. Decision-making situations can be divided into determined (making decisions under conditions of certainty), random (in risk conditions), and conflict (in conflict conditions, game theory). In the decision-making process, the following phases can be distinguished: recognition (collection of information), modeling (construction of a decision-making model), and deciding (selection of the best decision). The implementation of the objective requires the use of available resources (material, financial, time, human, etc.) in the form of a strategy.

The surgeon often makes decisions in conditions of external uncertainty (lack of sufficient knowledge about the operating environment) and internal (lack of knowledge and experience). Using advisory and training systems can reduce the area of uncertainty.

In situations of uncertainty, the decision-making procedure may be heuristic (open) or algorithmic (closed). Unlike an algorithm that provides a detailed description of steps, the heuristics provide useful hints. In situations without a clearly defined pattern of conduct and in unspecified conditions of the work environment, decisions should be of a creative nature. Computerized advisory systems and controlled, programmed navigation reduces the occurrence of a surgical error (through the analysis of diagnostic information and current real assessments of the operation field, the algorithm is introduced).

Erroneous decisions are based on the illusions of correlation and determination, as well as maintaining the first assessments and consequences. It should be noted that during a teleoperation the surgeon is away from the patient, but also from his or her team. The decisions made, especially in the situation of a threat, are influenced by many features, e.g., behavior, sensations, gestures of team members, temperature, smell. On the other hand, the possibility of moving away the elements disrupting the surgeon's work can play a positive role for the results (time and effect of the operation). However, to use the unique possibility of focusing only on the operated object, balanced movements, and the implementation of tasks, it is necessary to provide and adequately present the information necessary to conduct—including those that appear during the procedure outside the observation field.

Human decision-making is based on one of the two thought patterns: Cartesian (cause-and-effect thinking based on classic logic, thinking precedes action: "I think, I decide") or Darwinian (trial and error method, action precedes decisions: "I check effect of the decision made"). When it is possible (for example for neurosurgical operation), due to the predictable behavior and geometry of the object, based on diagnostics, the plan of the trajectory and tasks of the tool is made. After determining the spatial fit of the robot relative to the patient's body, tasks can be performed automatically to a large extent.

The Robin Heart project assumes the use of complete motion sequences encoded in the processor and actuated with microjoysticks, corresponding to functionalities such as sewing or tissue separation.

Surgical telemanipulators use both standards. The role of individual experience and the features associated with the selection of appropriate decisions is evidenced by the fact that in all statistical reports on the use of various surgical techniques, the most reliable ones are those referring to who (or what center) performed the operation. Similar relationships are observed in other team activities in which the role of a leader plays a decisive role. Thus, the surgeon's work bears not only the features of executive perfection (craftsman), but to a large extent, also managerial (conductor, trainer).

We are currently observing rapid progress in the development of artificial intelligence that will lead to the independence of robots. The surgeon-operator, however, will control surgical robots while there is no proper sensor. People make decisions based on information obtained through the senses and based on their own experience (learning process). There are currently no force sensors on the market with the appropriate sensitivity and strength to be mounted in surgical instruments. However, the door for the introduction of automation and AI is already opening: in March 2018, Corindus Vascular received 510 (k) clearance from the U.S. Food and Drug Administration (FDA) for the first automated robotic movement designed for its CorPath GRX platform. Called "Rotate on Retract" (RoR), the proprietary software feature is the first automated robotic movement in the technIQ Series for the CorPath GRX platform. It allows the operator to navigate quickly to a targeted lesion by automatically rotating the guidewire upon joystick retraction (Corindus, 2018).

5.3 Ergonomics

Ergonomics introduces a human factor to the design of machines, buildings, and objects in the human environment. It can be argued that, due to difficulties in optimizing the ergonomic work of the surgeon, only medical robots allow the introduction of all aspects of the surgeon's ergonomic workplace. In 1916, Frank Gilbreth, a pioneer in the field of time and motion study, stated that "…surgeons could learn more about motion study, time study, waste elimination, and scientific management from the industries than the industries could learn from the hospitals" (Gilbreth, 1916).

Efforts to create a more user-friendly operating room environment require the rethinking of traditional concepts of architecture, asepsis, and staffing. Berguer, in a review of ergonomics in surgery, divided the issue into the following categories:

a. visualization;

b. manipulation;

c. posture;

d. mental and physical workload; and

e. the operating room environment (Berguer, 1999).

The introduction of the possibility of long-distance surgery into the surgical practice will redefine the problem of locality (space, distance) and time (signal delays). At present, there are ongoing processes to reconcile the legal aspects of risk and responsibility sharing.

The creation of the Robin Heart Shell console project preceded the research of the method (kinematics, dynamics) of performing surgical operations using various types of tools (force measurements and motion analysis, trajectory, recorded with the help of several cameras). The result of these tests is the optimization of the haptic devices described above and the workstations (e.g., proper angles of setting monitors and access to all elements of the robot's control system). Mechanical problems of ergonomics are obvious; however, the issues of software ergonomics are worth special attention. Medical robots bring this completely new aspect to the responsibilities of a surgical equipment designer.

Bernstein is regarded as one of the founders of the modern theory of movement behavior, who in his work on the Structure of Movements (1947) made its foundations. The beginning of the movement is, according to Bernstein, possible after imagining the goal and constructing the program of action. The introduction of a computer between the surgeon's hand and the tool makes it possible to apply software limitations and improve the movement and tasks performed by the tools. Together with the operational planning system and advisory programs that can operate on line, it creates a system that limits risks, can be improved (AI), and can set standards for specific procedures. One of the most difficult challenges is the optimization of the motion transfer system from the operator's hand to the movement of the surgical tool (on-line orders for performed tasks) (Nawrat, 2011).

Bidard reviewed and described the design work of the input device for tele-surgery (Bidard et al., 2005). In the work of Mucha (Mucha et al., 2015, 2018a), the currently used user interfaces were reviewed, and the results of the Zabrze team's work on the technical solutions for the transfer of haptic information to Robin Heart robots were summarized.

5.4 Software ergonomics

A surgical robot is now more a mechatronic tool than an IT (information technology) tool. However, the challenges of appropriate decision-making flexibility and precision (especially in the absence of professional staff in the conditions of demographic problems) indicate that the direction of development toward autonomous (partly or fully) medical robots is the most up-to-date. The idea of spaceflight and establishing bases in distant objects from Earth and the need to secure medical service has also returned.

The principles of ergonomics should be applied when creating software—ensuring the effectiveness, efficiency, and satisfaction of the employee. These three properties make up the software utility. Let us mention a few principles (called heuristics) of software ergonomics: feedback (for each activity there should be a reaction or system information); availability (providing tools and information according to the user's needs); simple and natural dialogue; application of the user's language (language of symbols from the user's environment); reducing the load on short-term memory (no need to memorize a lot of information); confirmation of activities (information on the effect of activities); and elimination of errors. The software created as part of the Robin Heart project meets these principles.

The issues discussed are the key to building the proper telemanipulator control system and software. Man, the operator of a surgical robot, can be treated as an element of the control system (we take into account both the processing of information in the brain and its dependent motor coordination of precise control—motion control, position, speed, and strength—through the robot interfaces) connected by the system IT and operation of the robot electromechanical system with an executive tool. In terms of artificial organs—and so you can treat the robot (as a prosthesis, prolonging the surgeon's hand)—it is a hybrid organ (because it is necessary for proper operation to use cells, natural organs, simply human-operator) (Nawrat, 2011).

5.5 How to improve the decision making system of the surgeon—The robot's operator

The surgeon, knowing the patient's history (interview), documented diagnostic data, knowledge and professional experience, risk and chances, or potential prognosis of changes and progression of the disease, selects the optimal treatment strategy for the patient, including specific surgical interventions. However, when undertaking surgical procedures, it is usually necessary to modify the tactics by reacting (intelligence) to the existing image of the operation field and the emerging effects of introduced modifications of organs and tissues by means of surgical instruments, materials, devices, and implants.

Considering all of the above-mentioned areas of knowledge and uncertainty, Robin Heart's robotic development strategy assumes the introduction of a standard based on:

1. education (practical master-student type and theoretical type access to the knowledge base, clinical case database);
2. planning (based on virtual space technologies such as our virtual operating room);
3. training (both on physical trainers containing the model of surgical scenes and animal tissue as well as virtual and computer);
4. verification of progress (based on facts, i.e., measurable, objectified, traits of practical skills);
5. system of direct advising during on-line operation (based on facts: diagnostic data and results of simulations of the effects of surgery—e.g., blood pressure and blood flow during the operation of the cardiovascular system such as by pass or replacement of the heart valve).

From a practical point of view, however, the ergonomics of software and mechanical design as well as sensory information (operation field image and touch information) are of the greatest importance to the success of the telemanipulator implementation, convenience, and comfort of the workplace, i.e., user interfaces. For the surgeon, the most important in this regard is the haptic controller responsible for transferring decisions with the help of a hand with a specific motion and a task to perform for surgical instruments.

6 Conclusion

The success of robots in surgery depends on many factors: effective, precise mechatronic tools and an efficient, ergonomic control system. Lack of tactile information and force feedback impairs the surgeon's efficiency during surgery and increases the risk of making a mistake.

The Robin Heart family of Polish robots has a chance of becoming a commonly used telemedical system facilitating the performance of some parts of operations in minimally invasive, precise manner, safe for the patient and the surgeon (Nawrat et al., 2002; Nawrat and Kostka, 2006, 2008; Nawrat, 2008, 2011).

The lack of force feedback is one of the main barriers in the progress and widespread application of robotic surgery. The main tasks of the surgical robot control are the mapping and analysis of the movements of the surgeon operator (position/velocity and possibly other physical parameters), and also facilitating arm movement by providing control signals to the actuators. Additionally, it is desirable to reverse transfer the force/touch information to the person handling the tools. These signals can help the operator to make immediate correcting actions during the operation: cutting, separation, handle and move tissues, to care vascular clamping, to tie a knot, to recognize the type of tissue (pathology, calcification), to manipulate between different elements of internal organs without the risk of harming neighboring tissue, and also to sense collision of arms/or tools by automatic recognition.

Currently used endoscopic and robotic surgical instruments are rigid and straightforward, so they have limited reach. The introduction of adjustable stiffness and geometry tools can have a fundamental impact on extending endoscopic surgery capabilities to many patients in need.

A surgical robot is today an IT tool rather than a mechanical one. Man, the operator of a surgical robot, can be treated as an element of the control system connected by the IT system and operation of the robot electromechanical system with the executive tool (we take into account both the processing of information in the brain and its dependent motor coordination of precise control via robot interfaces).

Surgical robots are a way to introduce standardization and reduce invasiveness, while ensuring proper operation safety. Robots (cobots) interacting with people change the way of performing many professions. Robots create the possibility of standardization and constant improvement of quality through learning (AI) and communication with a professional information network (professional databases and management system) as well as with other medical devices (diagnostic, therapeutic, rehabilitation), and also elements of hospital infrastructure.

Robots are created to multiply human freedom. They reduce our workload, add strength, precision, efficiency, they bring back to society people who lost their freedom due to diseases and organ failures (artificial organs, exoskeletons). That is why it is really worth developing robotics.

Acknowledgments

The Robin Heart Project was supported by KBN 8 T11E 001 18 and projects: PW-004/ITE/02/2004, R1303301 and R13 0058 06/2009, and NCBR—The National Centre for Research and Development—Grants: R1303301 and R13 0058 06/2009, Robin PVA—no 178576, TeleRobin—no 181019 and many sponsors. The project of flexible tool supported in part by the European Commission within the STIFF-FLOP FP7 European project FP7/ICT-2011-7-287728 and grants of the ENIAC "INCITE" Project No. 621278.

The presented research and implementation works were financed from funds for Polish and European science, which in total do not exceed, so far, the purchase of one da Vinci robot. The team of the FRK participated in them and, at various stages, employees of several universities and enterprises—I would like to show everyone all the gratitude and respect. Especially for patients—for their faith—that we will make it.

References

Alexander III, A.D., 1974. Impacts of telemanipulation on modern society. In: 1st CISM-IFTomm Symposium on Theory and Practice of Robots and Manipulators, Udine. CISM Courses and Lectures No. 201, vol. II. Springer Verlag, Vien, New York.

Anvari, M., 2007. Remote telepresence surgery. The Canadian experience. Surg. Endosc. 21 (4), 537–541.

Anvari, M., McKinley, C., Stein, H., 2005. Establishment of the world's first telerobotic remote surgical service: for provision of advanced laparoscopic surgery in a rural community. Ann. Surg. 241 (3), 460–464.

Arezzo, A., Mintz, Y., Allaix, M.E., Arolfo, S., Bonino, M., Gerboni, G., Brancadoro, M., Cianchetti, M., Menciassi, A., Wurdemann, H., Noh, Y., Althoefer, K., Fras, J., Glowka, J., Nawrat, Z., Cassidy, G., Walker, R., Morino, M., 2017. Total mesorectal excision using a soft and flexible robotic arm: a feasibility study in cadaver models. Surg. Endosc. 31 (1), 264–273. https://doi.org/10.1007/s00464-016-4967-x (Epub 2016 June 23: 1–10).

Argenziano, M., Oz, M.C., Kohmoto, T., et al., 2003. Totally endoscopic atrial septal defect repair with robotic assistance. Circulation 108 (Suppl. 1), II191–194.

Berguer, R., 1999. Surgery and ergonomics. Arch. Surg. 134, 1011–1016.

Bidard, C., Brisset, J., Gosselin, F., 2005. Design of a high-fidelity haptic device for telesurgery. In: Proc. IEEE ICRA 2005, Barcelona, Spain, pp. 205–210.

Bowersox, J.C., Shah, A., Jensen, J., Hill, J., Cordts, P.R., Green, P.S., 1996. Vascular applications of telepresence surgery: initial feasibility studies in swine. J. Vasc. Surg. 23, 281–287.

Chitwood, W.R., Wixon, C.L., Elbeery, J.R., Moran, J.F., Chapman, W.H.H., Lust, R.M., 1997. Video-assisted minimally invasive mitral valve surgery. J. Thorac. Cardiovasc. Surg. 114 (5), 773–780.

Corindus Info, 2018. https://www.therobotreport.com/corindus-corpath-grx-clearance-japan/.

de Canniere, D., Wimmer-Greinecker, G., Cichon, R., Gulielmos, V., Van Praet, F., Seshadri-Kreaden, U., Falk, V., 2007. Feasibility, safety, and efficacy of totally endoscopic coronary artery bypass grafting: multicenter European experience. J. Thorac. Cardiovasc. Surg. 134, 710–716.

Falk, V., Diegeler, A., Walther, T., Jacobs, S., Raumans, J., Mohr, F.W., 2000. Total endoscopic off-pump coronary artery bypass grafting. Heart Surg. Forum 3 (1), 29–31.

Gao, C.Q., 2014. Overview of robotics cardiac surgery. In: Gao, C.Q. (Ed.), Robotic Cardiac Surgery. Springer Science + Business Media, Dordrecht, pp. 1–13. https://doi.org/10.1007/978-94-007-7660-9_1.

Gao, C.Q., Yang, M., Xiao, C., Wang, G., Wu, Y., Wang, J., Li, J., 2012. Robotically assisted mitral valve replacement. J. Thorac. Cardiovasc. Surg. 143 (4), 64–67.

Gilbreth, F.B., 1916. Motion study in surgery. Can. J. Med. Surg. 40, 22–31 (Google Scholar).

Humphries, M., 2019. China Performs First 5G Remote Surgery Matthew Humphries. (January 15). https://www.pcmag.com/news/365992/china-performs-first-5g-remote-surgery.

Kirkpatrick, T., LaGrange, C., 2016. Cases & Commentaries WebM&M. Robotic Surgery: Risks vs. Rewards (February 2016). https://psnet.ahrq.gov/webmm/case/368/Robotic-Surgery-Risks-vs-Rewards-.

Kroczek, K., Kroczek, P., Nawrat, Z., 2017. Medical robots in cardiac surgery—application and perspectives. Kardiochir. Torakochi. Pol. 14 (1), 1–5.

Lis, K., Lehrich, K., Mucha, Ł., Rohr, K., Nawrat, Z., 2015. Robin Heart PortVisionAble—idea, design and preliminary testing results. In: Proceedings of the 10th International Workshop on Robot Motion and Control. Poznan University of Technology, Poznan, Poland, pp. 176–181 (July 6–8).

Lis, K., Lehrich, K., Mucha, M., Nawrat, Z., 2017. Concept of application of the light-weight robot Robin Heart ("Pelikan") in veterinary medicine: a feasibility study. Med. Weter. 73 (2), 88–91. ISSN: 0025-8628.

Loulmet, D., Carpentier, A., d'Attellis, N., et al., 1999. Endoscopic coronary artery bypass grafting with the aid of robotic assisted instruments. J. Thorac. Cardiovasc. Surg. 118, 4–10.

Małota, Z., Nawrat, Z., Sadowski, W., 2018. Banchmarking for surgery simulators. In: Konstantinova, J., Wurdemann, H., Shafti, A., Shiva, A., Althoefer, K. (Eds.), Soft and Stiffness-Controllable Robotics Solutions for Minimally Invasive Surgery: The STIFF-FLOP Approach. River Publishers Series in Automation, Control and Robotics, England, ISBN: 978-87-93519-72-5, pp. 309–323.

Marescaux, J., Rubino, F., 2004. The Zeus telerobotic system. In: Ballantyne, G.H., Marescaux, J., Giulianotti, P.C. (Eds.), Primer of Robotic & Telerobotic Surgery. Lippincott Williams & Wilkins, pp. 61–65.

Mehmanesh, H., Henze, R., Lange, R., 2002. Totally endoscopic mitral valve repair. J. Thorac. Cardiovasc. Surg. 123 (1), 96–97.

Mohr, F.W., Falk, V., Diegeler, A., Walther, T., Gummert, J.F., Bucerius, J., Jacobs, S., Autschbach, R., 2001. Computer-enhanced "robotic" cardiac surgery: experience in 148 patients. J. Thorac. Cardiovasc. Surg. 121 (5), 842–853.

Mucha, Ł., Rohr, K., Lis, K., Lehrich, K., Kostka, P., Sadowski, W., Krawczyk, D., Kroczek, P., Małota, Z., Nawrat, Z., 2015. Postępy budowy specjalnych interfejsów operatora robota chirurgicznego Robin Heart. Med. Robot. Rep. 4, 49–55.

Mucha, Ł., Lis, K., Lehrich, K., Nawrat, Z., 2018a. RobinHand haptic device. In: Konstantinova, J., Wurdemann, H., Shafti, A., Shiva, A., Althoefer, K. (Eds.), Soft and Stiffness-Controllable Robotics Solutions for Minimally Invasive Surgery: The STIFF-FLOP Approach. River Publishers Series in Automation, Control and Robotics, England. ISBN: 978-87-93519-72-5, pp. 289–305.

Mucha, Ł., Lis, K., Krawczyk, D., Nawrat, Z., 2018b. The design of a functional STIFF-FLOP robot operator's console. In: Konstantinova, J., Wurdemann, H., Shafti, A., Shiva, A., Althoefer, K. (Eds.), Soft and Stiffness-Controllable Robotics Solutions for Minimally Invasive Surgery: The STIFF-FLOP Approach. River Publishers Series in Automation, Control and Robotics, England, ISBN: 978-87-93519-72-5, pp. 221–228.

Nawrat, Z., 2001. Perspectives of computer and robot assisted surgery for heart assist pump implantation. Planning for robotically assisted surgery. In: Darowski, M., Ferrari, G. (Eds.), Assessment and Mechanical Support of Heart and Lungs. In: Lecture Notes of the ICB SeminarsPolska Akademia Nauk, Warszaw, pp. 130–149.

Nawrat, Z., 2008. Medical robots in cardiac surgery. Kardiochir. Torakochir. Pol. 5 (4), 440–447.

Nawrat, Z., 2011. Robot chirurgiczny—projekty, prototypy, badania, perspektywy. (Rozprawa habilitacyjna), Katowice.

Nawrat, Z., 2012. The Robin Heart story. Med. Robot. Rep. 1, 19–21.

Nawrat, Z., Kostka, P., 2006. Polish cardio-robot "Robin Heart". System description and technical evaluation. Int. J. Med. Rob. Comput. Assisted Surg. 2, 36–44.

Nawrat, Z., Kostka, P., 2008. Robin Heart—perspectives of application of mini invasive tools in cardiac surgery. In: Bozovic, V. (Ed.), Medical Robotics. I-Tech Education and Publishing, Vienna, pp. 265–291.

Nawrat, Z., Koźlak, M., 2007. Robin Heart system modelling and training in virtual reality. J. Automat. Mob. Rob. Intel. Syst. 1 (2), 62–66.

Nawrat, Z., Podsędkowski, L., Mianowski, K., Wróblewski, P., Kostka, P., Baczyński, M., Małota, Z., Granosik, G., Jezierski, E., Wróblewska, A., Religa, Z., 2002. Robin Heart in 2002—actual state of polish cardio-robot project. In: RoMoCo 2002, IEEE, pp. 33–39.

Nawrat, Z., Rohr, K., Fürjes, P., Mucha, Ł., Lis, K., Radó, J., Dücső, C., Földesy, P., Sadowski, W., Krawczyk, K., Kroczek, P., Szebényi, G., Soós, P., Małota, Z., 2016. Force feedback control system dedicated for Robin Heart surgical robot. Procedia Eng. 168, 185–188. Available online at: www.sciencedirect.com, Elsevier, http://www.sciencedirect.com/science/article/pii/S1877705816335238.

Nawrat, Z., Mucha, Ł., Lis, K., Lehrich, K., Rohr, K., 2018a. EP3146930 A Surgical Robot's Tool Arms Assembly EP 15186289.3 22.09.2015. Decision 1.03.2018 (printed 06.06.2018).

Nawrat, Z., Mucha, Ł., Lis, K., Lehrich, K., Rohr, K., 2018b. EP 15179357.7 Tool manipulator for a training and testing medical device 31.07.2015 (printed 06.06.2018).

Podsędkowski, L., 2003. Forward and inverse kinematics of the cardio-surgical robot with non-coincident axis of the wrist. In: SyRoCo 2003, pp. 525–530.

Radó, J., Dücső, C., Földesy, P., Barsony, I., Rohr, K., Mucha, Ł., Lis, K., Sadowski, W., Krawczyk, K., Kroczek, P., Małota, Z., Szebényi, G., Santha, H., Nawrat, Z., Fürjes, P., 2018. Biomechanical tissue characterisation by force sensitive smart laparoscope of Robin Heart surgical robot. In: MDPI Proceedings EUROSENSORS 2018 Conference. CC by: http://creativecommons.org/licences/by/4.0/.

Sant'Anna, R.T., Prates, P.R.L., Sant'Anna, J.R.M., Prates, P.R., Kalil, R.A.K., Santos, D.E., Nesralla, I.A., 2004. Robotic systems in cardiovascular surgery. Rev. Bras. Cir. Cardiovasc. 19(2) http://www.scielo.br/scielo.php?pid=S0102-76382004000200012&script=sci_arttext&tlng=en.

Satava, R.M., 2011. Future direction in robotic surgery. In: Rosen, J., Hannaford, B., Satava, R.M. (Eds.), Surgical Robotics. System application and Visions. Springer New York, Dordrecht, Heidelberg, London https://doi.org/10.1007/978-1-4419-1126-1https://books.google.pl/books?id=2cswZL0L7iIC&pg=PA3&lpg=PA3&dq=richard+satava+robotics&source=bl&ots=I8iZexMB9C&sig=ACfU3U36NGPV67O0wIOwhLJaYrMJQEvzAw&hl=pl&sa=X&ved=2ahUKEwjqpv73-cDhAhVqk4sKHbm6DP0Q6AEwDHoECAcQAQ#v=onepage&q=richard%20satava%20robotics&f=false.

Srivastava, M.C., Taylor, B., Zimrin, D., Vesely, M.R., 2014. Hybrid coronary revascularization. In: Changqing, G. (Ed.), Robotic Cardiac Surgery. © Springer Science + Business Media, Dordrecht, pp. 135–140. https://doi.org/10.1007/978-94-007-7660-9_1.

Thurstone, L.L., 1938. Primary Mental Abilities. University of Chicago Press, Chicago.

Torracca, L., Ismeno, G., Alfieri, O., 2001. Totally endoscopic computer-enhanced atrial septal defect closure in six patients. Ann. Thorac. Surg. 72 (4), 1354–1357.

Williams, S.B., Prado, K., Hu, J.C., 2014. Economics of robotic surgery: does it make sense and for whom? Urol. Clin. North Am. 41 (4), 5916. https://doi.org/10.1016/j.ucl.2014.07.013. PMID 25306170. 2014.

Further reading

Intuitive Web, 2018. www.intuitive.com.

CHAPTER FOUR

Real-time object detection and manipulation using biomimetic musculoskeletal soft robotic grasper addressing robotic fan-handling challenge

Iyani N. Kalupahana[a,b], Godwin Ponraj[a], Guoniu Zhu[a], Changsheng Li[a] and Hongliang Ren[a,c]

[a]Department of Biomedical Engineering, National University of Singapore, Singapore
[b]Department of Electronic and Telecommunication Engineering, University of Moratuwa, Moratuwa, Sri Lanka
[c]NUS (Suzhou) Research Institute (NUSRI), Suzhou, China

1 Introduction

Advancement of technologies has enabled robotics to enter into the domestic and social setting in addition to their existing industrial presence. However, unlike the organized environment of an industry, the operating area of a social robot is often unstructured. Real-time accurate visual feedback is required for precisely identifying the object and to determine the best way to pick up the object. Identifying objects with respect to the task at hand from a random environment has always been a challenge for robotic manipulation. Once identified, the end effector has to efficiently handle the object to perform the required task. Industrial manipulators are often designed specifically for a particular task, for example, a gripper to pick and place objects in a supply chain or a drilling platform to handle various machining tools. A generic anthropomorphic gripper which can do more than pick and place tasks will benefit social robots more, since their usage is increasing and extending into several application domains. In this chapter, a viable solution is presented for one such application.

Initiated by IROS 2018 (IEEE/RSJ International Conference on Intelligent Robots and Systems) which was held at Madrid, Spain (October 1–5, 2018), the fan robotic challenge task is to pick up a Spanish fan from a random position on the table, to open and close it, and to place it back on the

Control Systems Design of Bio-Robotics and Bio-mechatronics with advanced applications
https://doi.org/10.1016/B978-0-12-817463-0.00004-6

table. The solution presented in this chapter is the only one demonstrated in the conference, which operates in a way similar to a human hand. The whole setup is established on a PC platform with Python development environment. The PyKinect2 API is used to acquire the visual data from Kinect v2 which consists of RGB camera, three-dimensional (3D) depth sensor, and multiarray mic. Object recognition from the background is done with the algorithm of color detection in thresholding from the output of RGB camera (resolution 1920 × 1080). Canny edge detection and noise reduction are performed to improve the sensitivity of the results. A method of classifying the same color, different paced objects is recognized by analyzing the contour area and depth distance from IR camera (depth camera with resolution of 512 × 424). Center of mass is calculated from the image moments of the selected contour. Even though pixel coordinates have been mapped to a real-world coordinate system using linear relationship, real world to robotic coordinate system relationship is much complex. A database was used for the analysis to find the relationship between the relevant coordinate systems.

A hybrid robotic gripper consisting of a rigid tendon-driven anthropomorphic hand with flexible soft layer enhancements was used for handling the fan. The rigid framework is 3D printed using PLA and the soft layers are made of stretchable EcoFlex™ 00-30 material. The tendons are fixed to servo motors to actuate the movement of the fingers. Torque-based control is used to drive the tendons for adjusting the grab strength where as a faster position-based control is reserved for the initial grab and the final release operations. The fan is opened due to a combination of gravity and inertial forces. While closing the fan, the gripper rotates 90 degrees in a swift motion similar to the way a human hand would operate. A UR5 robotic arm is used to move the gripper to the initial position to pick up the fan, through the desired open-close trajectory and to return the fan back to the table.

Overall, the performance of the completed task is about 98% in accuracy and the results are more robust than stereo images. Besides, the proposed methodology is less in cost comparing to other techniques with SURF, STICK; less in complexity with other software such as MATLAB; less in computational power than using neural networks. The remaining chapter is organized as follows. Section 2 summarizes related works which focus on similar object identification/classification tasks and a brief introduction on soft robotic grippers involved in manipulation tasks and tactile sensors. Section 3 describes the proposed object identification methodology in detail whereas Section 4 explains about the gripper used and the sequential steps followed during the manipulation of the fan. The final section concludes with remarks and future research possibilities.

2 Related work

2.1 Vision perception

Even though the visual perception system is task specific, we propose this method for robust and more general task-specific systems to increase its functionality. A similar concept has been used for the Amazon Picking Challenge of several years (Zeng et al., 2017; Jonschkowski et al., 2016; Roshni and Kumar, 2017a). Research works have been done on this approach by running a convolutional neural network trained by a large amount of data. Prescanned 3D models were used to get the 6D object pose. Deep reinforcement learning algorithm with bioinspired structures was used to design self-sufficient agents that can learn from observations in its environment (Bhagat et al., 2019). An unsupervised machine learning algorithm can be developed to detect known objects (Barbosa et al., 2017). Currently, imitation learning algorithms are used instead of mathematical models due to countless degrees of freedom (DOF). Since the scope of this chapter is limited to one object, instead of training a network, one can detect the object from feature extraction and color detection techniques. An intensity-based approach was used by Nadeau et al. (2015) to extract a moving object from the background. It was used for the 3D pose estimation task later in this chapter. This is not only limited to pick up the object but also for the opening of Spanish fan as to how human hand operates.

Point Cloud library, an open source library for algorithms of 3D point cloud processing, has been used for implementing numerous techniques in various applications (Cousins and Rusu, 2011; Alcantara et al., 2018). If multiple objects with the same color are present in the background, CIE-Lab color space and depth images can be taken to neglect other objects and identify the object of interest (Hernandez-Lopez et al., 2012). Object detection is done using more than one active camera in a dynamic environment (Celik et al., 2017). An object is detected based on the shape and a bounding box is created. Coordinates of the bounding box are used to calculate the centroid of the object and tracking is done with active motors via Arduino-MATLAB implementation. A quick object detection technique is used in autonomous vehicles or computer-assisted driving systems (Xitao and Zhang, 2010). Camera shot angle, camera focal length, and virtual square parameters are used in solving the coordination of an object from two simultaneous images. It is performed within a minimum time in the virtual square which reduced the image size. Adaptive color thresholding is used for detection applications in robotics and embedded systems (Soans et al., 2017). Image pyramid technique is used to reduce the size of an image, following color separating, finding the edge of the image by Sobel and thinning and locating the center of the image using Hough Transform (Palungsuntikul and Premchaiswadi, 2010).

Classification of features can be applied in binary style for object detection as it will make the application faster comparing to traditional technique (Dornaika and Chakik, 2010; Nakashima and Yabuta, 2018; Chelva et al., 2016). Binary support vector machine classifier is used to classify image features into object features and nonobject features. Features are extracted based on the continuous color model and discontinuous color model of image intensity patterns with adaptive color difference signals (Ukida et al., 2012). Pixel values can be calculated on canny edge detection or B edge detection in the matching process (Yadav and Singh, 2016; Mittal et al., 2019). Once the object is identified, its image moments can be used to calculate the center of mass of the selected contour (Balakrishnan and Jaya, 2014; Roshni and Kumar, 2017b). In previous works, the largest area contour algorithm and depth segmentation techniques have been used to recognize an object from same colored multiple objects and to solve object pose uncertainty (Fan et al., 2018).

2.2 Robotic gripper

Grasping and manipulating an object has been one of the most common tasks to be done by a robot. Depending on the complexity of the task itself, it can be done with two grippers or just one. In this chapter, we use a single robotic gripper to grasp, open and close a fan as a human hand would do. Several research works have been conducted to investigate the dynamics and capabilities of multifingered robotic hands with tendon-driven systems. For example, Catalano et al. (2014) explored the design and control strategies for the Pisa/IIT soft hand, and Dollar and Backus (2016) described a three-fingered prismatic gripper with passive rotational joints. Tendon-driven designs involve in better flexibility and dexterity comparing to concentric tube designs (Li et al., 2017b). For real-time position and shape sensing in tendon-driven robots, an electromagnetic sensor placed at the distal end of the robot to obtain positional and directional information based on Bezier curve (Song et al., 2015). This method is used for unknown payloads as it does not rely on a mechanical model.

Soft robots have proven themselves to be compliant in object-handling tasks (Finio et al., 2013; Banerjee et al., 2018c; Chen et al., 2018; Tse et al., 2018) in terms of conforming to the object surface, unstructured environment, and unknown objects. Soft origami robots can also work in unstructured environments (Chan et al., 2018; Banerjee et al., 2018b). Montmorillonite (MMT) hybrid nanocoatings were applied in the skin of soft robotic gripper to increase the stretchability in MMT/elastomer bilayer and it can endure direct flame contact without ignition (Chang et al., 2018). Soft robotic manipulator made of hyperelastic silicone rubber can produce safe and comfortable interactions with the subjects (Sun et al., 2016). It is widely used in minimally invasive surgery and applications that require closer inspection and operation. Hydrogel actuators and sensors are further used in soft robots especially for drug delivery, surgery, and biorobotics (Banerjee et al., 2018a; Ponraj et al., 2018; Banerjee and Ren, 2017).

Combination of the soft actuator with tendon actuation based on flexible shafts is owned softness and three DOF (Tan et al., 2018a). A soft actuator is proposed with fused-deposition modeling technique and kinematics and statics performance are analyzed under different design and fabrication settings, different shapes of cross-sections, infill patterns, infill densities, etc. The technique was tested with a robotic hand, multisegment continuum robot and miniaturized drilling device. An interval-based framework of inherent uncertainties and pose evaluation for soft robots with flexible shafts were analyzed with a general model that extracted from prior kinematics models (Tan et al., 2018b). By bringing together the soft and rigid structures, a hybrid gripper with better object conformation and simple control can be realized. Such grippers could be advantageous in grasping tasks. Especially, envelop grasps can easily be performed with a hybrid gripper due to soft layers without external interactions (Ilievski et al., 2011). Besides, hybrid grippers have further been designed to grasp and hold the object under external disturbance (McKenzie et al., 2017; Mizushima et al., 2018).

The flexible cable-driven robotic grasper is made with Lego-like modules and length can be adjusted by adding or removing additional modules that are connected by magnets without the need for rerouting or breaking the cables (Li et al., 2017a). It uses the automatically reshaping method based on the motors current during the operation. Nonsingular fast terminal sliding mode algorithm is used for fast positioning and force tracking in the multilateral teleoperation system with reconfigurable multifingered robot (Sun et al., 2018). It supports as an integrated perceptron for the operator in the multifingered robot. Another flexible, soft robotic manipulator was implemented with six ionic polymer metal composite segments and used statistics machine learning algorithms to plan the motion paths by learning from demonstration method (Wang et al., 2016). It overcomes the difficulty in path planning due to redundant DOF.

Calibration of robotic systems can be categorized as robot-world calibration, tool-flange (hand-eye) calibration, and kinematics calibration. Kinematics calibration is performed in several ways. The model that based on a minimal product of exponentials using only position measurements is accurate compared to the traditional model which uses position and orientation (Wu et al., 2015). The position of the robot base frame with respect to the world frame is analyzed under robot-world calibration. Calibration problem is framed as $AX = YB$ (Tan et al., 2018c) where A is partially known matrix and it is solved based on nonlinear optimization evolutionary computations. Hand-eye calibration is solved using $AX = XB$ equation.

2.3 Tactile sensing

As the domain of robotic applications expands day by day, more challenges are being discovered with respect to object interaction and manipulation. In addition to the shape and size of the objects, it is additionally beneficial for the robot to understand the contact force dynamics, texture, and other physical properties.

Resistive sensors, capacitive sensors, optical sensors, ultrasonic-based sensors, piezoelectric sensors, magnetism-based sensors, and MEMS tactile sensors are few of the available types of tactile sensors. Tactile sensors specific to robotic applications have to be secured on to the robot's surface seamlessly without affecting the functionality of the robot. For example, a bulky force sensor, at the tip of a robotic gripper, to measure contact force, might impede in grasping or object-handling tasks. Hence, as an additional requirement, tactile sensors for robotics have to be thin, flexible, and/or stretchable. Resistance and capacitance-based sensors are the two most widely preferred options for flexible and stretchable applications.

The robotic gripper used in the tasks detailed in this chapter is equipped with 4×4 piezoresistive fabric flexible-based tactile sensors at the fingertips. Fabric-based piezoresistive tactile sensors are selected due to their simple design, ease of fabrication, and low cost. The tactile sensor is used to identify and monitor object-handling tasks by the gripper.

3 Object detection and hand-eye calibration

Object identification is prevailing based on the imaging stream data (Guan et al., 2015; Nadeau et al., 2015; Shi et al., 2018; Ren et al., 2011, 2017; Ren and Dupont, 2012) (or RBGD modality), analysis of depth data, estimation of object position and orientation (Guo et al., 2019), and the eye–hand calibration for a variety of applications. An overview of the vision perception task of object identification to centroid identification in UR5 coordinate system is shown in Fig. 1.

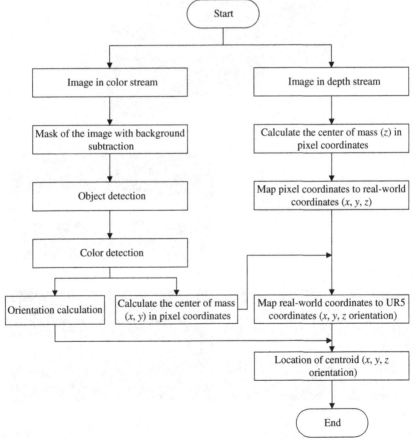

Fig. 1 Overview of the vision perception task of object identification to centroid identification in UR5 coordinates.

3.1 Object identification using color stream data

The object is identified using color detection algorithm by choosing lower and upper threshold values. Values should be selected accurately as it affects the sensitivity of the identification with the background (Fig. 2E). It is very sensitive to the changes in the lighting as well. Mask is created to save an image between the selected threshold window and it is used for the algorithms mentioned in the following sections (Fig. 2C). A wide range of edges

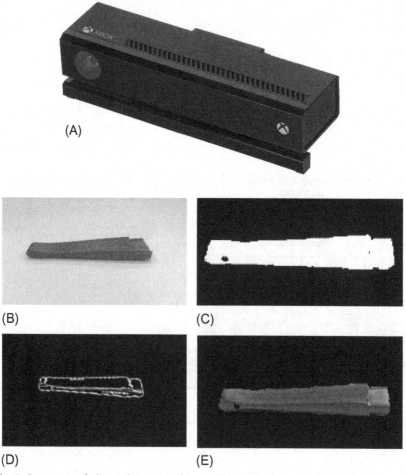

(A)

(B)   (C)

(D)   (E)

Fig. 2 Segments of object detection of Spanish fan. (A) Kinect v2, (B) an object in real space, (C) mask of the object under threshold value, (D) edge detected results, and (E) color object detection in success.

in the image can be detected using the multistage algorithm of Canny edge detection. Noise reduction is performed with a 5 × 5 Gaussian filter as the first step of the algorithm. It is used to filter with Sobel kernel in the horizontal and vertical direction which results in the first derivation of the horizontal (G_x) and vertical direction (G_y). Then, the edge gradient (G) and direction (θ) can be calculated as

$$G = \sqrt{G_x^2 + G_y^2} \tag{1}$$

$$\theta = \arctan\left(\frac{G_y}{G_x}\right) \tag{2}$$

Nonmaximum suppression and hysteresis thresholding are performed to the results from the gradient. Finally, strong edges of the image can be obtained through this method (Fig. 2D).

The algorithm was extended to find the contours of the captured view in color stream. Contour is a curve that joints all the continuous points, which is based on the same intensity or color. It can be calculated from the NumPy (Python library) array of coordinates in boundary points. Several contours may exist in the image as it captures the color detected results and the output of canny edge detection. A method is needed to implement for finding the contour of the target object with the others.

Based on the contours, the image moment can be calculated as a weighted average of the intensities in image pixels. It is further used to find special properties to the image such as area, the center of mass and information on its orientation. Then, areas of all the contours have been calculated and a NumPy array is created from them. The maximum value is selected from it and is considered as the object that is required to be identified from the background. Calculation of the moment in the selected contour is performed and the values are stored in another NumPy array. Center of mass is further calculated from the moments that have been obtained previously (Fig. 3B). Raw moments can be calculated as

$$M_{pq} = \int_{-\infty}^{\infty} \int_{-\infty}^{\infty} x^p y^q f(x, y)\, dx\, dy \tag{3}$$

where $p, q = 0, 1, 2, \ldots$, $f(x, y)$ is a 2D continuous function.

Coordinates (C_x, C_y) of the center of mass are defined as

$$C_x = \frac{M_{10}}{M_{00}}, \quad C_y = \frac{M_{01}}{M_{00}} \tag{4}$$

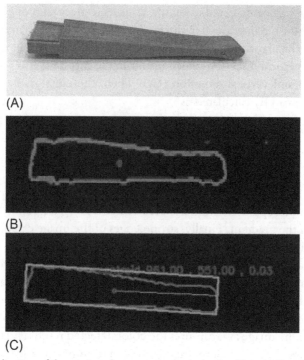

(A)

(B)

(C)

Fig. 3 Visualization of the center of mass and orientation in a Spanish fan. (A) Real-time object, (B) identified center of mass, and (C) center of mass with the suitable opening direction of the fan (orientation).

3.2 Depth data

The C_z coordinate is required to obtain the complete 3D coordinates of the center of mass. It is collected from the depth data in Kinect v2. Near mode is enabled in Kinect v2 before proceeding to collect data from the depth sensor. It is more sensitive to closer objects that lie between 50 and 200 cm. Each pixel in the depth frame gives the pixel's distance (in millimeters) from the Kinect sensor w.r.t. reference point. Since the RGB, and the IR camera are not located at the same location, the stereo vision problem arises, that is, coordinate values in one image are not the same in another image. To tackle this issue, image registration is conducted, which aligns RGB color stream values with depth values. Then, we can obtain the depth value of the centroid corresponding to the coordinates (x, y), which completes the 3D coordinates of the center of mass.

3.3 Orientation estimation

An accurate estimation of the orientation of the fan is needed for the robotic arm to approach the fan before grasping it. The orientations are expressed as the angle (α) of the line passing through the center of mass and the midpoint of the two narrow edges of the contour with respect to the XY-plane, YZ-plane, and XZ-plane of the Kinect. It will result in roll, pitch, and yaw angles. An exact position is required by the UR5 for grasping the Spanish fan as it concerns in opening and closing. Grasping the Spanish fan near to the center of mass is the best option that can be used to open the fan in a less force (Fig. 3C). Standard distance equations in trigonometry were used to calculate the angle α. This orientation estimation based on the Kinect data has a measurement accuracy of 1 degree.

3.4 Hand-eye calibration

The ROBOTIS Dynamixel SDK and the python-urx API are used in the control of the robotic gripper and robotic arm, respectively. The hand–eye calibration is performed between the UR5 and the Kinect v2 to establish the mapping relationship between the respective coordinates of the two devices. Chessboard pattern was used for the calibration. With the hand–eye calibration, the visual data read from the Kinect v2 can be directly sent to the UR5 for motion generation (Remy et al., 1997; Horaud and Dornaika, 1995).

Six DOF includes positional data frame (x, y, z) as well as from orientational data frame (R_x, R_y, R_z). Let A be the transformation between the two poses (positions and orientations for short) of the Kinect frame (x_pixel, y_pixel, z_pixel, $R_x$pixel, $R_y$pixel, $R_z$pixel) and let B be the transformation between the two poses of the hand frame (x_robo, y_robo, z_robo, $R_x$robo, $R_y$robo, $R_z$robo). Let X be the transformation between the robot frame and the Kinect frame. A, B, and X are related by the formula given by the following homogeneous equation. A is a 4 × 4 matrix.

$$AX = XB \tag{5}$$

$$A \equiv \begin{pmatrix} R_A & t_A \\ 0 & 1 \end{pmatrix} \tag{6}$$

where R_A is a 3 × 3 rotation matrix formulated from (R_x, R_y, R_z) and $t_A \equiv (x, y, z)^T$ is a 3D column vector.

$$R_A = R_x(\alpha) R_y(\beta) R_z(\gamma) \tag{7}$$

where yaw, pitch, and roll angles are represented by α, β, and γ, respectively.

$$R_x(\alpha) \equiv \begin{pmatrix} 1 & 0 & 0 \\ 0 & \cos(\alpha) & -\sin(\alpha) \\ 0 & \sin(\alpha) & \cos(\alpha) \end{pmatrix} \tag{8}$$

$$R_y(\beta) \equiv \begin{pmatrix} \cos(\beta) & 0 & \sin(\beta) \\ 0 & 1 & 0 \\ -\sin(\beta) & 0 & \cos(\beta) \end{pmatrix} \tag{9}$$

$$R_z(\gamma) \equiv \begin{pmatrix} \cos(\gamma) & -\sin(\gamma) & 0 \\ \sin(\gamma) & \cos(\gamma) & 0 \\ 0 & 0 & 1 \end{pmatrix} \tag{10}$$

In Kinect system, matrix A is obtained from calibrating the fixed chessboard pattern with its associated calibration frame. Consider, A_1 and A_2 as the transformations from calibration frame to Kinect frame at two different positions (Fig. 4). A can be derived as

$$A \equiv A_2 \cdot A_1^{-1} \tag{11}$$

If B_1 and B_2 are considered as transformations from hand frame to base frame in positions 1 and 2, B can be derived as

$$B \equiv B_2^{-1} \cdot B_1 \tag{12}$$

Let Y be transformation from hand frame to calibration frame. X can be calculated in the position 1 as follows:

$$X \equiv A_1 \cdot Y \tag{13}$$

Applying Eqs. (11), (12), into Eq. (5)

$$A_2 \cdot Y \equiv A_1 \cdot YB \tag{14}$$

After the camera is calibrated to its extrinsic parameters (A_i, to the position ith), matrix M_i is decomposed into intrinsic and extrinsic parameters of a 3×4 matrix.

$$M_i \equiv C \cdot A_i \tag{15}$$

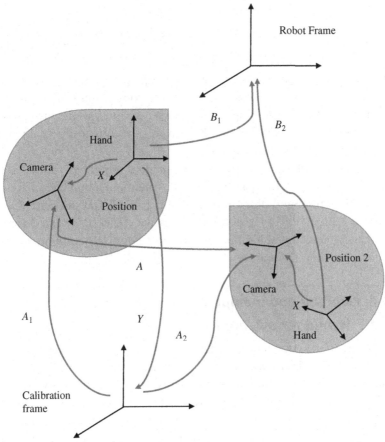

Fig. 4 Calibration diagram w.r.t. different frames, transformation between frames and two positions is marked.

$$M_i \equiv \begin{pmatrix} \alpha_u & 0 & u_0 & 0 \\ 0 & \alpha_v & v_0 & 0 \\ 0 & 0 & 1 & 0 \end{pmatrix} \begin{pmatrix} R_A^i & t_A^i \\ 0 & 1 \end{pmatrix} \qquad (16)$$

The parameters α_u, α_v, u_0, and v_0 describe the affine transformation between the camera frame and the image frame. By combining Eqs. (14), (15)

$$M_2 \cdot Y \equiv M_1 \cdot YB \qquad (17)$$

Projection of point P onto the image can explained as

$$u \equiv \frac{m_{11}x + m_{12}y + m_{13}z + m_{14}}{m_{31}x + m_{32}y + m_{33}z + m_{34}} \quad v \equiv \frac{m_{21}x + m_{22}y + m_{23}z + m_{24}}{m_{31}x + m_{32}y + m_{33}z + m_{34}} \quad (18)$$

where u and v are the image coordinates of the p projection of P and m_{ij} are the coefficients of M_1. x, y, and z represent the coordinates of P in the calibration frame. Previous two equations describe the line passing through the center of projection and through p. It is given according to the calibration frame as well as to the Kinect frame.

Closed-form solutions were implemented as separable solutions or simultaneous solutions. In separable solution, rotation is found first and positional components found later (Shiu and Ahmad, 1989; Tsai and Lenz, 1989; Chou and Kamel, 1991; Horaud and Dornaika, 1995; Park and Martin, 1994). In simultaneous solutions, rotation and positional components are found at same time (Daniilidis, 1999; Andreff et al., 1999).

4 Planning and manipulation

4.1 Hybrid robotic gripper

The hybrid robotic gripper used in the experiments has a two-part structure. The first part is a rigid tendon-driven modular anthropomorphic gripper which acts as a skeleton for the whole device (Fig. 5A). It has five fingers similar to a human hand and each finger has two links and one DOF. The thumb has an additional DOF orthogonal to the existing one to implement the abduction/adduction movement. The hand further has a single common joint in place of the metacarpals of the four fingers adding one more DOF. The links are designed in such a way that the internal joint angles do not exceed 180 degrees. The tendons are attached to servo motors and each of the seven DOF can be controlled individually by actuating the respective tendons. The modular design, kinematics, and the detailed explanation about the underactuated control of the gripper can be better understood from Li et al. (2017a) and Sun et al. (2018).

The second part is made up of flexible soft layers which can be attached to or removed from individual fingers by means of stretchable silicone attachment bands. It has a layered architecture specifically designed to enhance the grasp and to assist in object-handling tasks of the gripper. A 4×4 multitaxel piezoresistive fabric-based tactile sensor provides the necessary tactile feedback. It is encapsulated inside a thin soft layer at the rear side of the structure such that upon integration. It is sandwiched between the soft

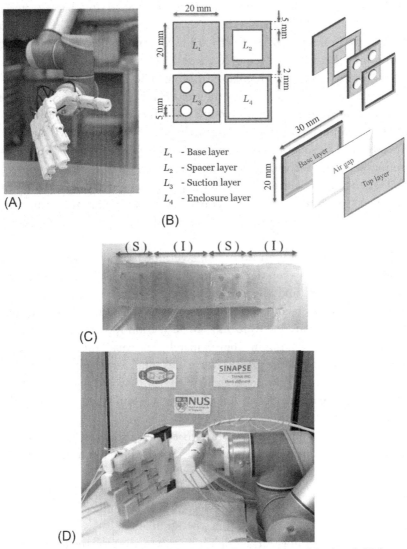

Fig. 5 The hybrid robotic gripper. (A) Rigid tendon-driven modular hand; (B) layered architecture of the soft layer; (C) actual soft layer; and (D) the combination of the rigid-soft parts.

enhancement layers and the rigid links of the gripper. Due to the scope of the chapter, the tactile sensor is not discussed in detail here. A detailed explanation on the structure, working principle, related electronics, and calibration of such tactile sensors, can be perceived our previous works (Kirthika et al., 2017; Ponraj et al., 2017; Ponraj and Ren, 2018).

The soft layers are made out of a soft yet strong and stretchable material EcoFlex™ 00-30 (Smooth-On, Inc., Macungie, PA, USA). The two-part precursors (EcoflexA and EcoflexB) are mixed in 1:1 ratio by weight and poured into the molds of required shapes. The mixture is cured at 80°C for 2 h to form a rubbery stretchable material. The dimensions of the structures are custom made to fit over the rigid hand. The soft layers comprise alternative inflation and suction modules which can be pneumatically actuated by vacuum pumps controlling their positive and negative pressure, respectively. The suction module is used to enhance the grip and the inflation module is for increasing the contact between the hand and the object. While the tendons control the overall gripping motion and the orientation of the hand, the soft layers provide contact modulation to improve the grip. The dimensions and layered architecture of the suction and the inflation modules are as given in Fig. 5B.

The suction module consists of four layers the base layer, the spacer layer to introduce gaps between the base and the suction layers, the suction layer with suction holes, and the enclosure layer to establish proper contact boundary with an object and to reduce air leaks. The layers are attached together using silicon glue to form a single suction module of dimensions 20 mm × 20 mm × 6 mm approximately. A small tube of diameter 1 mm was inserted into the spacer layer as a channel for the pneumatic actuation. When negative pressure is applied, the air between the suction layer and the object will be sucked out to create vacuum thus increasing the contact grip.

The inflation module consists of two layers of thickness 2 mm sealed together at the boundaries. The dimensions of the inflation modules (without actuation) are 20 mm × 30 mm × 4 mm approximately. Similar to the suction module, a small tube is inserted between the layers to facilitate actuation. When positive pressure is applied, the air is pumped into the space between the two layers thus inflating the module.

The suction and inflation modules are arranged alternatively and glued together to form a single soft finger (Fig. 5C). Additionally, thin soft strips are attached to the modules as extensions and loops to facilitate easy mounting over the rigid hand. The modules are arranged such that the suction module comes over the link of the rigid hand whereas the inflation module comes over the joint (Fig. 5C). The rationale behind such a design is to maximize the contact area between the robot and the object. During object-handling tasks using a rigid hand, the links of the gripper will come in contact with the object whereas the joints will remain untouched. By incorporating an inflation module over the joints, the gap between the gripper and object can be filled providing maximum contact similar to a human handling an object.

Each of the suction and the inflation modules were actuated using 12-V vacuum pumps individually. The normal operating pressure of the inflation module was in the range of 37–44 kPa whereas the suction module operated at −60 kPa. The variation in the operating range of the positive pressure is due to the fact that varying degrees of inflation is needed depending upon the space to be filled between the gripper and the object.

4.2 Planning and control

Corresponding to the seven DOF of the gripper, seven servo motors (Dynamixel MX-64RRobotis Dynamixel MX actuator, Lake Forest, CA, USA) are employed as the actuators, where each servo motor takes responsibility for one DOF. Torque control and position control strategies are used in the servo motor control during the fan manipulation. The opening and closing of the fan are performed by a combination of the robotic arm motion and the gripper manipulation. A python code was developed that moves the UR5 arm along a predefined trajectory following several waypoints while the gripper opens, closes, and adjusts the grip of specific fingers to manipulate the fan. Gravity and inertial force from the movement cause the fan to open and close. The opening and closing strategies of fan manipulation are shown in Fig. 6.

Once the position and orientation of the fan were estimated from Kinect data, the UR5 robotic arm is moved to an initial position from the desired approach angle based on the hand–eye calibration. The fingers are positioned in close proximity to the fan (Fig. 6A) using position control mode. A firm grip is made using a torque control mode, where the fingers close around the object till the target torque is reached. Comparing with a position control mode, a torque control strategy is more flexible and adaptive. Rather than specifying a target position, it drives the servo motors to achieve a target torque thereby making it adaptive to objects of various sizes. The readings from the tactile sensor at the fingertip of the index finger provide confirmation that the gripper now firmly holds the fan (inset from Fig. 6B). It is to be noted that the tactile sensor here was not included as an active part of the feedback control loop, but rather used as a passive observatory device to monitor the task.

The fan is then lifted up slightly above the surface of the table (Fig. 6B). Then the target torque value of the thumb is slightly increased such that an uneven force is applied on one side of the fan. This causes the fan blades to

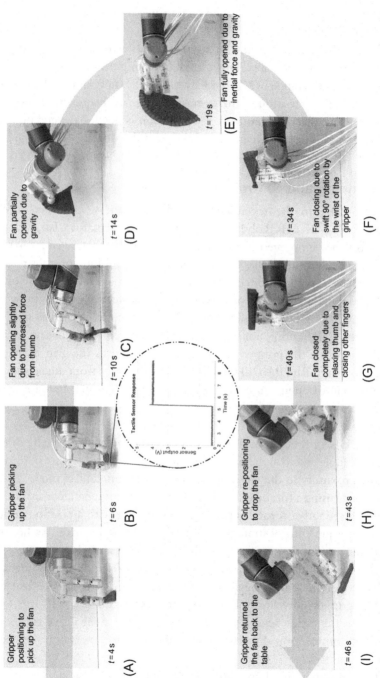

Fig. 6 Sequential steps in opening and closing strategies of the Spanish fan (clockwise from top to left). (A) Gripper moving toward the fan; (B) pick the closed fan from the table. *Inset* shows response from the tactile sensor in the index fingertip when the gripper holds the fan; (C) adjusting grab strength to begin opening the fan; (D) partial opening of the fan further with gravity; (E) fully opened fan due to gravity and inertial force; (F) partially closed fan due to inertia from fast wrist rotation; (G) fully closed fan fully with grab strength adjustment and finger movements; (H) positioning the gripper to return the fan; and (I) fan placed back on the table.

slide across each other and open slightly (Fig. 6C). With an upward movement of the UR5 arm, the fan opens further due to the effect of gravity and inertial force (Fig. 6D). Simultaneously, the grip on the fan is adjusted by relaxing the target torques of the index and middle fingers while the thumb keeps pushing one end of the fan toward the palm. A deliberate jolt from the UR5 finally opens the fan completely. A fully open fan positioned parallel to the background wall can be seen in Fig. 6E. A combination of gravity and the inertial force due to the fast movement of the UR5 arm is the reason for the opening of the fan. The opening strategy is loosely based on how a human will open a fan using a single hand.

A similar series of events are used to close the fan. First, the wrist of the robotic arm is rotated 90 degrees clockwise with respect to its z-axis. Due to the inertial force of this rotary motion, the fan closes to almost three-fourth of its size (Fig. 6F). It is closed completely by relaxing the thumb grip gradually and simultaneously closing the rest of the fingers to bring the other end of the fan blades together (Fig. 6G). The UR5 is now repositioned such that the gripper is ready to place the fan back at the table. One can see that the fan is now held between the dorsal side of the thumb and the ventral side of the other fingers (Fig. 6H). Except for this final finger configuration, the closing strategy is similarly based on human hand movement. Finally, the gripper opens and places the fan on the table by opening the thumb (Fig. 6I).

The entire series of tasks are based on rigorous revisions of trials to identify suitable waypoints for the UR5 trajectory. A careful combination of UR5 and gripper manipulation has made the successful accomplishment of this task possible. During IROS 2018, the task was performed in our lab in National University of Singapore, Singapore and telecasted live online to the conference site in Madrid, Spain. While there were other teams, from all over the world, with dedicated industrial nonhumanoid gripper mechanisms, which can complete the task much faster, our solution presented in this chapter was the only one which performed the task single-handed similar to human hand motion.

5 Data collection

Experimental data are collected by placing the object (Spanish fan) at an angle in between -75 and 75 degrees to the x-axis and recorded the position of the center of mass and orientation to y-axis and z-axis as discussed under Sections 3.1–3.3. It produces six DOF data reference to Kinect coordinate system. The UR5 arm is moved to the desired target so that the end effector (gripper) is ready to pick up the object. The position and orientation of the UR5 wrist are now recorded from the Polyscope robot user interface supported to UR5. This gives a 6D position vector containing x-axis, y-axis, and z-axis coordinates and roll (R_X), pitch (R_Y), and yaw (R_Z) angles. The roll, pitch, and yaw angles define the orientation in which the robot will approach the object. The position data recorded from the console is precise with an accuracy of 0.1 mm. Linear and nonlinear regression models are fitted to make the relationship between the Kinect coordinate system and robot coordinate system (Fig. 7). Graphs are generated to six DOF in a robot coordinate system w.r.t. the independent variable orientation.

6 Conclusion and future work

This chapter presents an overview of real-time object detection and manipulation architecture for fan robotic challenge which is initiated by IROS 2018. A visual perception system is developed based on the Kinect v2 sensor and computer vision methods. Color and depth images from Kinect are collected for visual tasks. The proposed algorithm achieves at the performance rate of grasping the object in planned position, holding the fan, opening, and closing and positioning it back on the table. This method is more robust than stereo vision images and the cost is less with other techniques such as SURF, SIFT, SICK, etc. Soft layers in hybrid robotic gripper improve the grip in the task and tendons control the total gripping motion. As hybrid gripper, it realizes the better object conformation and simple control even under environmental disturbance. Torque control strategy used in this application is more flexible and adaptive to grasp different sizes of objects in different environments.

P-value is the level of marginal significance within the hypothesis test probability of occurrence of a given event. A smaller P-value is stronger evidence to reject the null hypothesis and agree with the alternative hypothesis. The null hypothesis H_0 states that there is no significant relationship between angle and the variable given in the table. From the observed P-values ($<.001$), H_0 can be neglected and can be concluded that there is a significant relationship with those two variables. Although the value of .027 is less than that of .05, a significant relationship between the variable is quite poor (Table 1). R^2 value is the proportion of the variance for a dependent variable to an independent variable in the regression model. Correlation explains the strength of the relationship between an independent and dependent variable. Higher R^2 value (between 85% and 100%) indicates that the dependent

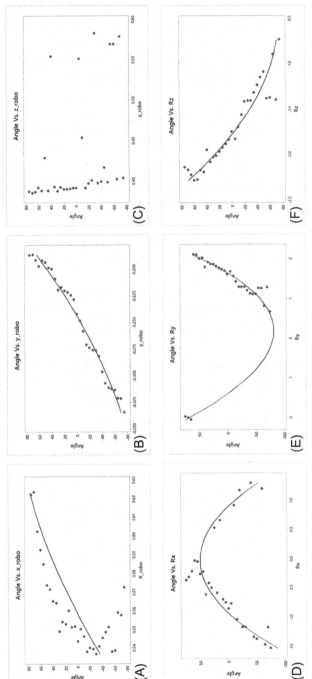

Fig. 7 The angle variations that are represented with y-axis (−75 to 75 degrees) to different coordinate systems that are represented in x-axis of each graph. (A) x-axis in UR5 coordinate system; (B) y-axis in UR5 coordinate system; (C) z-axis in UR5 coordinate system; (D) R_{Xrobo} axis in UR5 coordinate system; (E) R_{Yrobo} axis in UR5 coordinate system; and (F) R_{Zrobo} axis in UR5 coordinate system.

Table 1 Statistical data table of Kinect and robot frame, and R^2 values.

X_axis	P-value	R^2 (%)
x_robo	<.001	48.18
y_robo	<.001	98.18
z_robo	.027	15.80
R_xrobo	<.001	89.63
R_yrobo	<.001	94.56
R_zrobo	<.001	92.56

variable is relatively in line with the independent variable. If the R^2 value is low (70% or less), it does not follow the movement of the independent variable. y_robo, R_xrobo, R_yrobo, R_zrobo fit the regression model in the range of 85%–100% R^2.

z_robo have comparably low R^2 values to fit with the model. Hence, their values acquired based on the model might not be accurate. Since R^2 value does not sufficiently verify the accuracy of the chosen model, we follow the residual plots for further analysis. Residual plots determine whether a regression model is appropriate for the given data. As patterns can be observed (Fig. 8) from residual points, nonlinear models are appropriated. Regression models were difficult to plot on the independent variables of x_robo, y_robo, and z_robo as its points are randomly distributed on the axis. Z coordinate of the gripper and Kinect can be varied as those are not parallel and belongs to an intermediate state of another axis as well.

The program can be developed to recognize simple equipment that is easy to manipulate by human hand. Using the same procedure with developing algorithms, we are able to navigate that equipment through robot hand. In the future, advanced techniques like Dex-Net 2.0 deep learning robust robotic grasp will be investigated to detect an object from multiple objects for more complex tasks. As a follow up of the current work, the tactile sensor array will further be exploited by analyzing the data during the transition stages to meticulously control the entire process based on its feedback. Frequency domain analysis of tactile data could further be used for advanced slip detection and slip prevention controls.

Acknowledgments

This work is supported by Office of Naval Research Global under grant ONRG-NICOP-N62909-15-1-2029, NUSRI China Jiangsu Provincial Grant BE2016077 awarded to Dr. Hongliang Ren.

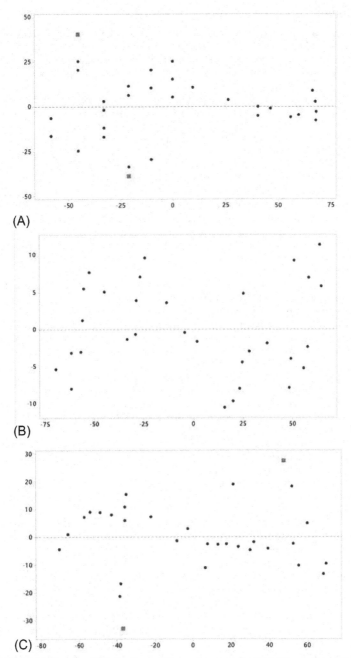

Fig. 8 Residual plots vs. fitted values. (A) x_robo coordinates, (B) y_robo coordinates, and (C) R_z*robo* coordinates.

References

Alcantara, G.K.L., Evangelista, I.D.J., Malinao, J.V.B., Ong, O.B., Rivera, R.S.D., Ambata, E.L.U., 2018. Head detection and tracking using OpenCV. In: 2018 IEEE 10th International Conference on Humanoid, Nanotechnology, Information Technology, Communication and Control, Environment and Management (HNICEM), pp. 1–5.

Andreff, N., Horaud, R., Espiau, B., 1999. On-line hand-eye calibration. In: Second International Conference on 3-D Digital Imaging and Modeling (Cat. No. PR00062), pp. 430–436.

Balakrishnan, G., Jaya, P., 2014. Contour based object tracking. Int. J. Comput. Sci. Inf. Technol. 5 (3), 4128–4130.

Banerjee, H., Ren, H., 2017. Optimizing double-network hydrogel for biomedical soft robots. Soft Robot. 4 (3), 191–201. https://doi.org/10.1089/soro.2016.0059.

Banerjee, H., Aaron, O.Y.W., Yeow, B.S., Ren, H., 2018a. Fabrication and initial cadaveric trials of bi-directional soft hydrogel robotic benders aiming for biocompatible robot-tissue interactions. In: 2018 3rd International Conference on Advanced Robotics and Mechatronics (ICARM), IEEE, pp. 630–635.

Banerjee, H., Pusalkar, N., Ren, H., 2018b. Single-motor controlled tendon driven peristaltic soft origami robot. J. Mech. Robot. 10 (6), 064501. https://doi.org/10.1115/1.4041200.

Banerjee, H., Tse, Z., Ren, H., 2018c. Soft robotics with compliance and adaptation for biomedical applications and forthcoming challenges. Int. J. Robot. Autom. 33 (1), 1–20. https://doi.org/10.2316/journal.206.2018.1.206-4981.

Barbosa, C., Santana, O., Silva, B., 2017. An unsupervised machine learning algorithm for visual target identification in the context of a robotics competition. In: 2017 Latin American Robotics Symposium (LARS) and 2017 Brazilian Symposium on Robotics (SBR), pp. 1–6.

Bhagat, S., Banerjee, H., Tse, Z.T.H., Ren, H., 2019. Deep reinforcement learning for soft, flexible robots: brief review with impending challenges. Robotics 8 (1), 4.

Catalano, M.G., Grioli, G., Farnioli, E., Serio, A., Piazza, C., Bicchi, A., 2014. Adaptive synergies for the design and control of the Pisa/IIT SoftHand. Int. J. Robot. Res. 33 (5), 768–782.

Celik, Y., Altun, M., Günes, M., 2017. Color based moving object tracking with an active camera using motion information. In: 2017 International Artificial Intelligence and Data Processing Symposium (IDAP), pp. 1–4.

Chan, Y.H., Tse, Z., Ren, H., 2018. Kirigami-inspired flexible and soft anthropomorphic robotic hand for rehabilitation and prosthetic purposes. Med. Biol. Eng. Comput. 10 (8), 453–467. https://doi.org/10.1007/s11517-017-1695-x.

Chang, T.H., Tian, Y., Wee, D.L.Y., Ren, H., Chen, P.-Y., 2018. Crumpling and unfolding of montmorillonite hybrid nanocoatings as stretchable flame retardant skin. Small 14 (21), 1800596. https://doi.org/10.1002/smll.201800596.

Chelva, M.S., Halse, S.V., Ratha, B.K., 2016. Object tracking in real time embedded system using image processing. In: 2016 International Conference on Signal Processing, Communication, Power and Embedded System (SCOPES), pp. 1840–1844.

Chen, F., Xu, W., Zhang, H., Wang, Y., Cao, J., Wang, M.Y., Ren, H., Zhu, J., Zhang, Y., 2018. Topology optimized design, fabrication and characterization of a soft cable-driven gripper. IEEE Robot. Autom. Lett. 3 (3), 2463–2470. https://doi.org/10.1109/LRA.2018.2800115.

Chou, J.C.K., Kamel, M., 1991. Finding the position and orientation of a sensor on a robot manipulator using quaternions. Int. J. Robot. Res. 10 (3), 240–254. https://doi.org/10.1177/027836499101000305.

Cousins, S., Rusu, R.B., 2011. 3D is here: point cloud library (PCL). In: Robotics and Automation (ICRA), pp. 1–4.

Daniilidis, K., 1999. Hand-eye calibration using dual quaternions. Int. J. Robot. Res. 18 (3), 286–298. https://doi.org/10.1177/02783649922066213.

Dollar, A.M., Backus, S.B., 2016. An adaptive three-fingered prismatic gripper with passive rotational joints. IEEE Robot. Autom. Lett. 1 (2), 668–675.

Dornaika, F., Chakik, F., 2010. Efficient object detection and matching using feature classification. In: 2010 20th International Conference on Pattern Recognition. pp. 3073–3076.

Fan, S., Gu, H., Zhang, Y., Jin, M., Liu, H., 2018. Research on adaptive grasping with object pose uncertainty by multi-fingered robot hand. Int. J. Adv. Robot. Syst. 15(2). https://doi.org/10.1177/1729881418766783.

Finio, B., Shepherd, R., Lipson, H., 2013. Air-powered soft robots for K-12 classrooms. In: 2013 IEEE Integrated STEM Education Conference (ISEC), pp. 1–6.

Guan, J., Zhang, W., Gu, J., Ren, H., 2015. No-reference blur assessment based on edge modeling. J. Vis. Commun. Image Represent. 29, 1–7. https://doi.org/10.1016/j.jvcir.2015.01.007.

Guo, J., Shi, C., Ren, H., 2019. Ultrasound-assisted guidance with force cues for intravascular interventions. IEEE Trans. Autom. Sci. Eng. 16 (1), 253–260. https://doi.org/10.1109/TASE.2018.2817644.

Hernandez-Lopez, J.J., Quintanilla-Olvera, A.L., López-Ramírez, J.L., Rangel-Butanda, F.J., Ibarra-Manzano, M.A., Almanza-Ojeda, D.L., 2012. Detecting objects using color and depth segmentation with Kinect sensor. Procedia Technol. 3, 196–204. https://doi.org/10.1016/j.protcy.2012.03.021.

Horaud, R., Dornaika, F., 1995. Hand-eye calibration. Int. J. Robot. Res. 14 (3), 195–210. https://doi.org/10.1177/027836499501400301.

Ilievski, F., Mazzeo, A.D., Shepherd, R.F., Chen, X., Whitesides, G.M., 2011. Soft robotics for chemists. Angew. Chem. Int. Ed. 123 (8), 1930–1935.

Jonschkowski, R., Eppner, C., Höfer, S., Martín-Martín, R., Brock, O., 2016. Probabilistic multi-class segmentation for the Amazon Picking Challenge. In: 2016 IEEE/RSJ International Conference on Intelligent Robots and Systems (IROS), pp. 1–7.

Kirthika, S.K., Vedhagiri, G.P.J., Ren, H., 2017. Fabrication and comparative study on sensing characteristics of soft textile-layered tactile sensors. IEEE Sens. Lett. 1 (3), 1–4. https://doi.org/10.1109/LSENS.2017.2708425.

Li, C., Gu, X., Ren, H., 2017a. A cable-driven flexible robotic grasper with Lego-like modular and reconfigurable joints. IEEE/ASME Trans. Mechatron. 22 (6), 2757–2767. https://doi.org/10.1109/tmech.2017.2765081.

Li, Z., Wu, L., Yu, H., Ren, H., 2017b. Kinematic comparison of surgical tendon-driven manipulators and concentric tube manipulators. Mech. Mach. Theory 107, 148–165. https://doi.org/10.1016/j.mechmachtheory.2016.09.018.

McKenzie, R.M., Barraclough, T.W., Stokes, A.A., 2017. Integrating soft robotics with the robot operating system: a hybrid pick and place arm. Front. Robot. AI 4, 39. https://doi.org/10.3389/frobt.2017.00039.

Mittal, M., Verma, A., Kaur, I., Kaur, B., Sharma, M., Goyal, L.M., Roy, S., Kim, T., 2019. An efficient edge detection approach to provide better edge connectivity for image analysis. IEEE Access. 7, 33240–33255. https://doi.org/10.1109/ACCESS.2019.2902579.

Mizushima, K., Oku, T., Suzuki, Y., Tsuji, T., Watanabe, T., 2018. Multi-fingered robotic hand based on hybrid mechanism of tendon-driven and jamming transition. In: 2018 IEEE International Conference on Soft Robotics (RoboSoft), pp. 376–381.

Nadeau, C., Ren, H., Krupa, A., Dupont, P.E., 2015. Intensity-based visual servoing for instrument and tissue tracking in 3D ultrasound volumes. IEEE Trans. Autom. Sci. Eng. 12 (1), 367–371. https://doi.org/10.1109/TASE.2014.2343652.

Nakashima, T., Yabuta, Y., 2018. Object detection by using interframe difference algorithm. In: 2018 12th France-Japan and 10th Europe-Asia Congress on Mechatronics, pp. 98–102.

Palungsuntikul, P., Premchaiswadi, W., 2010. Object detection and keep on a mobile robot by using a low cost embedded color vision system. In: 2010 Eighth International Conference on ICT and Knowledge Engineering, pp. 70–76.

Park, F.C., Martin, B.J., 1994. Robot sensor calibration: solving AX=XB on the Euclidean group. IEEE Trans. Robot. Autom. 10 (5), 717–721. https://doi.org/10.1109/70.326576.

Ponraj, G., Ren, H., 2018. Estimation of object orientation using conductive ink and fabric based multilayered tactile sensor. In: 2018 IEEE International Conference on Robotics and Automation (ICRA), pp. 1–7.

Ponraj, G., Kirthika, S.K., Thakor, N.V., Yeow, C., Kukreja, S.L., Ren, H., 2017. Development of flexible fabric based tactile sensor for closed loop control of soft robotic actuator. In: 2017 13th IEEE Conference on Automation Science and Engineering (CASE), pp. 1451–1456.

Ponraj, G., Kirthika, S.K., Lim, C.M., Ren, H., 2018. Soft tactile sensors with inkjet-printing conductivity and hydrogel biocompatibility for retractors in cadaveric surgical trials. IEEE Sensors J. 18 (23), 9840–9847. https://doi.org/10.1109/JSEN.2018.2871242.

Remy, S., Dhome, M., Lavest, J.M., Daucher, N., 1997. Hand-eye calibration. In: Proceedings of the 1997 IEEE/RSJ International Conference on Intelligent Robot and Systems. Innovative Robotics for Real-World Applications. IROS '97, vol. 2, pp. 1057–1065.

Ren, H., Dupont, P.E., 2012. Tubular enhanced geodesic active contours for continuum robot detection using 3D ultrasound. In: ICRA2012, IEEE International Conference on Robotics and Automation, 14–18 May, St. Paul, pp. 2907–2912.

Ren, H., Vasilyev, N.V., Dupont, P.E., 2011. Detection of curved robots using 3D ultrasound. In: IROS 2011, IEEE/RSJ International Conference on Intelligent Robots and Systems, pp. 2083–2089.

Ren, H., Anuraj, B., Dupont, P., 2017. Varying ultrasound power level to distinguish surgical instruments and tissue: toward intracardiac robotic surgery. Med. Biol. Eng. Comput. 10 (8), 453–467. https://doi.org/10.1007/s11517-017-1695-x.

Roshni, N., Kumar, T.K.S., 2017a. Pick and place robot using the centre of gravity value of the moving object. In: 2017 IEEE International Conference on Intelligent Techniques in Control, Optimization and Signal Processing (INCOS), pp. 1–5.

Roshni, N., Kumar, T.K.S., 2017b. Pick and place robot using the centre of gravity value of the moving object. In: 2017 IEEE International Conference on Intelligent Techniques in Control, Optimization and Signal Processing (INCOS), pp. 1–5.

Shi, C., Luo, X., Guo, J., Najdovski, Z., Fukuda, T., Ren, H., 2018. Three-dimensional intravascular reconstruction techniques based on intravascular ultrasound: a technical review. IEEE J. Biomed. Health Inform. 22 (3), 806–817. https://doi.org/10.1109/JBHI.2017.2703903.

Shiu, Y.C., Ahmad, S., 1989. Calibration of wrist-mounted robotic sensors by solving homogeneous transform equations of the form AX=XB. IEEE Trans. Robot. Autom. 5 (1), 16–29. https://doi.org/10.1109/70.88014.

Soans, R.V., Hegde, A., Singh, C., Kumar, A., 2017. Object tracking robot using adaptive color thresholding. In: 2017 2nd International Conference on Communication and Electronics Systems (ICCES), pp. 790–793.

Song, S., Li, Z., Ren, H., Yu, H., 2015. Shape reconstruction for wire-driven flexible robots based on Bezier curve and electromagnetic positioning. Mechatronics 29 (99), 28–35. https://doi.org/10.1016/j.mechatronics.2015.05.003.

Sun, Y., Song, S., Liang, X., Ren, H., 2016. A miniature soft robotic manipulator based on novel fabrication methods. IEEE Robot. Autom. Lett. 1 (2), 617–623. https://doi.org/10.1109/LRA.2016.2521889.

Sun, D., Gu, X., Li, C., Liao, Q., Ren, H., 2018. Multilateral teleoperation with new cooperative structure based on reconfigurable robots and type-2 fuzzy logic. IEEE Trans. Cybernet. 1–15. https://doi.org/10.1109/TCYB.2018.2828503.

Tan, N., Gu, X., Ren, H., 2018a. Design, characterization and applications of a novel soft actuator driven by flexible shafts. Mech. Mach. Theory 122, 197–218. https://doi.org/10.1016/j.mechmachtheory.2017.12.021.

Tan, N., Gu, X., Ren, H., 2018b. Pose characterization and analysis of soft continuum robots with modeling uncertainties based on interval arithmetic. IEEE Trans. Autom. Sci. Eng. 1–15. https://doi.org/10.1109/TASE.2018.2840340.

Tan, N., Gu, X., Ren, H., 2018c. Simultaneous robot-world, sensor-tip, and kinematics calibration of an underactuated robotic hand with soft fingers. IEEE Access. 6 (1), 22705–22715. https://doi.org/10.1109/ACCESS.2017.2781698.

Tsai, R.Y., Lenz, R.K., 1989. A new technique for fully autonomous and efficient 3D robotics hand/eye calibration. IEEE Trans. Robot. Autom. 5 (3), 345–358. https://doi.org/10.1109/70.34770.

Tse, Z.T.H., Chen, Y., Hovet, S., Ren, H., Cleary, K., Xu, S., Wood, B., Monfaredi, R., 2018. Soft robotics in medical applications. J. Med. Robot. Res. 1–18. https://doi.org/10.1142/s2424905x18410064.

Ukida, H., Terama, Y., Ohnishi, H., 2012. Object tracking system by adaptive pan-tilt-zoom cameras and arm robot. In: 2012 Proceedings of SICE Annual Conference (SICE), pp. 1920–1925.

Wang, H., Chen, J., Lau, H.Y.K., Ren, H., 2016. Motion planning based on learning from demonstration for multiple-segment flexible soft robots actuated by electroactive polymers. IEEE Robot. Autom. Lett. 1 (1), 391–398. https://doi.org/10.1109/LRA.2016.2521384.

Wu, L., Yang, X., Chen, K., Ren, H., 2015. A minimal POE-based model for robotic kinematic calibration with only position measurements. IEEE Trans. Autom. Sci. Eng. 12 (2), 758–763. https://doi.org/10.1109/TASE.2014.2328652.

Xitao, Z., Zhang, Y., 2010. An image-based object detection method using two cameras. In: 2010 International Conference on Logistics Engineering and Intelligent Transportation Systems, pp. 1–5.

Yadav, S., Singh, A., 2016. An image matching and object recognition system using webcam robot. In: 2016 Fourth International Conference on Parallel, Distributed and Grid Computing (PDGC), pp. 282–286.

Zeng, A., Yu, K.-T., Song, S., Suo, D., Walker, E., Rodriguez, A., Xiao, J., 2017. Multiview self-supervised deep learning for 6D pose estimation in the Amazon Picking Challenge. In: 2017 IEEE International Conference on Robotics and Automation (ICRA), IEEE, pp. 1386–1393.

CHAPTER FIVE

Formal verification of robotic cell injection systems

Adnan Rashid, Osman Hasan and Iram Tariq Bhatti
School of Electrical Engineering and Computer Science, National University of Sciences and Technology, Islamabad, Pakistan

1 Introduction

Biological cell injection is a method used for the insertion of small amount of substances, such as biomolecules, sperms, genes, and proteins, into the suspended or adherent cells. It is widely used in gene injection (Kuncova and Kallio, 2004), drug development (Nakayama et al., 1998), intracytoplasmic sperm injection (ISCI) (Yanagida et al., 1999), and in vitro fertilization (IVF) (Sun and Nelson, 2002). For example, in IVF, the sperm is injected into matured eggs for the treatment of infertility. Similarly, drug development involves the injection of drugs into a cell and the observation of its implication at the cellular level.

Robotic cell injection systems can automatically perform the task of cell injection as opposed to the conventionally used manual and semiautomated injection procedures, which require trained operators and involve time-consuming processes and also have low success rates. The most important parameters of a robotic cell injection system are coordinate frames, capturing the orientation and movement of its various components such as injection manipulator, digital cameras, sensors and microscope, and the force controlling the injection pipette (Huang et al., 2009a). A slight error in the orientation and movement of these components may result in injection into an undesired part of the cell. Similarly, a slight excess of force may damage the membrane of the cell (Huang et al., 2006) or an insufficient force may not be able to pierce the cell (Faroque and Nizam, 2016). Thus the accuracy of the orientation and movement of these fundamental components and the injection force are vital for the reliability of the overall system. Therefore, the robotic cell injection system designs need to be analyzed and verified quite carefully to ensure that these requirements are exhibited by the final systems.

Control Systems Design of Bio-Robotics and Bio-mechatronics
with advanced applications
https://doi.org/10.1016/B978-0-12-817463-0.00005-8

A robotic cell injection system is generally categorized into three types, namely 2-degree of freedom (DOF), 3-DOF, and 4-DOF, based on the DOF of the cell injection manipulator that is mounted on the motion stage and controls the motion of the injection pipette. The first step in the analysis of a robotic cell injection system is to model the coordinate frames corresponding to the orientations of its various components, such as the injection manipulator, cameras, and images. This model allows us to capture the movement and thus the positions of these components during the process of cell injection. Moreover, the relationship between these coordinates provides the relative positions of these components, which is an essential part of a successful cell injection procedure. Next, in order to perform the process of cell injection, the motion planning of the injection pipette is modeled using some force control algorithms, such as the contact-space-impedance force (Sun and Liu, 1997; Huang et al., 2009a) and the image-based torque controllers (Huang et al., 2006). These controllers capture the overall dynamics of the system and are mainly responsible for successful cell injection procedure and smooth functionality of the overall system.

Conventionally, the robotic cell injection systems have been analyzed using paper-and-pencil methods. However, these manual analysis techniques are prone to human error and also are not scalable for analyzing complex models like the robotic cell injection systems. Moreover, in some cases, all the required assumptions are not documented in the mathematical analysis, which may lead to erroneous designs and analysis. Similarly, the computer-based simulations and numerical techniques have been used for analyzing these systems. However, due to the involvement of the continuous–time (differential equation-based) models of the system in the analysis and the limited amount of computer memory and computational resources, the system is analyzed for a certain number of test cases only and thus the absolute accuracy cannot be achieved. Computer algebra systems, such as Mathematica (Mathematica, 2019), have also been used for analyzing these systems (Nethery and Spong, 1994). However, the symbolic algorithms residing in the core of these systems are unverified (Durán et al., 2013), which compromises the accuracy of these analyses. Due to the safety-critical nature of robotic cell injection systems, the above-mentioned conventional methods cannot be trusted as they are either prone to error or incomplete, which may lead to an undetected error in the analysis that may in turn lead to disastrous consequences.

Formal methods (Hasan and Tahar, 2015) are computer-based mathematical analysis techniques that can overcome the above-mentioned inaccuracies. Primarily, these techniques involve the development of a mathematical model of a system and verification of its properties using computer-based mathematical reasoning. There are mainly two types of formal methods, that is, probabilistic model checking (Clarke et al., 1999) and higher-order-logic theorem proving (Harrison, 2009), that have been used in this context. Probabilistic model checking involves the development of a state-space-based probabilistic model of the underlying system and the formal verification of its intended properties that are specified in temporal logic. It has been used by Sardar and Hasan (2017) for analyzing the robotic cell injection systems. However, the formal model involves the discretization of the differential equations modeling the dynamics of these systems, which compromises the accuracy of the corresponding analysis. Higher-order-logic theorem proving (Harrison, 2009) is an interactive verification method that can overcome these limitations. It primarily involves developing a mathematical model of the system based on higher-order logic and verifying its properties based on deductive reasoning. Given the high expressiveness of higher-order logic, it can truly capture the behavior of differential equations, which is not possible in the probabilistic model checking-based analysis. Rashid and Hasan (2018) have recently used this technique for the formal verification of the same robotic cell injection system. This chapter presents an overview of these efforts, which includes the formal verification of these systems using probabilistic model checking and higher-order-logic theorem proving. Finally, it highlights the strengths and weaknesses of these techniques for analyzing robotic cell injection systems.

The rest of the chapter is organized as follows: Section 2 provides some of the related work that has been done for analyzing the robotic cell injection systems. We provide a brief overview of the formal methods, that is, model checking and theorem proving in Section 3. Section 4 presents the formal verification of the robotic cell injection system using probabilistic model checking. We present the formal analysis of the robotic cell injection system using higher-order-logic theorem proving in Section 5. We provide some discussion about the analyses, presented in Sections 4 and 5, while highlighting their strengths and weaknesses in Section 6. Finally, Section 7 concludes the chapter.

2 Related work

Huang et al. (2006) have proposed a vision-based force control framework for the robotic cell injection systems. The authors have used biomembrane point-load model for the measurement of the injection force using visual cell deformation feedback. They have also developed a two-dimensional impedance force control strategy for the process of the robotic cell injection. Similarly, Huang et al. (2007) have presented a vision-based impedance force control algorithm for analyzing the three-dimensional cell injection systems. The authors have used xy coordinate frame deformation of the cell and microscope for the measurement of the total cell deformation during the process of cell injection. Later, the same authors have proposed a prototype cell injection system for automatic batch injection of suspended cells (Huang et al., 2009b). A force sensor is used to measure the real-time injection force applied to the cells during the process of cell injection. More-over, a force control algorithm is developed and also used for controlling the motion of the injection pipette during an out-of-plane cell injection. How-ever, all contributions presented earlier are based on conventional tech-niques and cannot be trusted, considering the safety-critical nature of the robotic cell injection systems.

Formal methods, such as probabilistic model checking and higher-order-logic theorem proving, have also been used for the formal analysis of the robotic cell injection systems. Sardar and Hasan (2017) have used probabi-listic model checking (Clarke et al., 1999), that is, a state-based formal method, to formally analyze the robotic cell injection systems. However, their methodology involves the discretization of the differential equations that model the dynamics of these systems, which compromises the accuracy of the corresponding analysis. Moreover, the analysis also suffers from the inherent state-space explosion problem (Clarke et al., 2012). Recently, Rashid and Hasan (2018) have proposed to use higher-order-logic theorem proving for formally analyzing the robotic cell injection systems. The authors have formalized various coordinate frames and formally verified their interrelationship. Moreover, they have also formally analyzed the dynamical behavior of these systems and the motion planning of the injec-tion pipette used in the process of cell injection. The main focus of this chapter is to present the efforts that have been made in these formal method-based analyses of the robotic cell injection systems.

3 Formal methods

3.1 Probabilistic model checking and PRISM

Model checking (Clarke et al., 1999) is one of the major formal techniques that is commonly used for analyzing concurrent systems, such as network and routing protocols. The model checking-based analysis involves the

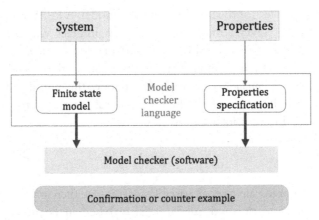

Fig. 1 Model checking.

construction of a state-space model of the given system and specification of the intended properties of the system in temporal logic (Pnueli, 1977), as depicted in Fig. 1. Next, both the model and properties are given to the model checker, which explores the state space exhaustively and automatically verifies the given system based on these properties. In the case of failure of a property, the model checker generates an error trace, which helps a user to identify and rectify the error found in the system's model. For larger systems, this technique is subject to the problem of state-space explosion (Baier and Katoen, 2008) caused by the limited availability of computer memory and other computational resources. The abstraction of the model or usage of the efficient bounded and symbolic model checking techniques are generally used to overcome this problem.

PRISM (Kwiatkowska et al., 2011) is a probabilistic model checker that is widely adopted for the modeling and analysis of the systems exhibiting probabilistic behaviors. It involves the development of a state-space model of the underlying system by assigning probabilities to its various transitions. Next, the intended properties, captured in temporal logic, are verified for the state-space model, developed in the previous step. Based on the nature of the underlying system, that is, discrete time or continuous time, it supports various types of probabilistic models such as discrete-time Markov chains (DTMCs) (Kulkarni, 2016), continuous-time Markov chains (Kulkarni, 2016), probabilistic timed automata (Puterman, 2014), and Markov decision process (Segala and Lynch, 1995).

The formal model of a system is developed using a state-based language, that is, *PRISM language*, which is based on Alur's reactive modules formalism. *PRISM language* provides several fundamental components, that is, *modules* and *variables*, for the development of a model. A PRISM model generally consists of several modules, which can interact with each other. A module contains a set of local variables and guarded *commands*. The values of these variables at a time instant provide the state of the modules. Moreover, the guarded *commands* capture the behavior of the modules. The global state of the overall system is identified by the local states of all modules. The syntax of a PRISM command is given below:

```
[] guard -> prob_1 : update_1 + ...+ prob_n : update_n;
```

The guard is a predicate over all variables. If it is true then it allows the transition of states and performs the update of states based on the initially assigned probabilities of the transitions.

The PRISM model of a system can be formally verified using the properties specified in probabilistic temporal logics, such as probabilistic computation tree logic (PCTL (Hansson and Jonsson, 1994) and PCTL* (Aziz et al., 1995)), continuous stochastic logic (Aziz et al., 1996), probabilistic linear temporal logic, and nonprobabilistic temporal logic computational tree logic (Orgun and Ma, 1994), which are formulas in temporal logic and support specifications based on temporal, path, logical, probability, and reward operators. The supported temporal operators include next (X), always (G), and eventually (F). Similarly, the path operators are there exists (E) and forall (A), whereas negation (!), disjunction (|), conjunction (&), and implication (=>) are the logical operators. The probability operators include probability (P), maximum probability (Pmax), and minimum probability (Pmin). Similarly, the reward operator involves reward (R). Table 1 presents the property specification, the temporal and path operators, and their graphical representation.

3.2 Theorem proving and HOL Light

Theorem proving (Harrison, 2009) involves constructing a mathematical model of the given system and the intended properties of the system based on an appropriate logic as shown in Fig. 2. Next, both the model and properties are presented as theorems to the theorem prover, which uses deductive reasoning to develop their proofs based on well-known axioms and hypothesis and thus verifies the given system based on these properties. Based on the decidability or undecidability of the underlying logic, that is, propositional or higher-order logic, theorem proving can be automatic or interactive,

Table 1 Property specification using the temporal and path operators.

Property	Graphical representation

E F p

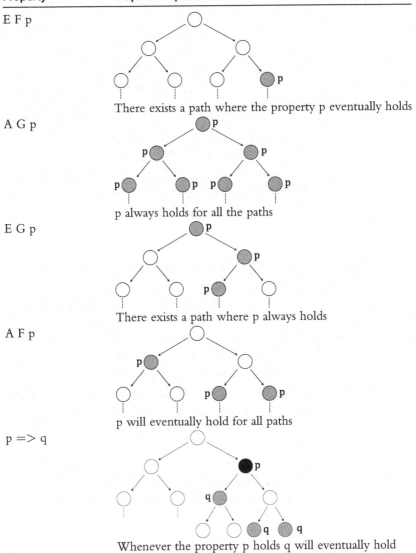

There exists a path where the property p eventually holds

A G p

p always holds for all the paths

E G p

There exists a path where p always holds

A F p

p will eventually hold for all paths

p => q

Whenever the property p holds q will eventually hold

respectively. Every theorem prover comes with a set of axioms and inference rules, which, along with the already verified theorems, are the only ways to prove the new theorems. This purely deductive feature ensures soundness, that is, every sentence proved in the system is actually true.

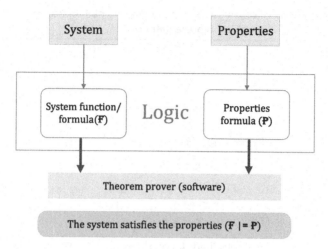

Fig. 2 Theorem proving.

HOL Light (Harrison, 1996a) is a higher-order-logic proof assistant that ensures secure theorem proving using the Objective CAML (OCaml) language, which is a variant of the strongly typed functional programming language ML (Paulson, 1996). HOL Light users can interactively verify theorems by applying the available proof tactics and proof procedures. A HOL Light theory consists of types, constants, definitions, and theorems. HOL Light theories are built in a hierarchical fashion and new theories can inherit the definitions and theorems of their parent theories. HOL Light provides an extensive support for the analysis based on Boolean algebra, real arithmetics, multivariable calculus, and vectors (Harrison, 2013). There are many automatic proof procedures (Harrison, 1996b) available in HOL Light, which help the user in concluding a proof more efficiently.

Table 2 illustrates some symbols, that is, their HOL Light and standard representations, and their meanings, which have been very often used in this chapter.

4 Model checking-based analysis of robotic cell injection systems

4.1 Robotic cell injection systems

A robotic cell injection system is generally composed of three modules, namely executive, sensory, and control modules as depicted in Fig. 3. The executive module comprises positioning table, working plate, and

Table 2 HOL Light symbols.

HOL Light symbols	Standard symbols	Meanings
/\	And	Logical *and*
\/	Or	Logical *or*
~	Not	Logical *negation*
==>	→	Implication
<=>	=	Equality in Boolean domain
!x.t	$\forall x.t$	For all $x{:}t$
?x.t	$\exists x.t$	There exists $x{:}t$
λx.t	$\lambda x.t$	Function that maps x to $t(x)$
num	{0, 1, 2, ...}	Natural numbers data type
real	All real numbers	Real data type
SUC n	$(n + 1)$	Successor of natural number
&a	$\mathbb{N} \to \mathbb{R}$	Typecasting from natural to real numbers
abs x	$|x|$	Absolute function
EL n l	*element*	nth element of list l

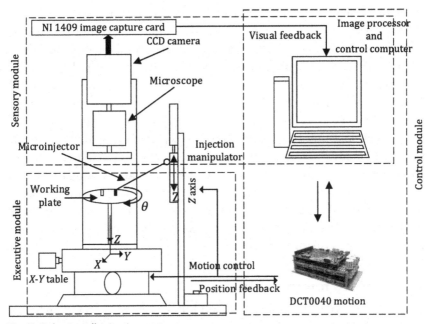

Fig. 3 Robotic cell injection systems.

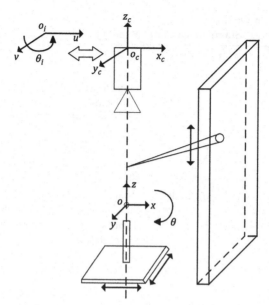

Fig. 4 Configuration of the robotic cell injection systems.

the injection manipulator. The cells that need to be injected are placed on a working plate, which is mounted on a positioning table ($XY\theta$-axis) and the injection manipulator is mounted on Z-axis as shown in Fig. 3.

The sensory module consists of a vision system that has four parts, namely charge–coupled device (CCD) camera, peripheral component interconnect (PCI) image capture, optical microscope, and a processing card. The CCD camera is used to capture the process of the robotic cell injection using a PCI image capture. The control module comprises a host computer and a DCT0040 motion control system. Fig. 4 depicts the configuration of a robotic cell injection system. The axis $o - xyz$ represents the stage (table and working plate) coordinate frame, where o is the origin of these coordinates representing the center of the working plate and z is along the optical axis of the microscope. Similarly, $o_c - x_cy_cz_c$ represents the camera coordinate frame where o_c is the center of the microscope. The coordinate frame in image plane is represented by $o_i - uv$, where o_i is the origin and the axis uv is perpendicular to the optical axis.

4.2 Proposed formal model

Probabilistic model checking (Baier and Katoen, 2008) was used by Sardar and Hasan (2017) for the formalization of the robotic cell injection systems.

This model incorporates various random factors that are vital for the analysis of these systems due to their safety-critical nature. Probabilistic model checking has been widely utilized for analyzing a variety of systems exhibiting the probabilistic or random behavior that belong to various domains, such as randomized distributed algorithms, security, communication and routing protocols, biological systems, etc. The proposed model supports quantitative analysis while capturing the real-world factors of random nature and thus helps in determining the efficiency of these systems.

4.2.1 Proposed modeling approach and formalization

The force control module of the robotic cell injection systems is modeled as a closed-loop control system since it involves a feedback loop from the output to the input of the underlying system and thus can provide a better performance by minimizing the peak error as compared to the open-loop system. Fig. 5 depicts the model for the control of the injection pipette's trajectory in the X-coordinate. The term plant is used for the robotic cell injection system to distinguish it from the rest of the controller. The variable X_d represents the desired position of the pipette, whereas the position of the pipette, given by the encoder, is denoted by X_n. The difference of both these positions, that is, $e_x = X_d - X_n$ denotes the trajectory error in X-axis. The disturbances (internal and external) are modeled as additive noise. The injection pipette controllers adjust the driving motor's input torque τ_x using the error e_x and the current value of the external force f_{ex} applied to the actuators. The resulting torque τ_n is provided to the plant, which is further used to compute the actual position of the injection pipette X. The measurement noise due to various factors, such as encoder noise, calibration error, fabrication variation, is modeled as additive noise to X, providing X_n, which is fed back to the controller (Sardar and Hasan, 2017).

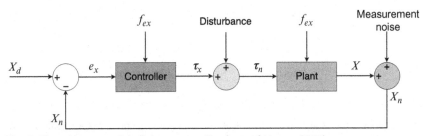

Fig. 5 Closed-loop model of the system (Sardar and Hasan, 2017).

Our proposed formalization of the robotic cell injection systems is based on DTMC and involves the discretization of the implementation of the actual system. In the case of the robotic cell injection systems, there is a specific time interval for capturing the images of the injected cell using CCD camera. Also, it requires time to process the values of radius a and depth w_d of the dimple created. These values are used to estimate the force feX, which is fed to the controller for the computation of the torque Tau_x, as depicted in Fig. 6. After adding disturbances (internal and external), the resulting torque Tau_n is provided to the plant's motor. This noise also corrupts the actual position of the pipette X_cur resulting in encoder measurement X_n. Finally, the error ex in the trajectory of the pipette is computed and fed back to the force controller (Sardar and Hasan, 2017). We consider

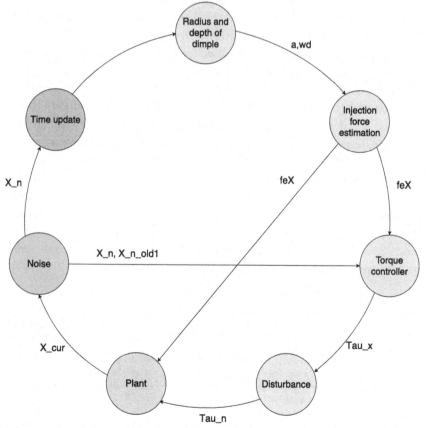

Fig. 6 Finite state-space model of the robotic cell injection system (Sardar and Hasan, 2017).

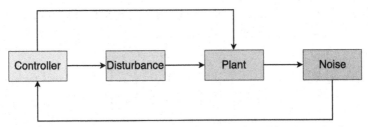

Fig. 7 PRISM modules and their interconnections (Sardar and Hasan, 2017).

one time unit as the total time taken during the process of robotic cell injection. This process is repeated at each time instant for various tasks, such as prepiercing, piercing, injection, and injector pulling out periods. These tasks are repeated for each of the new cells and this process continues in this manner to provide the injection for the whole batch of cells. Our PRISM model comprises four main modules: plant, controller, noise, and disturbance, and their interaction is shown in Fig. 7. The plant module models the dynamics of the plant. The control module implements the image-based force controller capturing the movement of the pipette. The noise and disturbance modules provide our formalization of the internal and external disturbance, as well as the measurement noise (Sardar and Hasan, 2017).

4.3 Formalization of the plant

The main modules and the configuration (orientation of various coordinate frames) of the robotic cell injection systems are depicted in Figs. 3 and 4. The image-stage coordinate frame interrelationship is mathematically expressed as

$$
\begin{bmatrix} u \\ v \end{bmatrix} = \begin{bmatrix} f_x \cos\alpha & f_x \cos\alpha \\ -f_y \sin\alpha & f_y \cos\alpha \end{bmatrix} \begin{bmatrix} X \\ Y \end{bmatrix} + \begin{bmatrix} f_x d_x \\ f_y d_y \end{bmatrix} \tag{1}
$$

where $f_x = \lambda/\delta_u$ and $f_y = \lambda/\delta_v$ are the display resolutions of the vision system in x- and y-directions, respectively. Here, λ represents the magnification factor of the microscope objective. Similarly, δ_u and δ_v denote the actual distances between the two adjoining pixels in the CCD sensor u-v coordinate frame. The variables dx and dy denote distances between the stage and the camera coordinate frames in x- and y-directions, respectively. Also, α represents the angle between the stage and the camera coordinate frames.

The dynamics of the 2-DOF motion stage, based on Lagrange's equation, is mathematically expressed as

$$
\begin{bmatrix} m_x + m_y + m_p & 0 \\ 0 & m_y + m_p \end{bmatrix} \begin{bmatrix} \dfrac{d^2x}{dt} \\ \dfrac{d^2y}{dt} \end{bmatrix} + \begin{bmatrix} 1 & 0 \\ 0 & 1 \end{bmatrix} \begin{bmatrix} \dfrac{dx}{dt} \\ \dfrac{dy}{dt} \end{bmatrix} = \begin{bmatrix} \tau_x \\ \tau_y \end{bmatrix} - \begin{bmatrix} fex^d \\ fey^d \end{bmatrix} \quad (2)
$$

where m_p, m_x, and m_y are the masses of the working plate and the x and y positioning tables, respectively. Similarly, τ_x and τ_y model the x and y components of the input torque to the driving motor, respectively. Similarly, fex^d and fey^d present the x and y components of the desired external force applied to the actuators during the process of cell injection.

Using the values obtained from the real-world experimental setup in Eq. (2), we obtain the following second-order differential equation after simplification (Sardar and Hasan, 2017):

$$
0.022180\ddot{X} + (2.465e - 0.03)\dot{X}^2 - 0.0479\dot{X} = 1.5 \times 10^{-7}(\tau_x - f_{ex}) - 1.146 \quad (3)
$$

The earlier differential equation is discretized to its corresponding differential equation by using finite difference method (LeVeque, 2007) having backward difference operators. We utilize the first-order approximation to avoid the overhead cost generated by *higher-order difference* approach. Representing the value of the variable X at time instant t_i as X_i, the first- and second-order derivatives of X using the backward difference method are mathematically expressed as Sardar and Hasan (2017)

$$
\dot{X}_i = \frac{X_i - X_{i-1}}{\Delta t} \quad (4)
$$

$$
\ddot{X}_i = \frac{X_i - 2X_{i-1} + X_{i-2}}{(\Delta t)^2} \quad (5)
$$

where Δt denotes the step size for time. Similarly, X_{i-1} and X_{i-2} represent the value of X at time instants t_{i-1} and t_{i-2}, respectively. Using these values in Eq. (3) and ignoring the higher-order terms, the solution for X_i, for $\Delta t = 1$ after simplification and rounding off the values to six decimal places, is given as (Sardar and Hasan, 2017)

$$
X_i = -77.143817X_{i-1} + 39.510075X_{i-2}
$$
$$
- 5.926511 \times 10^{-6}(\tau_x - f_{ex}) + 45.278546 \quad (6)
$$

The earlier equation is formalized in PRISM as follows (Sardar and Hasan, 2017):

```
[ ] guard -> (X_cur' = ceil (p1 * X_old1 + p2 * X_old2 -
                             p3 * (TauX - feX) + p4));
```

where X_cur represents the current position of the injection pipette along the X-axis. Similarly, X_old1 and X_old2 capture the pipette's position along the X-axis at time instants t_{i-1} and t_{i-2}, respectively. Moreover, p1, p2, p3, and p4 are defined as constants in PRISM and their values are obtained from Eq. (6). Moreover, TauX represents the per-unit torque input to the driving motors and feX is the external force applied to the actuators during the process of the robotic cell injection. Since PRISM does not support rational numbers for state variables, we utilized the PRISM function ceil to round the final result to the nearest integer value. Moreover, the guard represents the condition for sequencing using a global variable count.

4.3.1 Formalization of the controller

The biomembrane point load model (Sun et al., 2003) is used for the vision-based estimation of the cell injection force F, which is mathematically expressed as

$$F = \frac{2\pi E h w_d^3}{a^2(1-\gamma)} \frac{3 - 4\zeta^2 + \zeta^4 + 2\ln\zeta^2}{(1-\zeta^2)(1-\zeta^2 + \ln\zeta^2)^3} \quad (7)$$

where E and h represent the membrane elastic modulus and thickness of the biomembrane, respectively. Similarly, w_d denotes the depth of the dimple resulted due to injection of the pipette and a represents the radius of the dimple after injection of the pipette. γ is the Poisson ratio and $\zeta = c/a$, where c is the radius of the pipette. Eq. (7) is formalized in PRISM as Sardar and Hasan (2017)

```
[ ] guard -> (Force' = min (ceil ((2 * pi * EM * h * pow (wd,3)) *
    (3 - 4 * pow (c/a,2) + pow (c/a,4) + 2 * log (pow (c/a,2),e)) /
    ((pow (a,2) * (1 - gamma)) * (1 - pow (c/a,2)) *
    pow ((1 - pow (c/a,2) + log (pow (c/a,2),e)),3))),
    Force_max))
```

where the variable Force captures the injection force and EM denotes the membrane elastic modulus. Similarly, e and pi are defined as constants in PRISM. The PRISM functions pow (i,j) and log (i,j) compute i to the power of j and the log of i to the base j, respectively. Similarly, the function min (i,j) accepts two values i and j and returns the minimum out of those values. It is utilized to restrict the values of the applied force under its upper bound Force_max.

To reproduce the image processing results using our proposed approach, a and w_d are modeled differently in the four different time zones: prepiercing, piercing, injection, and pulling out of the injection pipette. During the prepiercing phase, the values of a and w_d increase by a nondecreasing factor due to the increasing velocity, whereas their values decrease during the piercing phase. Similarly, during the injection phase, they remain constant. Finally, the values decrease initially and then increase during the pulling out period.

4.3.2 Formalization of the random factors

Two main random factors have been incorporated into our PRISM model, which are classified as either disturbance or measurement noise. Disturbance can be further categorized into types: internal and external disturbances (Sardar and Hasan, 2017). The internal disturbances include plant uncertainties, such as electromagnetic effects of the components of the system, variation in parameters of the process, and distortion due to nonlinear elements (Dorf and Bishop, 2011). External disturbances mainly occur due to the environmental effects, such as temperature and electromagnetic effects of components in the surrounding. For example, the high temperature may cause the image degradation (PULNiX, 2017), which may introduce an error in the values of the parameters a and w_d of the biological cell. The measurement noise is caused by the sensor error, which includes calibration error, fabrication variation, and the lifetime of the sensor (Dorf and Bishop, 2011). The amount of the noise is characterized by the noise-to-signal ratio (Levine, 1999). Our PRISM models of the disturbance and measurement noise are depicted in Figs. 8 and 9. Since the disturbance is generally greater than

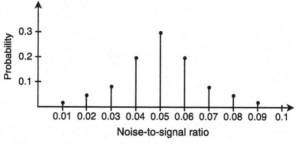

Fig. 8 Model of the disturbance (Sardar and Hasan, 2017).

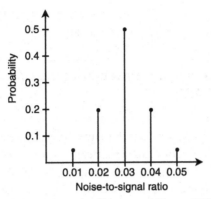

Fig. 9 Model of the measurement noise (Sardar and Hasan, 2017).

the measurement noise, the noise–to–signal ratios of 0.1–0.9 are used for disturbance, with the peak value at 0.05, as shown in Fig. 8. We implemented this in PRISM as follows (Sardar and Hasan, 2017):

```
[] guard -> 0.02:(Tau_n' = ceil (Tau_x + 0.01 * Tau_x)) +
            0.05:(Tau_n' = ceil (Tau_x + 0.02 * Tau_x)) +
            0.08:(Tau_n' = ceil (Tau_x + 0.03 * Tau_x)) +
            0.2:(Tau_n' = ceil (Tau_x + 0.04 * Tau_x)) +
            0.3:(Tau_n' = ceil (Tau_x + 0.05 * Tau_x)) +
            0.2:(Tau_n' = ceil (Tau_x + 0.06 * Tau_x)) +
            0.08:(Tau_n' = ceil (Tau_x + 0.07 * Tau_x)) +
            0.05:(Tau_n' = ceil (Tau_x + 0.08 * Tau_x)) +
            0.02:(Tau_n' = ceil (Tau_x + 0.09 * Tau_x));
```

where Tau_x and Tau_n represent the per-unit torques from the controller and after the addition of distortion.

The noise–to–signal ratios of 0.1–0.5 are used for measurement noise, with peak value at 0.03, as shown in Fig. 9. This is implemented in PRISM as follows (Sardar and Hasan, 2017):

```
[] guard -> 0.05:(X_n' = ceil (X_cur + 0.01 * X_cur)) +
            0.2:(X_n' = ceil (X_cur + 0.02 * X_cur)) +
            0.5:(X_n' = ceil (X_cur + 0.03 * X_cur)) +
            0.2:(X_n' = ceil (X_cur + 0.04 * X_cur)) +
            0.05:(X_n' = ceil (X_cur + 0.05 * X_cur));
```

where X_cur denotes the position of the pipette on the X-axis and X_n represents its position after adding the measurement noise of the encoder.

5 Theorem proving-based analysis of robotic cell injection systems

This section provides the theorem proving-based formal analysis of the robotic cell injection systems, which mainly includes the higher-order-logic formalization of various coordinate frames and their interrelationship. It also includes the formal verification of the solutions of the differential equations capturing the continuous dynamics of the 2-DOF robotic cell injection systems using HOL Light theorem prover. Moreover, we also present the formal modeling of the torque and force controllers and the formal verification of their implication relationship.

In order to facilitate the understanding of the chapter, the formal analysis of the robotic cell injection system is presented using a mix Math/HOL Light notation.

5.1 Formalization of the coordinate frames and their interrelationship

The coordinate frames of a robotic cell injection system, such as camera, image, and stage coordinates, are generally modeled as two-dimensional coordinates. These coordinates are modeled in HOL Light as follows (Rashid and Hasan, 2018):

Definition 5.1. Two-dimensional coordinates

$$\vdash_{def} \forall x\ y\ t.\ \text{two\_dim\_coordin}\ x\ y\ t = \begin{bmatrix} x_{(t)} \\ y_{(t)} \end{bmatrix}$$

where x: $\mathbb{R} \to \mathbb{R}$ and y: $\mathbb{R} \to \mathbb{R}$ are functions of time modeling the respective axes and t is a variable representing the time.

Next, we model a matrix providing a rotation from the stage coordinate frame ($o - xyz$) to the camera coordinate frame ($o_c - x_c y_c z_c$), and the two-dimensional vector representing the displacement between the origins of both these frames (Rashid and Hasan, 2018):

Definition 5.2. (Rotation matrix and displacement vector).

$$\vdash_{def} \forall\ \alpha.\ \text{rotat\_matrix}\ \alpha = \begin{bmatrix} \cos\alpha & \sin\alpha \\ -\sin\alpha & \cos\alpha \end{bmatrix}$$

$$\vdash_{def} \forall dx\ dy.\ \text{displace\_vector}\ dx\ dy = \begin{bmatrix} dx \\ dy \end{bmatrix}$$

where $\alpha\alpha$ denotes the angle between the two frames. Similarly, the variables dx and dy denote distances between the two coordinate frames in x- and y-directions, respectively.

The camera, image, and stage coordinate frames ensure the correct orientation and movement of various components of a robotic cell injection system, like, injection manipulator, microscope, stage frame, etc., and are mainly responsible for a reliable operation of the system. Therefore, the verification of the relationship between these coordinate frames is of utmost importance. The camera–stage coordinate frame interrelationship is formally verified by the following HOL Light theorem (Rashid and Hasan, 2018):

Theorem 5.1. *Camera-stage coordinate frame interrelationship*

$\vdash_{thm}$ ∀xc yc x y α dx dy t.

[A1]: 0 < dx ∧ [A2]: 0 < dy

$$\Rightarrow (\text{relat\_camera\_stage\_coordin}\; xc\; yc\; x\; y\; \alpha\; dx\; dy\; t \Leftrightarrow$$

$$\begin{bmatrix} xc(t) \\ yc(t) \end{bmatrix} = \begin{bmatrix} x(t)*\cos\alpha + y(t)*\sin\alpha + dx \\ -x(t)*\sin\alpha + y(t)*\cos\alpha + dy \end{bmatrix}$$

where the function relat_camera_stage_coordin *represents the camera-stage coordinate frame interrelationship. Assumptions* A1–A2 *ensure the nonnegativity of the distances* dx *and* dy*, respectively, and are design constraints for the relationship. The verification of the previous theorem is based on the properties of vectors and matrices alongside some real arithmetic reasoning. Now, to verify the image-camera coordinate frame interrelationship, we require modeling the display resolution matrix (Rashid and Hasan, 2018):*

Definition 5.3. Display resolution matrix

$\vdash_{def}$ ∀fx fy. display_resol_matrix fx fy $= \begin{bmatrix} fx & 0 \\ 0 & fy \end{bmatrix}$

where fx and fy are the display resolutions of the vision system in x- and y-directions, respectively. Next, the image-camera coordinate frame interrelationship is verified as

Theorem 5.2. *Image-camera coordinates interrelationship*

$\vdash_{thm}$ ∀xc yc u v t fx fy.

[A1]: 0 < fx ∧ [A2]: 0 < fy

$$\Rightarrow (\text{relat\_image\_camera\_coordin}\; xc\; yc\; u\; v\; t\; fx\; fy \Leftrightarrow$$

$$\left(\begin{bmatrix} u(t) \\ v(t) \end{bmatrix} = \begin{bmatrix} fx*xc(t) \\ fy*yc(t) \end{bmatrix} \right)$$

where the function relat_image_camera_coordin *provides the image-camera coordinate frame interrelationship. Assumptions* A1–A2 *of Theorem 5.2 model the design constraints for the relationship, that is, the nonnegativity of* fx *and* fy*, respectively.*

Next, to verify the image-stage coordinate frame interrelationship, given in Eq. (1), we require modeling the transformation matrix:

Definition 5.4. Transformation matrix

$\vdash_{def}$ ∀fx fy α. transform_matrix fx fy $\alpha = \begin{bmatrix} fx*\cos\alpha & fx*\sin\alpha \\ -fy*\sin\alpha & fy*\cos\alpha \end{bmatrix}$

Now, the image-stage coordinate frame interrelationship (Eq. 1) is verified as

Theorem 5.3. *Image-stage coordinates interrelationship*

$\vdash_{thm}$ ∀x y u v t fx fy dx dy α xc yc. [A1]: 0 < dx ∧ [A2]: 0 < dy ∧
[A3]: 0 < fx ∧ [A4]: 0 < fy ∧
[A5]: two_dim_coordin u v t = display_resol_matrix fx fy **
                                          two_dim_coordin xc yc t ∧
[A6]: two_dim_coordin xc yc t = rotat_matrix α **
                         two_dim_coordin x y t + displace_vector dx dy
⇒ two_dim_coordin u v t = transform_matrix fx fy α **
                              two_dim_coordin x y t + $\begin{bmatrix} fx*dx \\ fy*dy \end{bmatrix}$

*where ** represents the operator for the multiplication of a matrix with a vector and vice versa. Assumptions* A1-A4 *present the design constraints for the image-stage coordinate frame interrelationship. Assumption* A5 *models the image-camera coordinates interrelationship. Similarly, Assumption* A6 *provides the camera-stage coordinate frame interrelationship. Finally, the conclusion captures the relationship between the image and stage coordinate frames. The proof process of Theorem 5.3 is mainly based on Theorems 5.1 and 5.2 along with some properties of the vectors and matrices. The verification of these relationships ensures the correct orientation and movement of various components of the robotic cell injection system, that is, injection manipulator, working plat, microscope, camera, etc., and is vital considering the safety-critical nature of the underlying system.*

Next, we model the dynamics of the robotic cell injection systems, which are generally modeled as a set of differential equations and formally verify the solution of these differential equations. For the sake of simplicity, we consider 2-DOF motion stage of the system, which considers the process of cell injection in the xy plane only. The dynamics of the cell injection system, that is, Eq. (2), is formalized in HOL Light as follows:

Definition 5.5. Dynamics of the 2-DOF motion stage

$\vdash_{def}$ ∀mx my mp x y t taux tauy fexd feyd.
    dynamics_2dof_motion_stage mx my mp x y t taux tauy fexd feyd ⇔
      mass_matrix mx my mp ** sec_order_deriv_stage_coordin x y t +
        posit_table_matrix ** fir_order_deriv_stage_coordin x y t =
          torque_vector taux tauy - desired_force_vector fexd feyd

where `mass_matrix` is the matrix containing the respective masses. The function `sec_order_deriv_stage_coordin` models the first-order derivative of the stage coordinates. Similarly, the function `posit_table_matrix` provides the diagonal matrix. The function `sec_order_deriv_stage_coordin` captures the second-order derivative of the stage coordinates. Similarly, the functions `torque_vector` and `desired_force_vector` present the vectors with their elements representing the x and y components of the applied torque and desired force, respectively.

Under the condition of the applied torque and force vectors equal to zero, the injection pipette does not touch the cells. Thus, the dynamics of the underlying system, that is, Eq. (2), is transformed as follows:

$$
\begin{bmatrix} m_x + m_y + m_p & 0 \\ 0 & m_y + m_p \end{bmatrix} \begin{bmatrix} \dfrac{d^2x}{dt} \\ \dfrac{d^2y}{dt} \end{bmatrix} + \begin{bmatrix} 1 & 0 \\ 0 & 1 \end{bmatrix} \begin{bmatrix} \dfrac{dx}{dt} \\ \dfrac{dy}{dt} \end{bmatrix} = \begin{bmatrix} 0 \\ 0 \end{bmatrix} \tag{8}
$$

The solution of the dynamics of the motion stage of the cell injection system, that is, Eq. (8), is verified as

Theorem 5.4. *Verification of solution of dynamics of motion stage*

$\vdash_{thm}$ ∀x y mx my mp taux tauy fexd feyd alpha x0 y0 xd0 yd0.

[A1]: 0 < mx ∧ [A2]: 0 < my ∧ [A3]: 0 < mp ∧

[A4]: x(0) = x0 ∧ [A5]: y(0) = y0 ∧ [A6]: $\frac{dx}{dt}$(0)= xd0 ∧

[A7]: $\frac{dy}{dt}$(0)= yd0 ∧ [A8]: $\begin{bmatrix} taux \\ tauy \end{bmatrix} = \begin{bmatrix} 0 \\ 0 \end{bmatrix}$ ∧ [A9]: $\begin{bmatrix} fexd \\ feyd \end{bmatrix} = \begin{bmatrix} 0 \\ 0 \end{bmatrix}$ ∧

[A10]: (∀ t. x(t) = (x0 + xd0 * (mx + my + mp)))

$\qquad\qquad$ - xd0 * (mx + my + mp) * $e^{\frac{-1}{mx+my+mp}t}$ ∧

[A11]: (∀ t. y(t) = (y0 + yd0 * (my + mp)) - yd0 * (my + mp) * $e^{\frac{-1}{my+mp}t}$)

⇒ dynamics_2dof_motion_stage mx my mp x y t taux tauy fexd feyd

Assumptions A1-A3 *present the nonnegativity of the masses* mx, my, *and* mp, *respectively. Assumptions* A4-A7 *model the initial conditions, that is, the values of the stage coordinates* x *and* y *and their first-order derivatives* $\frac{dx}{dt}$ *and* $\frac{dy}{dt}$ *at* t = 0. *Assumptions* A8-A9 *provide the condition that the torque and the force vectors are zero. Assumptions* A10-A11 *capture the values of the xy stage coordinates at any time* t. *Finally, the conclusion provides the dynamical behavior of the 2-DOF motion stage of the underlying system. The verification of Theorem 5.4 involves the properties of derivatives of the real-valued functions, transcendental functions, vectors, and matrices along with some arithmetic reasoning. We verify an alternate form of the image-stage coordinate frame interrelationship, which depends on the dynamical behavior of the motion stage (Definition 5.5) and is characterized as a vital property for the analyzing the robotic cell injection systems. For this purpose, we need to model the positioning table and inertia matrices:*

Definition 5.6. Positioning table and inertia matrices

$\vdash_{def}$ ∀fx fy α. posit_table_matrix_fin fx fy α =

    posit_table_matrix ∗∗ matrix_inv (transform_matrix fx fy α)

$\vdash_{def}$ ∀mx my mp fx fy α. inertia_matrix mx my mp fx fy α =

    mass_matrix mx my mp ∗∗ matrix_inv (transform_matrix fx fy α)

where the function `matrix_inv` takes a matrix $A:\mathbb{R}^{NM}$ and returns its inverse. Now, the alternate representation of the relationship of the image and the stage coordinate frames is verified in HOL Light as the following theorem:

Theorem 5.5. *Image-stage coordinates interrelationship*

$\vdash_{thm}$ ∀xc yc u v x y fx fy dx dy mx my mp taux tauy fexd feyd α.

[A1]: 0 < dx ∧ [A2]: 0 < dy ∧ [A3]: 0 < fx ∧ [A4]: 0 < fy ∧

[A5]: invertible (transform_matrix fx fy α) ∧

[A6]: (∀t. u real_differentiable atreal t) ∧

[A7]: (∀t. v real_differentiable atreal t) ∧

[A8]: (∀t. $\frac{du}{dt}$ real_differentiable atreal t) ∧

[A9]: (∀t. $\frac{dv}{dt}$ real_differentiable atreal t) ∧

[A10]: (∀t. relat_image_camera_coordin xc yc u v t fx fy) ∧

[A11]: (∀t. relat_camera_stage_coordin xc yc x y α dx dy t) ∧

[A12]: dynamics_2dof_motion_stage mx my mp x y t taux tauy fexd feyd

    ⇒ inertia_matrix mx my mp fx fy α ∗∗

                            second_order_deriv_image_coordin u v t +

posit_table_matrix_finfxfyα ∗∗ first_order_deriv_image_coordin u v t =

            torque_vector taux tauy - desired_force_vector fexd feyd

Assumptions A1-A4 *provide the design constraints for the image-stage coordinate frame interrelationship. Assumption* A5 *presents the condition about the existence of the inverse of the transformation matrix, that is,* (transform_matrix, *Definition 5.4) is invertible. Assumptions* A6-A9 *describe the differentiability condition for the image coordinates and their first-order derivatives. Assumptions* A10-A11 *model the image-camera and camera-stage coordinate frame interrelationships, respectively. Assumption* A12 *presents the dynamical behavior of the 2-DOF motion stage of the system. Finally, the conclusion of Theorem 5.5 provides the alternate form the relationship between the image and stage coordinate frames. The proof process of Theorem 5.5 is mainly based on the properties of the derivatives of the real-valued functions, vectors, and matrices along with some real arithmetic reasoning.*

5.2 Formalization of the motion planning of the injection pipette

The motion of the injection pipette is vital for the process of the robotic cell injection as a slight excessive force applied on the pipette may damage the membrane of the cell or an insufficient force may not be able to pierce the cell. The motion of the pipette is generally controlled by the force and the torque controllers, which are mainly responsible for controlling the applied injection force and the torque applied to the deriving motor. We formalize both these controllers and formally verify their implication relationship. The impendence force control for a robotic cell injection system is mathematically expressed as

$$m\ddot{e} + b\dot{e} + ke = f_e \tag{9}$$

where m, b, and k represent the desired impendence parameters. Similarly, f_e is the two-dimensional vector containing the x and y components of the applied force. Moreover, e, $\dot{e}$, and $\ddot{e}$ are the two-dimensional vectors capturing the position errors of the xy motion stage coordinate frame, its first-order and second-order derivatives, respectively, and are given as follows:

$$e = \begin{bmatrix} x_d \\ y_d \end{bmatrix} - \begin{bmatrix} x \\ y \end{bmatrix}, \quad \dot{e} = \begin{bmatrix} \dfrac{dx_d}{dt} \\ \dfrac{dy_d}{dt} \end{bmatrix} - \begin{bmatrix} \dfrac{dx}{dt} \\ \dfrac{dy}{dt} \end{bmatrix}, \quad \ddot{e} = \begin{bmatrix} \dfrac{d^2x_d}{dt} \\ \dfrac{d^2y_d}{dt} \end{bmatrix} - \begin{bmatrix} \dfrac{d^2x}{dt} \\ \dfrac{d^2y}{dt} \end{bmatrix} \tag{10}$$

where x and y are the actual axes and x_d and y_d are the desired axes of the stage coordinate frame. Now, the image-based torque controller for the xy stage coordinates is mathematically expressed as

$$\begin{bmatrix} \tau_x \\ \tau_y \end{bmatrix} = \begin{bmatrix} m_x + m_y + m_p & 0 \\ 0 & m_y + m_p \end{bmatrix} \begin{bmatrix} f_x\cos\alpha & f_x\sin\alpha \\ -f_y\sin\alpha & f_y\cos\alpha \end{bmatrix} \begin{bmatrix} \dfrac{d^2x_d}{dt} \\ \dfrac{d^2y_d}{dt} \end{bmatrix}$$

$$+ \begin{bmatrix} m_x + m_y + m_p & 0 \\ 0 & m_y + m_p \end{bmatrix} \begin{bmatrix} f_x\cos\alpha & f_x\sin\alpha \\ -f_y\sin\alpha & f_y\cos\alpha \end{bmatrix}$$

$$\times m^{-1}(b\dot{e} + ke - f_e) + \left(\begin{bmatrix} 1 & 0 \\ 0 & 1 \end{bmatrix} \begin{bmatrix} f_x\cos\alpha & f_x\sin\alpha \\ -f_y\sin\alpha & f_y\cos\alpha \end{bmatrix}^{-1} \right) \tag{11}$$

$$\times \begin{bmatrix} f_x\cos\alpha & f_x\sin\alpha \\ -f_y\sin\alpha & f_y\cos\alpha \end{bmatrix} \begin{bmatrix} \dfrac{dx}{dt} \\ \dfrac{dy}{dt} \end{bmatrix} + \begin{bmatrix} fex^d \\ fey^d \end{bmatrix}$$

Eq. (11) can be written in a compact form as

$$\vec{\tau} = MT \begin{bmatrix} \dfrac{d^2 x_d}{dt} \\ \dfrac{d^2 y_d}{dt} \end{bmatrix} + MTm^{-1}(b\dot{e} + ke - f_e) + NT \begin{bmatrix} \dfrac{dx}{dt} \\ \dfrac{dy}{dt} \end{bmatrix} + \vec{f_{ed}} \qquad (12)$$

where $\vec{\tau}$ and $\vec{f_{ed}}$ describe the torque and desired force vectors. Similarly, M, N, and T model the inertia, positioning table, and transformation matrices. Eq. (12) was wrongly presented in simulation-based analysis (Huang et al., 2006) as follows:

$$\vec{\tau} = M \begin{bmatrix} \dfrac{d^2 x_d}{dt} \\ \dfrac{d^2 y_d}{dt} \end{bmatrix} + Mm^{-1}(b\dot{e} + ke - f_e) + N \begin{bmatrix} \dfrac{dx}{dt} \\ \dfrac{dy}{dt} \end{bmatrix} + \vec{f_{ed}} \qquad (13)$$

In Eq. (13) (used in the simulation-based analysis; Huang et al., 2006), the transformation matrix (T) is missing, which involves the contributions of the applied force and the angle at which the injection pipette is pierced into the cell and its absence can lead to disastrous consequences, that is, excess substance injection, injection of the substance at the wrong location, cell tissues damage, etc. We caught this wrong interpretation of Eq. (12) in the simulation-based analysis during the verification of the implication relationship between the force and torque controllers. We verified the implication relationship between the impendence force control and the image-based torque controller (Eq. 12) in HOL Light as follows:

Theorem 5.6. *Relationship between force and torque controllers*

$\vdash_{thm}$ ∀xd yd x y t mx my mp fx fy α taux tauy fex fey fexd feyd m b k.

   [A1]: 0 < m ∧ [A2]: 0 < k ∧ [A3]: 0 < b ∧

   [A4]: invertible (transform_matrix fx fy α) ∧

   [A5]: force_controller xd yd x y t m b k fex fey ∧

   [A6]: dynamics_2dof_motion_stage mx my mp x y t taux tauy fexd feyd

      ⇒ torque_controller xd yd x y t mx my mp fx fy

            α taux tauy fex fey fexd feyd m b k

Assumptions A1-A3 *model the nonnegativity of the desired impendence parameters. Assumption* A4 *presents the existence of the inverse of the transformation matrix. Assumption* A5 *provides the impendence force controller (Eq. 9). Similarly, Assumption* A6 *describes the dynamical behavior of the 2-DOF motion stage. Finally, the conclusion provides the image-based torque controller (Eq. 11). The proof process of Theorem 5.6 is mainly based on the properties of derivatives of the real-valued functions, matrices, and vectors. This concludes our formalization of the motion planning of the injection pipette used in the process of the robotic cell injection.*

6 Discussions

The probabilistic model checking-based formalization of the robotic cell injection systems using PRISM, presented earlier, is based on DTMC and thus involves the discretization of the continuous dynamics (modeling differential equations) of these systems. Moreover, due to the assignment of probabilities to transitions of the state-based PRISM model, it incorporates various disturbances and measurement noises associated with the underlying system. However, the proposed framework only involves the development of the formal model and thus it lacks the property verification corresponding to this model, which can be done automatically. Moreover, it enables the formalization of 2-DOF cell injection system and cannot be used to reason about 3-DOF and 4-DOF robotic injection system.

In comparison to the model checking-based analysis, the higher-order-logic theorem proving-based approach allows us to model the dynamics of the cell injection systems involving differential and derivative (Eqs. 2, 9, 11) in their true form, whereas, in their model checking-based analysis (Sardar and Hasan, 2017), they are discretized and modeled using a state-transition system. Moreover, all the verified theorems are universally quantified and can thus be specialized to the required values based on the requirement of the analysis of the cell injection systems. However, due to the undecidable nature of the higher-order logic, the verification results involve manual interventions and human guidance. Moreover, it only provides the formalization of 2-DOF cell injection system and cannot be used to reason about 3-DOF and 4-DOF robotic injection system. Table 3 presents a comparison of various analysis techniques, summarizing their strength and weaknesses, for analyzing the robotic cell injection systems. This comparison is performed based on various parameters such as expressiveness, accuracy, and automation. For example, in model checking, we cannot truly model the differential equations, and their discretization results in an abstracted model, which makes it less expressive. Moreover, higher-order-logic theorem proving enables the verification in an interactive manner due to the undecidable nature of the underlying logic.

Table 3 Comparison of techniques for analyzing robotic cell injection systems.

	Paper-and-pencil proof	Simulation	Computer algebra system	Model checking	Theorem proving
Expressiveness	✓	✓	✓		✓
Accuracy	✓ (?)			✓	✓
Automation		✓	✓	✓	

7 Conclusions

Robotic cell injection involves the insertion of bimolecules, sperms, DNA, and proteins into a specific location of suspended or adherent cells and is widely used in drug development, cellular biology research, and transgenics. This chapter provides our probabilistic model checking and higher-order-logic theorem proving-based formalization of the 2-DOF robotic cell injection systems. Finally, a discussion provides the strengths and weaknesses of these analyses and thus enables a user to select the appropriate analysis technique based on a particular scenario.

References

Aziz, A., Singhal, V., Balarin, F., Brayton, R.K., Sangiovanni-Vincentelli, A.L., 1995. It usually works: the temporal logic of stochastic systems. In: LNCS. Computer Aided Verification, vol. 939. Springer, pp. 155–165.
Aziz, A., Sanwal, K., Singhal, V., Brayton, R., 1996. Verifying continuous time Markov chains. In: LNCS. Computer Aided Verification, vol. 1102. Springer, pp. 269–276.
Baier, C., Katoen, J.P., 2008. Principles of Model Checking. MIT Press.
Clarke, E.M., Grumberg, O., Peled, D., 1999. Model Checking. MIT Press.
Clarke, E.M., Klieber, W., Nováček, M., Zuliani, P., 2012. Model checking and the state explosion problem. In: LNCS. Tools for Practical Software Verification, vol. 7682. Springer, pp. 1–30.
Dorf, R.C., Bishop, R.H., 2011. Modern Control Systems. Pearson.
Durán, A.J., Pérez, M., Varona, J.L., 2013. The misfortunes of a mathematicians' trio using computer algebra systems: can we trust? CoRR. abs/1312.3270.
Faroque, M., Nizam, S., 2016. Virtual reality training for micro-robotic cell injection. Technical report. Deakin University, Australia.
Hansson, H., Jonsson, B., 1994. A logic for reasoning about time and reliability. Form. Asp. Comput. 6 (5), 512–535.
Harrison, J., 1996a. HOL Light: a tutorial introduction. In: LNCS Formal Methods in Computer-Aided Design. vol. 1166. Springer, pp. 265–269.

Harrison, J., 2009. Handbook of Practical Logic and Automated Reasoning. Cambridge University Press.

Harrison, J., 2013. The HOL Light theory of Euclidean space. J. Autom. Reason. 50 (2), 1–18.

Harrison, J., 1996b. Formalized Mathematics. Turku Centre for Computer Science.

Hasan, O., Tahar, S., 2015. Formal verification methods. In: Encyclopedia of Information Science and Technology. third ed. IGI Global Pub., pp. 7162–7170

Huang, H., Sun, D., Mills, J.K., Li, W.J., 2006. A visual impedance force control of a robotic cell injection system. In: Robotics and Biomimetics, IEEE, pp. 233–238.

Huang, H., Sun, D., Mills, J.K., Li, W.J., 2007. Visual-based impedance force control of three-dimensional cell injection system. In: Robotics and Automation, IEEE, pp. 4196–4201.

Huang, H., Sun, D., Mills, J.K., Li, W.J., Cheng, S.H., 2009. Visual-based impedance control of out-of-plane cell injection systems. Trans. Autom. Sci. Eng. 6 (3), 565–571.

Huang, H.B., Sun, D., Mills, J.K., Cheng, S.H., 2009. Robotic cell injection system with position and force control: toward automatic batch biomanipulation. IEEE Trans. Robot. 25 (3), 727–737.

Kulkarni, V.G., 2016. Modeling and Analysis of Stochastic Systems. CRC Press.

Kuncova, J., Kallio, P., 2004. Challenges in capillary pressure microinjection. In: Engineering in Medicine and Biology Society, vol. 2. IEEE, pp. 4998–5001.

Kwiatkowska, M., Norman, G., Parker, D., 2011. PRISM 4.0: verification of probabilistic real-time systems. In: LNCS. Computer Aided Verification, vol. 6806. Springer, pp. 585–591.

LeVeque, R.J., 2007. Finite Difference Methods for Ordinary and Partial Differential Equations: Steady-State and Time-Dependent Problems. vol. 98. SIAM.

Levine, W.S., 1999. Control System Applications. CRC Press, Boca Raton, FL.

Mathematica, 2019. https://www.wolfram.com/mathematica/. (Accessed January 2019).

Nakayama, T., Fujiwara, H., Tastumi, K., Fujita, K., Higuchi, T., Mori, T., 1998. A new assisted hatching technique using a piezo-micromanipulator. Fertil. Steril. 69 (4), 784–788.

Nethery, J.F., Spong, M.W., 1994. Robotica: a mathematica package for robot analysis. IEEE Robot. Autom. Mag. 1 (1), 13–20.

Orgun, M.A., Ma, W., 1994. An overview of temporal and modal logic programming. In: LNCS. Temporal Logic, vol. 827. Springer, pp. 445–479.

Paulson, L.C., 1996. ML for the Working Programmer. Cambridge University Press.

Pnueli, A., 1977. The temporal logic of programs. In: Foundations of Computer Science, IEEE, pp. 46–57.

PULNiX, 2017. TM-6701 AN progressive scan full-frame shutter camera. www.jai.com/SiteCollectionDocuments/Camera_Solutions_Datasheets/TM-Datasheets/Datasheet_TM-6701AN.pdf.

Puterman, M.L., 2014. Markov Decision Processes: Discrete Stochastic Dynamic Programming. John Wiley & Sons.

Rashid, A., Hasan, O., 2018. Formal analysis of robotic cell injection systems using theorem proving. In: LNCS Special Issue on the Theme of Design, Modeling and Evaluation of Cyber Physical SystemsSpringer. http://save.seecs.nust.edu.pk/pubs/2018/CyPhy_2017.pdf.

Sardar, M.U., Hasan, O., 2017. Towards probabilistic formal modeling of robotic cell injection systems. In: Models for Formal Analysis of Real Systems. Open Publishing Association, Uppsala, Sweden, pp. 271–282.

Segala, R., Lynch, N., 1995. Probabilistic simulations for probabilistic processes. Nordic J. Comput. 2 (2), 250–273.

Sun, D., Liu, Y., 1997. Modeling and impedance control of a two-manipulator system handling a flexible beam. In: IEEE International Conference on Robotics and Automation, vol. 2. IEEE, pp. 1787–1792.

Sun, Y., Nelson, B.J., 2002. Biological cell injection using an autonomous microrobotic system. Robot. Res. 21 (10–11), 861–868.

Sun, Y., Wan, K.T., Roberts, K.P., Bischof, J.C., Nelson, B.J., 2003. Mechanical property characterization of mouse zona pellucida. IEEE Trans. Nanobiosci. 2 (4), 279–286.

Yanagida, K., Katayose, H., Yazawa, H., Kimura, Y., Konnai, K., Sato, A., 1999. The usefulness of a piezo-micromanipulator in intracytoplasmic sperm injection in humans. Hum. Reprod. 14 (2), 448–453.

CHAPTER SIX

Identifying vessel branching from fluid stresses on microscopic robots

Tad Hogg
Institute for Molecular Manufacturing, Palo Alto, CA, United States

1 Introduction

Microscopic devices small enough to pass through the circulatory system are useful for biological research and medicine (Freitas, 1999; Martel, 2007; Monroe, 2009; Hill et al., 2008; Schulz et al., 2009). For instance, nanoparticles can precisely deliver drugs (Allen and Cullis, 2004). More elaborate applications could arise from devices with a full range of robotic capabilities, including sensing, computation, communication, and locomotion. These microscopic robots could sense a variety of signals, including chemicals on cells (Park et al., 2008), fluid shear (Korin et al., 2012), light, or temperature (Sershen et al., 2000).

Applications of microscopic robots could improve their precision if robots can identify type of tissue they are passing through. For circulating robots in the vasculature, especially useful identification methods are those available to the robot from properties measurable from within the vessels. For instance, the geometry of capillaries differs among organs (Augustin and Koh, 2017) and between normal and tumor vasculature (Nagy et al., 2009; Jain et al., 2014). Thus, robots able to determine vessel geometry could supplement other available information, such as chemicals, to more accurately identify different types of tissue.

An important aspect of vessel geometry is when a vessel splits into branches, or when several vessels merge into a larger one. Microscopic robots might detect branches acoustically (Freitas, 1999), although interpreting reflected signals could require significant computation in the presence of multiple reflections, scattering, and the small difference in acoustic impedance between the fluid and walls of tiny vessels. As a complementary

Control Systems Design of Bio-Robotics and Bio-mechatronics with advanced applications
https://doi.org/10.1016/B978-0-12-817463-0.00006-X

171

approach, this chapter describes and evaluates how robots could detect branches from changes in the patterns of fluid-induced stresses on their surfaces.

Specifically, this chapter first summarizes related work in microscopic devices. Then, the following two sections describe typical geometric parameters of tiny vessels and the stresses on the surface of an object moving with the fluid in such vessels. The next two sections show how a robot can use stress measurements to identify when it is passing vessel branches. Section 7 evaluates the performance of this approach. The remaining sections discuss possible extensions to more complicated scenarios. The "Appendix" section describes the scenarios used for training and testing, and the resulting the classifier used to identify branches.

2 Related work

Some existing devices (Koman et al., 2018; Jager et al., 2000; Martel et al., 2007; Sitti et al., 2015) demonstrate various robotic capabilities in small volumes, although they are generally too large to fit through capillaries. Smaller demonstrated devices include DNA nanodevices (Ke et al., 2018; Li et al., 2018; Thubagere et al., 2017), biohybrids with bacteria (Martel, 2014), and sequential logic within cells (Andrews et al., 2018), with more limited sensing and computational capabilities than larger counterparts. Building on these demonstrations, future microscopic robots could provide a wide range of capabilities in a volume small enough to pass through capillaries (Freitas, 1999).

For navigation, fluid stresses provide useful information at larger scales, for example, for fish (Bleckmann and Zelick, 2009) and underwater vehicles (Yang et al., 2006; Vollmayr et al., 2014). Micromachine sensors motivated by those of fish can detect relatively small changes in fluid motion (Kottapalli et al., 2014). These fluid stresses provide information about the environment within about a body length of the object in the fluid (Sichert et al., 2009), although extracting this information requires significant computation (Bouffanais et al., 2010). Another example of sensing fluid motion is the use of artificial whiskers, modeled on the geometry of seal whiskers, to estimate the direction, size, and speed of moving objects from their wakes (Eberhardt, 2016).

For microscopic objects in fluids, viscous rather than inertial forces dominate the flow (Dusenbery, 2009; Purcell, 1977). Viscous flows have a linear relation between stresses and velocities (Happel and Brenner, 1983), allowing simpler computations to interpret stresses than required for larger-scale flows. This simplicity enables stress-based navigation by microscopic robots (Hogg, 2018) in spite of their computational limits, both in terms of operating speed and memory. Moreover, viscous flow generally has gentle gradients, so the effect of boundaries extends relatively far into the fluid. This potentially allows a microscopic robot to use stresses to infer properties of the boundaries at larger distances, relative to its size, than is feasible for larger robots.

3 The geometry of microscopic vessels

This chapter examines spherical robots in moving fluids similar to water, with parameters given in Table 1. We numerically evaluate robot behavior in vessels with geometry comparable to that of short segments of capillaries. These vessels generally have radii of curvature of tens of microns (Pawlik et al., 1981), and when they split, they typically split into just two branches (Cassot, 2006). The branches have a larger total cross-section than the main vessel (Murray, 1926), leading to slower flows in the branches (Sochi, 2015).

For simplicity, we focus on planar vessel geometry. Specifically, incoming and outgoing axes of curved vessels are in the same plane. Similarly, branching vessels have the axes of the main vessel and the branches in the same plane.

Fig. 1 shows examples of the vessel geometries considered here. The fluid flow speed for these two cases is chosen so the robots have the same average stress magnitudes on their surfaces at the indicated position along each paths. Specifically, the maximum speed at the inlet is 1000 and 530 μm/s, for the branch and curve, respectively. At the position of the

Table 1 Typical parameters for fluids and microscopic robots

Density	ρ	10^3 kg/m^3
Viscosity	η	10^{-3} Pa s
Kinematic viscosity	$\nu = \eta/\rho$	10^{-6} m^2/s
Vessel diameter	d	5–10 μm
Maximum flow speed	u	200–2000 μm/s
Reynolds number	$\mathrm{Re} = ud/\nu$	<0.04
Robot radius	r	1 μm

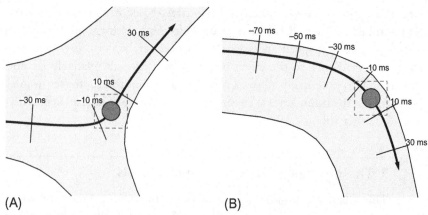

Fig. 1 Robot motion through (A) branched and (B) curved vessels. The *curved arrow* shows a portion of the path of the robot's center as it moves through the vessel with the fluid. The *ticks along the paths* indicate times, in milliseconds, before or after the location of the robot indicated by the *gray disk*. The *dashed rectangles* indicate the parts of the vessel shown in Fig. 2. The robot has radius 1 μm and the vessel inlets, to the left of the sections shown in the figure, both have diameters of 7.8 μm.

robot shown in the branch, the robot moves at 189 μm/s and rotates with angular velocity −34 rad/s, that is, clockwise. For the curve, these values are 186 μm/s and +39 rad/s.

For developing and testing a branch classifier based on stresses on the robot's surface, we create samples of robots in branch and curve vessels. Appendix A.1 describes the parameters for the vessel geometry, initial robot position, and fluid flow speed used to create these samples. These vessels are similar in size to the examples of Fig. 1.

4 Robot stresses and motion in vessels

We determine stresses on the robot surface by numerically evaluating the flow and robot motion in a segment of the vessel. For the vessel sizes, planar geometries and fluid speeds considered here, the motion and stresses can be approximated by two-dimensional quasistatic Stokes flow (Hogg, 2018). As examples, Fig. 2 shows the fluid velocity near the robot, in the section of the vessel indicated by the dashed rectangles in Fig. 1. In spite of the different vessel geometries, the flow near the robot is similar in the two cases.

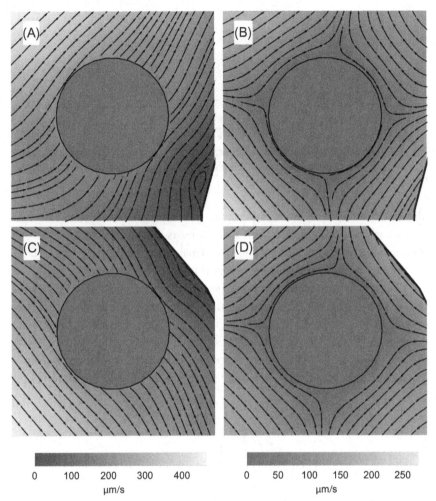

Fig. 2 Fluid flow near the robot for the scenarios shown in Fig. 1. *Arrows* show streamlines of the flow and *colors* show the flow speed. (A) Fluid velocity in the branch with respect to the vessel. Velocity is zero at the vessel wall, and matches the motion of the robot at its surface. (B) Fluid velocity in the branch with respect to the robot. Velocity is zero at the robot surface. (C) Fluid velocity in the curve with respect to the vessel. (D) Fluid velocity in the curve with respect to the robot. The legend at the bottom of each column applies to the two plots above it.

4.1 Stresses on robot surface

We denote the stresses on a robot's surface by $\mathbf{s}(\theta, t)$ for the stress vector at angle θ at time t. The angle θ specifies a location on a robot's surface, measured from an arbitrary fixed point on the robot called its "front," as

Fig. 3 A robot, indicated as a *gray disk*, with $n = 30$ stress sensors, shown as points, spaced uniformly around its surface. The *green line* (*gray line in print version*) indicates the front of the robot.

illustrated in Fig. 3. A robot can estimate $\mathbf{s}(\theta, t)$ by measuring stresses at various locations on its surface with force sensors, and interpolating between these locations. By measuring forces normal and tangential to the surface, the sensors determine the stress vector (Hogg, 2018).

The stresses depend on the vessel geometry near the robot and the speed of the flow. However, this relationship is not unique: different geometries can produce similar stress patterns. Fig. 1 provides one such example. In these cases, Fig. 4 shows how stresses vary over the robot's surface, with the arrows indicating the stress vectors at points spaced uniformly around the surface. The patterns of stress on the surfaces are nearly the same. Thus,

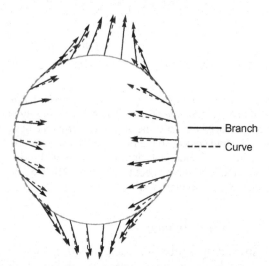

Fig. 4 Stress vectors on the robot surface for the indicated robot positions in the branch and curve scenarios of Fig. 1. The *arrow lengths* show the magnitude of the stresses, ranging up to 0.58 Pa, at locations of sensors on the robot surface.

in general, the pattern of stresses at a single instant does not reliably identify the geometry of the vessel near the robot.

4.2 Changing stress patterns

As a robot moves through a vessel with changing geometry, the stresses on its surface change. In many cases, the stresses change significantly as a robot moves through a branch, whereas changes are fairly small as a robot moves around a curve.

One measure of changing stresses is the correlation of the stress pattern at two times. Suppose $\mathbf{f}(\theta)$ and $\mathbf{g}(\theta)$ are two vector-valued functions of the angle θ around the robot's surface. The correlation between these vector fields is

$$\text{cor}(\mathbf{f}, \mathbf{g}) = \frac{1}{\|\mathbf{f}\|\|\mathbf{g}\|} \int_0^{2\pi} \mathbf{f}(\theta) \cdot \mathbf{g}(\theta)d\theta \tag{1}$$

with the norm $\|\mathbf{f}\| = \sqrt{\int_0^{2\pi} \mathbf{f}(\theta) \cdot \mathbf{f}(\theta)d\theta}$ and $\mathbf{f} \cdot \mathbf{g}$ denoting the inner product of the two vectors.

In our case, the vector fields are the stresses on the robot surface. As an example, the correlation between the surface stresses in the two cases shown in Fig. 4 is 0.98. Due to the normalization, the correlation is independent of the overall magnitude of the stresses. In particular, this means that the correlation does not depend on the fluid viscosity.

As noted earlier, the robot can estimate the stress field $\mathbf{s}(\theta, t)$ by interpolating surface stress measurements. A particularly useful method is interpolating the stress pattern from a few Fourier modes. The correlation between two stress patterns can be computed directly from the Fourier coefficients, avoiding the need to explicitly evaluate the integrals in Eq. (1) (Hogg, 2018). Specifically, we use the first six modes, which capture most of the variation in stress over the robot's surface for the cases considered here.

When viewed from a fixed location on the robot surface, for example, a specific sensor, the stress changes both due to changing vessel geometry and due to the robot's rotation caused by the fluid. This rotation is not relevant for identifying vessel geometry for the spherical robots considered here. Thus, to identify changes in geometry, the robot must remove the change due to its rotation. A simple way to do so is to maximize the correlation over all possible rotations of the robot between the two measurements used in the correlation. Specifically, this approach compares the stress at time t, $\mathbf{s}(\theta, t)$, with shifted versions of the stress at a prior time, $\mathbf{s}(\theta + \Delta\theta, t - \Delta t)$, and uses the maximum correlation over all shifts $\Delta\theta$. That is, the robot measures changes in the pattern of stress that are *not* due to its rotation by evaluating

$$c(t, \Delta t) = \max_{\Delta\theta} \ \text{cor}(\mathbf{s}(\theta, t), \mathbf{s}(\theta + \Delta\theta, t - \Delta t)) \tag{2}$$

for the correlation function of Eq. (1). Another application of this maximization is estimating the robot's angular velocity because the shift in angle, $\Delta\theta$, giving the maximum is an estimate of how much the robot has rotated between these two times (Hogg, 2018).

As an example, Fig. 5 shows the correlation between stress patterns separated by 10 ms as the robot moves through the branch and curved vessels of Fig. 1. For the curve, the stress pattern remains nearly the same, so correlations are close to one. For the branch, however, the correlation drops as the robot approaches the branch, about 30 ms before it reaches the position shown in Fig. 1. Later, the correlation drops again as the robot moves into one of the branches. Similarly, if the robot was moving in the opposite direction, that is, the flow was a merge of two small vessels into a larger one, the robot would encounter these drops in correlation in the opposite order and at times shifted by 10 ms as it compares its current stress pattern with the pattern it had encountered earlier along the reversed path.

For the flow speeds and vessel sizes considered here, $\Delta t = 10$ ms is a reasonable choice: over that time a robot can move a distance comparable to the

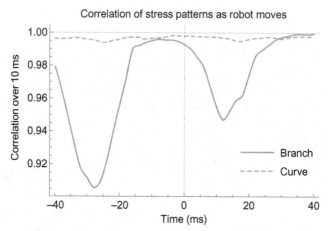

Fig. 5 Correlation $c(t, \Delta t)$ between stress patterns at times t and $t - \Delta t$, with $\Delta t = 10$ ms, as the robot moves along the paths shown in Fig. 1. The times correspond to the *tick marks* along the paths shown in Fig. 1, with $t = 0$ corresponding to the robot positions shown in that figure.

extent of branching or curving, but not so far as to completely pass the changing geometry. However, the precise value of Δt is not important. For example, comparing stresses separated by $\Delta t = 5$ or 20 ms is qualitatively similar to the behavior shown in Fig. 5. For definiteness, we use $\Delta t = 10$ ms in the following discussion.

5 Classifying vessel geometry

Fig. 5 suggests a robot could identify vessel branches by checking for when the correlation of stresses separated by a short time is sufficiently small. This procedure could reliably distinguish branches from curves if there is little overlap in the distributions of minimum correlations for these two geometries. Unfortunately, the behavior shown in Fig. 5 does not occur in all cases. Instead, the distributions of minimum correlation for curves and branches have considerable overlap, especially when the robot is near the center of the vessel.

Fig. 6 quantifies this difficulty by showing how the minimum correlation along a path depends on how close the robot is to the vessel wall before it reaches the curve or branch, that is, while still in a fairly straight portion of

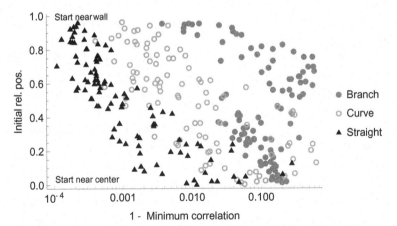

Fig. 6 Scatterplot of the minimum value of $c(t, \Delta t)$ along a path, for $\Delta t = 10$ ms, and initial relative position, defined by Eq. (3). The points distinguish robots moving through a branch, around a curve, or in a straight vessel. To highlight the differences among these geometries, the *horizontal axis* shows $1 - c$ on a logarithmic scale. Thus, situations where the stress pattern changes only slightly over time Δt, that is, correlations are close to 1, appear on the *left side* of the diagram.

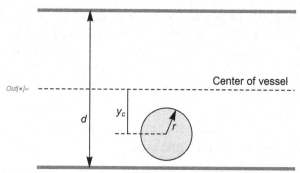

Fig. 7 Position of robot (*gray disk*) in a straight segment of a vessel, used to define the robot's relative position.

the vessel. As a measure of how close a robot is to the vessel wall, we use the robot's *relative position*, given by

$$\mathrm{rp} = \frac{|y_c|}{d/2 - r} \tag{3}$$

where y_c is the position of the robot's center relative to the central axis of the vessel, d is the vessel diameter, and r is the robot's radius, as illustrated in Fig. 7. Relative position ranges from 0, for a robot at the center of the vessel, to 1, for a robot just touching the vessel wall. A robot can accurately estimate its relative position while in a straight vessel segment from surface stresses (Hogg, 2018).

When the robot starts near the center of the vessel, Fig. 6 shows that curved paths have a wider range of minimum correlations than branches, and this range includes the values occurring in branches. Combined with smaller, and hence noisier, stresses for robots relatively far from the vessel wall (Hogg, 2018), this indicates correlation is not a reliable identifier of branches when robots start near the vessel center.

To improve identification for paths starting close to vessel center, a third piece of information is helpful, namely a summary of the stress pattern at the time the robot evaluates the correlation. That is, to identify branches, at time t, we use three pieces of information: the correlation $c(t, \Delta t)$, the relative position for the path that was determined prior to any significant change in the stress, and the current stress $s(\theta, t)$.

To create a vessel geometry classifier using this information, we generate a set of training samples. Specifically, for each training sample (created as described in Appendix A.1), we determine the time t along the path with the minimum correlation, and use the three pieces of information from that time along the path. This method is an example of off-line training. That is, we suppose the training samples have measurements from a completed path, that is, a path starting before the robot reaches the branch or curve, and continuing until the robot is well past those changes. With measurements along the whole path, the time of minimum correlation can be obtained and values at that time used for training. Using such training samples from both branch and curve vessels results in the logistic regression classifier for branches described in Appendix B.2.

6 Applying the classifier to identify branches

The classifier described earlier was trained with the minimum correlation along a path. A robot using the same method when applying the classifier would have to wait until it was sufficiently far past a changing geometry to be sure it had detected the minimum correlation along the path for that change. This off-line application of the classifier could be useful in reporting vessel geometry changes well after encountering them, for example, to provide a description of the path leading the robot to a target location.

We focus on the more demanding classification task of recognizing a branch near the time the robot encounters it. This online or real-time classification allows the robot to take action, while still near the branch. In this case, a robot uses the classifier by repeatedly evaluating Eq. (B.1) as it moves. During these evaluations, the correlation $c(t, \Delta t)$ is not necessarily the minimum correlation along the path: that is, the robot could encounter smaller values as it continues through the vessel. Thus, the robot using this method is extrapolating beyond the values used for training.

The classifier uses the robot's relative position in the vessel before it encounters a branch or significant curve. Thus, the robot must save its estimate of relative position, updating the value only while vessel geometry is not changing. A robot could determine when this steady behavior occurs by checking when the correlations between stresses at various delay times Δt are close to 1, and the pattern of stress on its surface is consistent with its presence in a straight vessel segment (Hogg, 2018). When the robot encounters changing geometry, it uses the saved estimate of its relative position when evaluating the classifier, that is, Eq. (B.1). This procedure applies to typical capillaries (Pawlik et al., 1981) where significant geometry changes are separated by tens of microns, a considerably larger distance than the size of the robots small enough to pass through those vessels.

6.1 Example

As an example of how the classifier applies to online classification, Fig. 8 shows the estimates of P_{branch} from Eq. (B.1) along the robot paths of Fig. 1. The branch path has higher values than the curve. This example indicates how a robot could use the classifier for online branch identification: consider a branch to be nearby whenever P_{branch} exceeds a predetermined threshold. For this example, a threshold around 0.8 distinguishes the branch from the curve.

In addition to identifying *whether* the robot passes a branch, Fig. 8 indicates *when* this method detects the branch. In this case for the branching vessel, P_{branch} becomes large about 30 ms before the robot reaches the location shown in Fig. 1. This corresponds to the robot entering the branch. The robot slows as it moves through the branch, leading to a period of large correlation for about 20 ms. As the robot leaves the branch, the stress pattern changes again, leading to a second minimum in the correlation (see Fig. 5) and another maximum in P_{branch}. In other cases, the robot moves more rapidly through the branch, so the $\Delta t = 10$ ms time difference used here gives a single minimum in the correlation and, correspondingly, a single peak in P_{branch}. A robot could distinguish these cases by checking for a second peak within a few tens of milliseconds. Over that amount of time, the second peak indicates the robot is leaving the branch rather than encountering a second branch. This is consistent with the typical spacing of branches in capillaries (see Section 3).

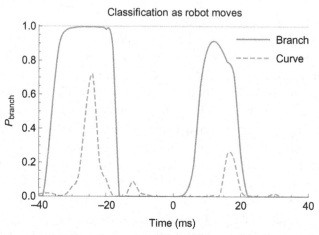

Fig. 8 Estimated probability of encountering a branch, P_{branch}, vs. time as the robot moves along the paths shown in Fig. 1, based on correlation $c(t, \Delta t)$ with $\Delta t = 10$ ms.

6.2 Selecting a threshold to identify branches

A suitable threshold to use for identifying branches with this classifier depends on the relative importance of false negatives (i.e., missing a branch) and false positives (i.e., incorrectly considering a curved vessel to have a branch). The importance of these errors depends on the application.

For example, if a robot with locomotion capability needs to move into a branch, it is better to recognize a branch before passing it, so the robot would only need to actively move a short distance to reach the desired branch. This contrasts with the situation of not recognizing the branch until well after the robot has passed it, in which case it would need a more larger distance and upstream against the flow of the fluid. In this case, the robot could use a relatively low threshold, thereby being fairly sure it will identify branches as it encounters them, although it may also, incorrectly, attempt to move into a branch when passing through some curved vessels. Such errors are more likely the earlier the robot needs to identify a branch because the flow well upstream of a branch is similar to that in vessel without branch.

Another action the robot could take upon detecting a branch is to move to the vessel wall near the branch and act as a beacon to other robots arriving at the branch. The beacon signal could, for example, direct subsequent robots into one branch or the other to ensure roughly equal numbers explore each branch in spite of the fluid flow favoring one branch over the other. More generally, branches could be useful locations to station robots for forming a navigation network (Freitas, 1999). With a limited number of robots to form this network and if it is sufficient to station robots at some rather than all branches, the robot could use a high threshold to ensure identified branches are very likely to actually be branches.

Other applications have less need for identifying branches, while robots are close to them, but instead collect information on the number and spacing of branches for later use, for example, as a map or diagnostic tool at larger scales than short segments of a single vessel. This applies to identifying vessel geometries that distinguish different organs or normal from cancer tissue (Nagy et al., 2009; Jain et al., 2014). In such cases, the emphasis could be on identifying all branches, favoring a relatively low threshold, and then verifying detected branches with subsequent measurements after the robot has moved past each possible branch, thereby reducing the false positives while keeping sensor information obtained near the branches from the original identification of those cases later verified to be branches.

Multiple robots spaced closely enough could communicate verified branch detections to robots upstream of the branch. With a message from a downstream robot that it verified its recent passage of a branch, a robot could lower its threshold in anticipation of that upcoming branch. The message could come from a downstream robot that entered a different branch of the splitting vessel than the robot receiving the message. For example, with acoustic communication robots could compare measurements over 100 μm or so (Hogg and Freitas, 2012), which is farther than the typical spacing between branches in capillaries described in Section 3.

6.3 Verification after passing a curve or branch

This chapter describes how a robot can use stresses to identify branches as it encounters them. This contrasts with off-line methods that collect time-stamped information before, during, and after a robot passes a branch, and later use the entire path history to identify if and when the robot passed branches. A possible combination of the two approaches is to use online classification to identify likely branches, record information about them, and later verify the branch detection when the robot has additional information after passing the possible branch.

Information a robot could use for this verification task includes changes in vessel diameter, relative position, and robot speed before and after passing the curve or branch. Typically, these changes are much larger after passing a branch than a curve. Such comparisons must wait until the robot is well past the branch or curve, when it is again in a nearly straight vessel segment, where stress-based navigation allows estimating these quantities (Hogg, 2018).

Specifically, when a vessel splits into branches, the branches have smaller diameter than the main vessel, but the combined cross-sections of the branches is larger than that of the main vessel (Murray, 1926). Thus, measuring vessel diameter before and after the branch gives a direct indication of branching.

The increased cross-sectional area after a vessel splits leads to slower flows in the branches (Sochi, 2015), whereas flow in a curve remains nearly constant. This suggests a robot could check for changes in its speed through the vessel before and after passing a branch or curve to verify the identification. However, slower fluid speed in the branch does not necessarily mean the robot's speed changes in the same proportion, because the robot's speed depends both on the speed of the flow and how close the robot is to the wall, that is, its relative position. Along with flow speed, the relative position can change as a robot passes a branch.

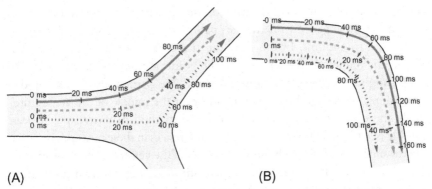

Fig. 9 Robot paths through (A) branched and (B) curved vessels. Each path is for a robot traveling by itself through the vessel with the fluid. The *ticks along the paths* indicate times, in milliseconds, from the start of each path when the maximum inlet flow speed is 1000 μm/s. Robots along paths near the center of the vessel move more rapidly than those near the walls, indicated by the wider spacing between successive ticks along the central paths. The vessel inlets both have diameters of 7.8 μm.

Fig. 9 illustrates this behavior. Robots close to the wall remain close upon entering a branch or moving around a curve. But robots entering a branch near the center of the vessel move to near the wall of the branch, and robots starting between the center and the wall move to near the middle of a branch. This change in relative position can distinguish branching from curved paths that are not close to the wall. Similarly, the change in speed is particularly large for branch paths when the robot is not near the wall, so its speed reflects the decreasing flow through the branches. When the robot is near the wall, in either a branch or curve, it moves relatively slowly and remains near the wall, giving relatively little change. By contrast, Fig. 10

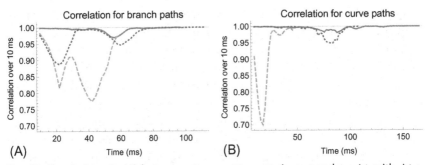

Fig. 10 Correlation $c(t, \Delta t)$ between stress patterns at times t and $t - \Delta t$, with $\Delta t = 10$ ms, as the robot moves along the (A) branched and (B) curved paths shown in Fig. 9. The times correspond to the *tick marks* along the paths. In each case, the path near the center of the vessel has the largest decrease in correlation.

shows that the minimum correlation of the stress patterns does not, by itself, distinguish these paths. In particular, paths close to the center of the vessels have small minimums. In this case, the curve has a smaller minimum than that of the branch, corresponding to the points at the lower right of Fig. 6.

Changes in relative position and speed are particularly useful for distinguishing curves and branches for paths near the middle of the vessels, precisely the cases where the correlation-based method in this chapter is least reliable (see Fig. 6). Moreover, the larger differences in these measures between curves and branches for paths near the center of the vessel could compensate to some extent for the lower accuracy of estimating position and speed from stresses when the robot is near the center of the vessel (Hogg, 2018).

Due to errors in estimating position, speed, and vessel diameter from stresses (Hogg, 2018), evaluating changes from a combination of these measures before and after passing a branch or curve is likely to be more robust than relying on a single method.

7 Classification performance

This section evaluates how the classifier performs using a set of test samples.

In the branch vessels considered here, the fluid flows from the larger vessel into the two branches. Thus, testing the classifier with these paths evaluates how well a robot recognizes when the vessel splits into two smaller vessels, with the robot moving into one of them. The situation for flow merging from smaller branches into a larger vessel is similar. This is because for a robot at a given location in a vessel, stresses on the robot surface are the same for either direction of the flow due to the reversibility of Stokes flow (Happel and Brenner, 1983). Thus, the stress measurements along a path are the same for paths through merging branches, but in the reverse order.

For merging vessels, the classification operates in the same way as for splitting vessels, but with a shift of time Δt in when the correlation is measured. For instance, for the robot at the location shown in Fig. 1A, the correlation $c(t, \Delta t)$ compares the stresses at the robot's indicated location with the stresses when the robot's center was at the point along the path indicated by the tick for -10 ms when using $\Delta t = 10$ ms. Conversely, a robot moving in the reverse direction along the path, that is, from the branch at the upper right into the main vessel at the left, would compare its stress with that 10 ms earlier on the reverse path, corresponding to the location of the tick for 10 ms on the forward path. Thus while stresses are the same along the forward and reversed paths, the comparison used for the correlation and the initial position in the vessel before encountering the branch are different for these two directions.

When the robot along the reverse path is at the position indicated by the tick for -10 ms, it would compare stresses at the same two times that the robot on the forward path does at the position indicated in the figure. So, a robot on the reverse path computes the same correlations as a robot on the forward path, but with a shift of Δt in time. In this comparison, the robots use the same correlation in the classifier. But because they are at different locations along the path when they evaluate that correlation, they would have different values for the other two values used in the classifier: the current stress and the initial relative position.

The classifier training only included the forward direction for each branch, that is, moving from a main vessel into one of the branches at a split. As a test of how well the classifier generalizes, we use both path directions for each branch test sample.

7.1 Accuracy

Since the choice of threshold depends on the application, instead of characterizing a classifier for a single choice of threshold, a better measure is the trade-off between true and false positives over the range of threshold values. At one extreme, a threshold equal to one means the robot never recognizes a branch (no true positives) but also never mistakenly considers a curve to be a branch (no false positives). At the other extreme, a threshold equal to zero means the robot always considers itself to be encountering branches: this identifies all actual branches, but also mistakenly recognizes curves and straight vessel segments as branches (high false positives). For a perfect classifier, there would exist an intermediate threshold allowing it to identify all branches without also mistaking other geometries for branches.

For the classifier developed in this chapter, Fig. 11 shows the performance as the threshold varies from 1 (at lower left) to 0 (at upper right). Specifically, for each threshold value, the figure evaluates P_{branch} with Eq. (B.1) along the path of each test sample. If P_{branch} exceeds the threshold anywhere along the path that sample is classified as a branched vessel. The true positives are the branch samples identified as branches using that threshold, and the false positives are the curve samples incorrectly identified as branches.

This classifier performs well: with suitable threshold, the classifier recognizes most branches with only a few mistakes, as indicated by the curve in

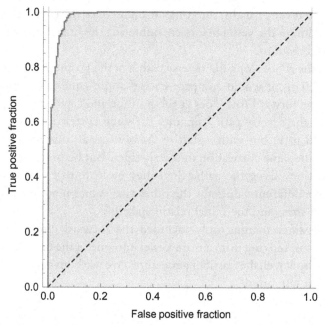

Fig. 11 Classification performance on test samples. The curve would follow the *dashed diagonal line* if the classifier did not discriminate between paths through branch and curve vessels.

Fig. 11 passing close to the ideal behavior of recognizing all true positives and none of the false positives (i.e., the upper-left corner of the figure). This trade-off curve includes both forward and reverse paths. Examining performance separately for each direction (i.e., splitting or merging vessels) shows similar curves. Thus, there is no penalty for not including reverse paths when training the classifier. This is an indication of the robustness of the information used for this classifier, and the simplicity from the reversibility of Stokes flow.

An overall performance measure from the trade-off of Fig. 11 is the area under the curve. In this case, the area is 98.6% of the total area. By comparison, the area would be 50% for a classifier that made no distinction between branch and curve, and 100% for a perfect classifier.

7.2 When branches are identified

In addition to how accurately this classifier detects branches, an important performance measure is *where* along a path the robot first detects a branch.

One of the features this classifier uses is the correlation between stresses at the robot's current location and those at the time $\Delta t = 10$ ms earlier. Typically, stresses change the most as the robot passes through the branch, and during 10 ms the robot does not move significantly past the branch. This means the classifier typically detects the branch, while the robot is within the branching section of the vessel.

Quantitatively, Fig. 12 shows this classifier generally identifies branches near the middle of the sample paths. This corresponds to when the robot is close to the branch, rather than well before or after passing the branch. That is, this classifier identifies branches, while the robot is close to them. Thus, a robot can use this stress-based classifier to identify when it is passing a branch, well before it passes significantly downstream of the branch. On the other hand, branches are not identified well before the robot reaches the branch. Thus, this classifier is not useful for applications requiring significant advance notice that the robot is approaching a branch.

This classifier can detect most branches (high true positive fraction) with only a few false positives (see Fig. 11). Thus, applications will likely use thresholds low enough to give true positive fraction above, say, 80% or so, corresponding to the right portion of Fig. 12. With this choice for the threshold, branch detection will generally occur at smaller fractions of the path length for forward paths than for reverse paths. This corresponds to

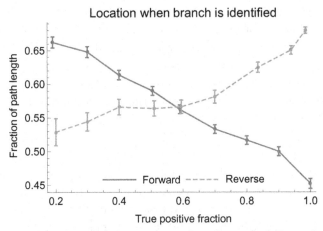

Fig. 12 Fraction of path length of the test sample paths at which P_{branch} first exceeds the detection threshold corresponding to the true positive fraction indicated on the horizontal axis. The *solid and dashed curves* are for forward and reverse branch paths, respectively. *Error bars* show the standard error of the means, indicated by the points.

a robot moving toward a vessel split detecting the branch just as the main vessel splits. Conversely, a robot moving in a vessel that merges with another to form a larger vessel will identify the branch just as the vessels merge. Quantitatively, the difference path fraction for these two cases shown in the figure (about 0.15 for thresholds giving true positives above 80%) corresponds to a difference of about 6 μm for the path lengths used here (see Appendix A.1). Thus, the difference in where a branch is first identified for splitting or merging vessels is just a few times the robot diameter.

7.3 Noise

Sensor measurements are subject to noise, so in practice robots will not have exact values for the stress-based quantities used by the classifier. Evaluating the effect of noisy measurements on classification accuracy is an important performance measure.

The classifier uses three pieces of information: correlation of two stress measurements separated by Δt, relative position estimated from stress while in a straight segment, and principal components from the latest stress used for correlation. These quantities involve stress measurements at different times, except that one of the stresses used for correlation is the same as used to determine principal components. Thus, a reasonable model of the noise is independent errors for the three values used as inputs by the classifier. This assumes noise at different times, separated by at least a few milliseconds, is uncorrelated, for example, as is the case for thermal noise.

As one example of the effect of independent noise, Fig. 13 shows the consequence of relative errors for the area under the trade-off curve for true and false positives shown in Fig. 11. For simplicity, the noise used in this figure takes each of the three inputs to the classifier to have a noisy value given by a random relative error drawn from a normal distribution with zero mean and standard deviation equal to the amount of relative noise indicated on the horizontal axis of the figure. This shows little change in classifier performance with up to about 10% noise. More generally, each of the three inputs to the classifier could have a different amount of noise, depending on how noise from stress measurements propagates to the calculated values of correlation, relative position, and principal components.

If noise is large enough to significantly degrade classifier performance, the robot could average the result of several subsequent evaluations. For instance, averaging stress measurements over several milliseconds can significantly reduce the effect of thermal noise on stress measurements

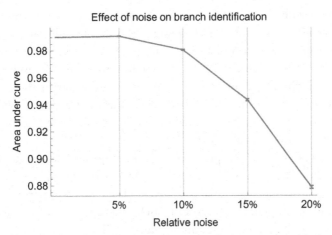

Fig. 13 Area under the trade-off curve as a function of relative noise added to classifier inputs. Each value is the average of 100 noisy versions of the measurements along paths of all the test samples. The error bars are the standard error of those averages, and are only slightly larger than the thickness of the curve connecting the points.

(Hogg, 2018). This averaging will also improve the stress-based inputs to the classifier, provided noise at different times is independent.

Averaging to reduce the effect of independent noise contrasts with systematic errors, which produce noise correlated over time or across different sensors. Examples of such systematic errors include a stress sensor giving consistently smaller readings than the others due to poor calibration, or if sensor responses change as the robot moves through the vessel due to sensors becoming covered by material adhering to the robot surface. Another source of systematic error is a robot's clock running at the wrong speed. That could, for instance, cause a robot to actually compare stresses separated by $\Delta t = 15$ ms, while using a classifier trained with correlations with $\Delta t = 10$ ms. Such systematic errors are not reduced by averaging so will need to be handled by designing stable sensors or providing a way to calibrate them while in use. Alternatively, if such errors differ among robots, for example, because they have somewhat different clock speeds, nearby robots could communicate the information they use for branch identification. Each robot could then average its own measurement with those of other robots to reduce the effect of these errors.

8 Discussion

This chapter describes how a robot moving with fluid in a small vessel can use the changing stress on its surface to identify when it encounters a branch. Repeatedly using this classifier, a robot can identify changes in the vessel geometry as it moves. For example, the robot could record the time and hence distance, with an estimate of its speed obtained, for example, from stress measurements (Hogg, 2018) between its initial approach and final passing of a branch. Thus, the robot can determine the size of the branching region. Over longer times, the robot could measure distances between successive branches, thereby estimating the larger-scale geometry of successive splits and merges of small vessels.

The classifier in this chapter uses only a few properties of the changing stress pattern. These properties are sufficient to identify branches for the scenarios considered here. It remains to be seen how well these properties perform in more complex situations. One such situation is nonplanar branching vessels (Sochi, 2015), which require evaluating 3D flows. Another situation is the fluid containing a variety of objects in addition to the robot. Such objects include cells that deform as they are pushed around curves or into branches, requiring more elaborate simulations (Hoskins et al., 2009), especially in somewhat larger vessels than considered here (Bagchi, 2007). In these or other more complex situations, the properties used in the classifier described here may not be sufficient for high accuracy. In such cases, classifiers could incorporate additional information available to the robot about its local environment. These additions include prior information about vessel structure and additional measurements.

For instance, the robot could combine the classifier's output with prior probabilities of branches. This prior could, for instance, consider typical distances between successive branches and how often vessels bend without branching. Values for the priors could come from studies of microvasculature in general or in specific organs (Augustin and Koh, 2017) or types of tissue (Nagy et al., 2009; Jain et al., 2014). Using priors specific to individual organs or tissues would require the robot to determine which organ or type of tissue it is in, for example, using chemical signatures or external localizing signals (Freitas, 1999). A robot could also adjust its priors based on its recent history, for example, using the frequency of verified branches it has encountered recently in small vessels. The robot can combine any prior probabilities it has with the result P_{branch} from the classifier using Bayes theorem. For example, if branches are rare compared to curves, the robot could increase the threshold it uses to identify a branch to reduce the number of false positives. This would adjust for the use of an equal number of curve and branch paths used for the training in this chapter, which corresponds to an equal likelihood for the robot to encounter a curve or branch.

The robot could use a wider variety of values to aid classification in more complex scenarios. For instance, the classifier of this chapter evaluates stress changes over a fixed interval of time, namely 10 ms. More generally, robots could use stress correlations over multiple time intervals, for example, using Δt values of both 10 and 20 ms. A mix of times could help identify vessel structure when flow speed varies over a larger range than considered here, for example, due to vessel blockages.

Another example of additional information the robot could use is a broader measure of how stress changes. The correlation measure used here measures the change in the *shape* of the stress pattern on the robot surface. As a robot moves, the stress magnitudes change as well. For instance, moving into a branch with slower flow typically decreases the stress, whereas moving closer to the wall increases it. Thus, another measure of changing stress is ratio of current and previous average stress magnitudes over the surface, which characterizes the change in the *magnitude* of stresses. For the cases considered here, the change is stress magnitude gives similar information as that obtained from the correlation, and does not noticeably improve classifier performance. Nevertheless, in other cases, changes in shape and magnitude may provide different information, thereby improving performance when used together.

In addition to using the information a robot obtains while passively moving with the fluid, robots could actively probe their environments. One such approach is making changes to the surface that affect stresses. For instance, a robot that can alter its shape (Castano et al., 2000) could measure how its shape affects its surface stresses to gain additional information on the nearby vessel geometry. Another possibility is for a robot with locomotion capability to alter its distance to the vessel wall before it encounters a branch or curve to increase the accuracy of classification or verification after passing the change. A robot could move closer to the center of the vessel so its distance to the wall will change even more when encountering a branch than a curve (see Fig. 9) thereby improving verification. Alternatively, it could move closer to the wall so the change in correlation is more distinct between branches and curves (see Fig. 6).

Beyond classifying the type of geometry (e.g., a branch), a robot might use stresses to estimate quantitative properties of how the geometry changes. This possibility arises from stresses depending on the geometric properties such as the branching angle when a vessel splits and the size of the branches, as well as the radius of curvature of a curved vessel. Estimating such properties requires determining how they influence the stress pattern as the robot moves, and then determining how to extract property values from these influences on measured stresses. Such estimates could also benefit from prior information on relations between branching angles and diameters (Thompson, 1992).

Communication among nearby robots could enhance these extensions. In particular, when multiple communicating robots pass through a vessel at around the same time, they could compare stresses at different positions in the vessel. These comparisons could include robots close to the wall or close to the middle, and robots approaching, near and past a branch. More broadly, robots with a variety of sensors could combine information from stresses with other measures, such as changing chemical concentrations or acoustic echoes (Freitas, 1999), to evaluate the changing geometry of their environments.

9 Conclusion

In summary, a robot can use fluid stresses on its surface to identify when it encounters branches in microscopic vessels. The classifier discussed here combines three types of information derived from stresses: how stress patterns change over a short time, the position of the robot in the vessel before it encounters a branch or curve, and the shape of the stress pattern. Together, these values accurately distinguish branches from curves for a typical range of geometries encountered with microscopic vessels in tissue. Moreover, the time over which stresses are compared, for example, 10 ms, is short enough that the robot moves just a few times its size during the evaluation. This allows the classifier to identify branches, while the robot is still near the branch.

Fabrication of microscopic robots with stress sensors and even relatively simple computers involve significant technological challenges. Prior to the feasibility of robot fabrication, theoretical studies, such as presented in this chapter, can quantify likely robot capabilities and their performance in various applications. These studies can be useful guides to the kinds of applications robots will be able to perform as their capabilities improve. Conversely, these studies quantify the performance requirements of the robot sensors and computers needed to address these applications.

Appendix

A.1 Samples of robot motion in small vessels

We determine the relationship between surface stresses and vessel geometry from a set of samples in known geometries. If robots could be fabricated, these samples could be obtained experimentally, for example, by measuring forces on microfluidic devices (Wu et al., 2010). Since this experimental approach is not yet feasible, we instead create samples from numerical solution of the flow with a robot in vessels with various geometries.

We create samples of robot paths for a range of vessel diameters and flow speeds corresponding to small blood vessels, with parameters given in Table 1. These samples are variations of the situation shown in Fig. 1 with parameters chosen uniformly at random according to:

- Maximum inlet fluid speed between 800 and 1000 μm/s.
- For curved vessels:
  - Vessel diameter between 6 and 13 μm.
  - Bend angle, between direction of inlet and outlet, between 25 and 75 degrees.
- For a vessel splitting into two branches:
  - Diameters of the branches, d_1 and d_2, between 6 and 10 μm.
  - The diameter of the main vessel, d, is determined from d_1 and d_2 according to Murray's law (Murray, 1926; Sherman, 1981; Painter et al., 2006), that is, $d^3 = d_1^3 + d_2^3$.
  - The two branch angles chosen uniformly between 25 and 75 degrees and −25 and −75 degrees, respectively.
- Vessel segment length extending 30 μm in each direction from the curve or branch.
- Initial position of robot's center: eight times the robot radius, that is, $8r = 8$μm, from the vessel inlet, and randomly positioned between the vessel's walls with minimum gap $0.2r$ between the robot surface and the wall.
- Robot orientation between 0 and 360 degrees.

From the initial robot position, we solve for the robot's motion through the vessel until it comes within 8 μm of an outlet. Boundary conditions on the flow are a parabolic velocity profile at the inlet, no-slip along the vessel wall and zero pressure at the outlet for a curve, or at both outlets for a branch.

This study used a total of 2000 samples created according to this procedure, with 1000 for each vessel type, that is, curve or branch. For each of these vessel types, 800 samples were used for training the classifier and the remaining 200 were used to test the classifier performance. The paths in these samples are about 40 μm in length. The fluid typically moves the robot along the path in around 100 ms.

B.2 Identifying branches from stress measurements

This section describes how the vessel branch classifier was trained and the parameters of the resulting model. The section also estimates the computational requirements for using the trained classifier.

B.2.1 Regression classifier for branch detection

This chapter identifies branches with a logistic regression based on three characteristics of the robot's stresses along its path through a vessel. First, the correlation between the current stress and that of a short time earlier. Second, the robot's relative position in the vessel evaluated during the robot's most recent passage through a nearly straight section of the vessel. The third characteristic is a measure of the shape of the robot's current stresses, specifically the first principal component of the Fourier coefficients of the stress pattern (Hogg, 2018).

The regression model for the probability of a branch, P_{branch}, is

$$P_{\text{branch}} = \frac{1}{1 + \exp\left(-b(\log(1-c), \text{rp}, p_1)\right)} \tag{B.1}$$

where c is the correlation between changing stress patterns, defined in Eq. (2), rp is the relative position, defined in Eq. (3), p_1 is the first principal component of the stress pattern, and

$$b(\text{lc}, \text{rp}, p_1) = \beta_0 + \beta_1 \text{lc} + \beta_2\, \text{rp} + \beta_{1,1}\, \text{lc}^2 + \beta_{2,2}\, \text{rp}^2 + \beta_3\, p_1$$

where the $\beta_{...}$ are the parameters, given in Table B.1, determined from the training samples.

Training this regression used stress measurements along the complete path of each sample to identify the time with minimum correlation along the path. This "off-line" training corresponds to the situation after robots complete their paths, so stress measurements all along the paths are available

Table B.1 Regression parameters for the probability a robot encounters a branch, Eq. (B.1)

Parameter	Value	Standard error
β_0	− 0.8	0.7
β_1	− 1.7	0.6
β_2	9.0	1.8
$\beta_{1,1}$	− 0.97	0.11
$\beta_{2,2}$	6.8	2.3
β_3	11.6	0.8

for training. Specifically, for each training sample, rp is the relative position at the start of the sample path where, by construction, the robot is in a straight vessel segment prior to reaching the curve or branch. Moreover, c is the minimum correlation along the path, that is, the minimum value of $c(t, \Delta t)$ along a path, for $\Delta t = 10$ ms. The branch and curve points in Fig. 6 are examples of these relative position and correlation values. Finally, the value of p_1 used for training is the principal component of the robot's stress at the time of the minimum correlation.

B.2.2 Computational requirements

An important metric for classifiers is the computational cost to train and use them. The training considered here is off-line, from stress measurements collected along a sample of paths. In this case, training could take place in a conventional computer rather than in the robots, and the resulting regression parameters stored in the robot's memory. Thus, training is not significantly constrained by the robots' onboard computational capabilities.

On the other hand, robots applying the classifier would use their onboard computer to repeatedly evaluate the trained classifier from their stress measurements. Microscopic robots are likely to have limited onboard computational capabilities. Thus, it is important to evaluate the computational cost for robots using the classifier, and the memory required to store the parameters obtained from the training.

The classifier used in this chapter involves a small set of parameters (see Table B.1), and recording several sets of stress measurements (represented by their low-frequency Fourier coefficients) to allow evaluating correlations. This information amounts to about 100 numbers, which could fit in a kilobyte of memory. The regression parameters are just a few additional numbers, so do not add significantly to the memory requirement.

The classifier uses simple functions of stress measurements. Specifically, Eq. (B.1) involves correlations, estimates of a robot's relative position, and the principal component of Fourier modes of the stresses. All these quantities can be computed from a few low-frequency Fourier coefficients of stress measurements on the robot's surface. A computer capable of 10^6 operations/s could evaluate these values every few milliseconds (Hogg, 2018).

Such computation is well beyond the demonstrated capabilities of current nanoscale computers, for example, DNA or RNA-based logic operations (Douglas et al., 2012; Green et al., 2017) or programmable microorganisms (Ferber, 2004). However, theoretical analyses indicate more elaborate molecular computers could be both small enough to fit within micron-size robots and readily exceed this estimate of the required computational performance (Drexler, 1992; Merkle et al., 2018). Although such computers cannot yet be manufactured, the same technology required to build the micron-scale robots considered in this chapter is also likely to be able to fabricate these computers. Thus, whenever these robots can be manufactured, they likely will have sufficient computing capability to apply the stress-based estimates described in this chapter. Nevertheless, early versions of such robots may not have sufficient computation to evaluate the classifier, but still have enough to encode stress measurements for communication. In that case, an alternate method for applying the classifier would be to send stress measurements to an external computer that would evaluate the classifier and return the value to the robot. This illustrates how microscopic robots could rely on onboard computation or communication, depending on which is easier to manufacture.

The classifier considered here relies on the correlation between stresses separated by $\Delta t = 10$ ms. Similar classifiers could be trained with other values of Δt. However, the change in stress and hence the correlation values depend on Δt: generally, stresses change more over longer times, leading to smaller correlations. Thus, to use a classifier trained with a specific value of Δt requires the robot measure correlations over approximately the same time interval. For the classifier considered here, this requires the robot have a clock with, roughly, millisecond precision over time intervals of tens of milliseconds. An alternative to onboard clocks is an external timing signal (Freitas, 1999). For example, timing signals could use sound waves, which travel at about 1500 m/s in water and tissue. Thus, a millisecond-period timing signal would travel 1.5 m, thereby providing a common millisecond time standard to robots operating within a few centimeters. Such an external timing signal would not only provide timing for individual robots, but also a global standard for a group of robots, allowing them to synchronize measurements.

Acknowledgment
The author thanks Robert A. Freitas Jr., Ralph C. Merkle, and James Ryley for helpful discussions.

References
Allen, T.M., Cullis, P.R., 2004. Drug delivery systems: entering the mainstream. Science 303, 1818–1822.
Andrews, L.B., Nielsen, A.A.K., Voigt, C.A., 2018. Cellular checkpoint control using programmable sequential logic. Science 361, 1217.

Augustin, H.G., Koh, G.Y., 2017. Organotypic vasculature: from descriptive heterogeneity to functional pathophysiology. Science 357, eaal2379.

Bagchi, P., 2007. Mesoscale simulation of blood flow in small vessels. Biophys. J. 92, 1858–1877.

Bleckmann, H., Zelick, R., 2009. Lateral line system of fish. Integr. Zool. 4, 13–25.

Bouffanais, R., Weymouth, G.D., Yue, D.K.P., 2010. Hydrodynamic object recognition using pressure sensing. Proc. R. Soc. A 467, 19–38.

Cassot, F., 2006. A novel three-dimensional computer assisted method for a quantitative study of microvascular networks of the human cerebral cortex. Microcirculation 13, 15–32.

Castano, A., Shen, W.M., Will, P., 2000. CONRO: towards miniature self-sufficient metamorphic robots. Auton. Robot. 8, 309–324.

Douglas, S.M., Bachelet, I., Church, G.M., 2012. A logic-gated nanorobot for targeted transport of molecular payloads. Science 335, 831–834.

Drexler, K.E., 1992. Nanosystems: Molecular Machinery, Manufacturing, and Computation. John Wiley, New York, NY.

Dusenbery, D.B., 2009. Living at Micro Scale: The Unexpected Physics of Being Small. Harvard University Press, Cambridge, MA.

Eberhardt, W.C., 2016. Development of an artificial sensor for hydrodynamic detection inspired by a seal's whisker array. Bioinspir. Biomim. 11, 056011.

Ferber, D., 2004. Microbes made to order. Science 303, 158–161.

Freitas Jr., R.A., 1999. Nanomedicine. In: Basic Capabilities, vol. I. Landes Bioscience, Georgetown, TX.

Green, A.A., 2017. Complex cellular logic computation using ribocomputing devices. Nature 548, 117–121.

Happel, J., Brenner, H., 1983. Low Reynolds Number Hydrodynamics, second ed. Kluwer, The Hague.

Hill, C., Amodeo, A., Joseph, J.V., Patel, H.R.H., 2008. Nano- and microrobotics: how far is the reality? Expert Rev. Anticancer Ther. 8, 1891–1897.

Hogg, T., 2018. Stress-based navigation for microscopic robots in viscous fluids. J. Micro-Bio Robot. 15, 59–67.

Hogg, T., Freitas Jr., R.A., 2012. Acoustic communication for medical nanorobots. Nano Commun. Netw. 3, 83–102.

Hoskins, M.H., Kunz, R.F., Bistline, J.E., Dong, C., 2009. Coupled flow-structure-biochemistry simulations of dynamic systems of blood cells using an adaptive surface tracking method. J. Fluids Struct. 25, 936–953.

Jager, E.W.H., Inganas, O., Lundstrom, I., 2000. Microrobots for micrometer-size objects in aqueous media: potential tools for single-cell manipulation. Science 288, 2335–2338.

Jain, R.K., Martin, J.D., Stylianopoulos, T., 2014. The role of mechanical forces in tumor growth and therapy. Annu. Rev. Biomed. Eng. 16, 321–346.

Ke, Y., Castro, C., Choi, J.H., 2018. Structural DNA nanotechnology: artificial nanostructures for biomedical research. Ann. Rev. Biomed. Eng. 30, 377–403.

Koman, V.B., 2018. Colloidal nanoelectronic state machines based on 2D materials for aerosolizable electronics. Nat. Nanotechnol. 13, 819–827.

Korin, N., 2012. Shear-activated nanotherapeutics for drug targeting to obstructed blood vessels. Science 337, 738–742.

Kottapalli, A., Asadnia, M., Miao, J., Triantafyllou, M., 2014. Touch at a distance sensing: lateral-line inspired MEMS flow sensors. Bioinspir. Biomim. 9, 046011.

Li, S., 2018. A DNA nanorobot functions as a cancer therapeutic in response to a molecular trigger in vivo. Nat. Biotechnol. 36, 258–264.

Martel, S., 2007. The coming invasion of the medical nanorobots. Nanotechnol. Percept. 3, 165–173.

Martel, S., 2014. Computer 3D controlled bacterial transports and aggregations of microbial adhered nano-components. J. Micro-Bio Robot. 9, 23–28.

Martel, S., 2007. Automatic navigation of an untethered device in the artery of a living animal using a conventional clinical magnetic resonance imaging system. Appl. Phys. Lett. 90, 114105.

Merkle, R.C., Freitas Jr., R.A., Hogg, T., Moore, T.E., Moses, M.S., Ryley, J., 2018. Mechanical computing systems using only links and rotary joints. ASME J. Mech. Robot. 10, 061006.

Monroe, D., 2009. Micromedicine to the rescue. Commun. ACM 52, 13–15.

Murray, C.D., 1926. The physiological principle of minimum work: I. The vascular system and the cost of blood volume. Proc. Natl. Acad. Sci. USA 12, 207–214.

Nagy, J.A., Chang, S.-H., Dvorak, A.M., Dvorak, H.F., 2009. Why are tumour blood vessels abnormal and why is it important to know? Br. J. Cancer 100, 865–869.

Painter, P.R., Eden, P., Bengtsson, H.-U., 2006. Pulsatile blood flow, shear force, energy dissipation and Murray's law. Theor. Biol. Med. Model. 3, 31.

Park, J.H., 2008. Magnetic iron oxide nanoworms for tumor targeting and imaging. Adv. Mater. 20, 1630–1635.

Pawlik, G., Rackl, A., Bing, R.J., 1981. Quantitative capillary topography and blood flow in the cerebral cortex of cats: an in vivo microscopic study. Brain Res. 208, 35–58.

Purcell, E.M., 1977. Life at low Reynolds number. Am. J. Phys. 45, 3–11.

Schulz, M.J., Shanov, V.N., Yun, Y., 2009. Nanomedicine Design of Particles, Sensors, Motors, Implants, Robots, and Devices. Engineering in Medicine and Biology, Artech House, Boston, MA.

Sershen, S., Westcott, S., Halas, N.J., West, J., 2000. Temperature-sensitive polymer-nanoshell composite for photothermally modulated drug delivery. J. Biomed. Mater. Res. 51, 293–298.

Sherman, T.F., 1981. On connecting large vessels to small: the meaning of Murray's law. J. General Physiol. 78, 431–453.

Sichert, A.B., Bamler, R., van Hammen, J.L., 2009. Hydrodynamic object recognition: when multipoles count. Phys. Rev. Lett. 102, 058104.

Sitti, M., 2015. Biomedical applications of untethered mobile milli/microrobots. Proc. IEEE 103, 205–224.

Sochi, T., 2015. Fluid flow at branching junctions. Int. J. Fluid Mech. Res. 42, 59–81.

Thompson, D.W., 1992. On Growth and Form. Cambridge University Press, Cambridge, MA.

Thubagere, A.J., 2017. A cargo-sorting DNA robot. Science 357, 112.

Vollmayr, A.N., 2014. Snookie: an autonomous underwater vehicle with artificial lateral-line system. In: Bleckmann, H. (Ed.), FlowSensing in Air and Water, vol. 20. Springer, pp. 521–562.

Wu, J., Day, D., Gu, M., 2010. Shear stress mapping in microfluidic devices by optical tweezers. Opt. Express 18, 7611–7616.

Yang, Y., 2006. Distant touch hydrodynamic imaging with an artificial lateral line. Proc. Natl. Acad. Sci. USA 103, 18891–18895.

CHAPTER SEVEN

Navigation and control of endovascular helical swimming microrobot using dynamic programing and adaptive sliding mode strategy

Mohammad Javad Pourmand[a] and Mojtaba Sharifi[b]
[a]School of Mechanical Engineering, Shiraz University, Shiraz, Iran
[b]Department of Mechanical Engineering, Sharif University of Technology, Tehran, Iran

1 Introduction

Microrobots and miniature robots have been recently designed, studied, and employed in different applications, including medical and biological operations. Several novel designs have been presented for microrobots; however, before introducing them it is needed to know about an untethered microagent. The untethered microrobotics is now the superior approach for delicate micromanipulations in inaccessible or hardly accessible environments with medical purposes. Exemplifications of these environment and areas are the circulatory system (Belharet et al., 2011), urinary system and prostate (Kristo et al., 2003), central nervous system (Duffner et al., 2003; Zaaroor et al., 2006), eye (Kummer et al., 2010; Yesin et al., 2006), ear (Nelson et al., 2010), and fetus (Flake, 2003). Microagents in these environments can be utilized for diversity of operations such as the biopsy (De Cristofaro et al., 2010), drug delivery (Khalil and Misra, 2017; Nelson et al., 2010), marking (Fluckiger and Nelson, 2007), occlusion (Jeong et al., 2016), ablation (Kim et al., 2016), and so on (Nelson et al., 2010). Accordingly, the key point in designing microrobots is insightful knowledge of the operation environment as its target. Most of the proposed microrobots have been designed to operate in human body (generically mammals' bodies); therefore, it is evident that these microrobots should swim inside body fluids in different parts. Relatively intricate and expensive local instruments (i.e., microsensors and microactuators) have made researchers prone to benefit from distant sensors, power supply, and actuators in miniature environments. The most promising distant navigation employed for microrobots is the magnetic navigation, used in the magnetically driven elastic flagellated propulsion (Evans and Lauga, 2010), magnetically driven rigid flagellated propulsion (Abbott et al., 2010), and OctoMag system (Kummer et al., 2010).

The idea of controlling a microrobot within an intravascular environment can be involved with parametric uncertainties and unmodeled dynamics (Folio and Ferreira, 2017; Arcese et al., 2013; Hund et al., 2017; Pourmand et al., 2018), although in a highly viscous environment, using a bioinspired flagellated propulsion can accomplish an effective locomotion (Evans and Lauga, 2010) that relatively increases the structured and unstructured uncertainties (in comparison with other methods). Moreover, pulsatile flow causes remarkable disturbances in the system that requires an effective robust control strategy. Accordingly, several controllers have been proposed for such conditions. For instance, the model predictive (Belharet et al., 2011), adaptive (Arcese et al., 2013), and H_∞ (Marino et al., 2014) control strategies are presented for microrobots, and they failed to perform an effective control in a realistic environment (Folio and Ferreira, 2017). Comparing magnetic resonance propulsion (MRP, particularly executed in the magnetic resonance imaging (MRI) device) and magnetically driven helical-flagellated propulsion (MHP) results in that MHP produces larger propulsive forces; however, it can elevate the uncertainty of the system. One of recently introduced methods to deal with the mentioned uncertainties is considering some of the unstructured uncertainties in the path planning of the robot inside endovascular system (Folio and Ferreira, 2017).

The sliding mode control (SMC) as a robust strategy was employed in uncertain nonlinear systems (Slotine and Li, 1991; Sharifi and Moradi, 2017). The basic issues of this controller are two interconnected phenomenon: the chattering phenomenon and high control effort (Utkin and Poznyak, 2013; Aghajanzadeh et al., 2018; Sharifi et al., 2017). The magnitude of a discontinuous SMC input is proportional to the amplitude of chattering; therefore, the adaptive siding mode control was introduced to reduce the magnitude of SMC gain and the chattering issue (Utkin and Poznyak, 2013; Torabi et al., 2017). Moreover, the adaptive sliding mode control is one of the best robust strategies for the nonlinear systems that their dynamic models cannot be linearly parametrized or the parametrization is complicated.

Microrobots have some other control issues due to the limitation of applicable equipment such as the precision of sensory data and the under-actuation characteristic. Furthermore, microrobots in the fields of bio-robotics (Nelson et al., 2010; Abbott et al., 2010) have currently encountered some considerations such as biocompatibility and a large number of uncertainties due to the interaction with biosystems. As a result, developed control models for varieties of these microrobots (Arcese et al., 2012) have been operated in the circulatory system involved with unmodeled dynamics (Belharet et al., 2011) and structured uncertainties (Arcese et al., 2012; Hund et al., 2017; Kenner, 1989).

With respect to these issues, the sliding mode control (SMC) method with robustness property against structured and unstructured uncertainties seems to be an effective strategy for the control of endovascular microrobots. Accordingly, a nonlinear adaptive sliding mode controller is designed for the first time in this work for a magnetically driven monohelical flagellated microrobot. To resolve the chattering issue and autonomously overcome uncertainties, an adaptation law is presented for online adjustment of the controller gain. The global stability and tracking convergence of the proposed control method is proven via the Lyapunov theorem. In addition, a three-dimensional (3D) optimal path planning is employed based on the dynamic programming (DP) for generating the desired trajectory of the microrobot. Some simulations are executed to evaluate the objective achievement and the controller performance.

2 3D optimal path planning

Path planning in 2D environment has been widely used and discussed for microrobots; however, a 3D path planning is necessary for the endo-vascular environment. Due to the lack of direct measurement of the micro-agent velocity using currently available imaging devices, an appropriate feedback controller has to be devised instead of previous approaches needed the velocity signal. Firstly, it is considered that we have a 3D image of the vascular system obtained from MRI, computed tomography (CT), or any other imaging devices. In the domain $C_D \subset \mathbb{R}^3$ that is in the permissible space of the microrobot operation, any path starts from $p(0) \in C_D$ and ends at $p(1) \in C_D$ can be expressed by

$$p : l \in [0, 1] \rightarrow p(l) \in C_D \tag{1}$$

Now, the control effort is defined as

$$w(\boldsymbol{v}(\boldsymbol{p}(l)), \boldsymbol{f}_m(\boldsymbol{p}(l))) = \arg\max\left(\boldsymbol{f}_m(\boldsymbol{p}(l)) \cdot \boldsymbol{v}(\boldsymbol{p}(l))\right), \qquad (2)$$

where f_m and $\boldsymbol{v}$ are the control force and micororbot velocity, respectively, at each location $\boldsymbol{p}(l)$ of the path (Folio and Ferreira, 2017).

Assumption. The induced magnetic force is controllable in any direction and the flow velocity is not directly measurable since conventional imaging devices cannot provide such data. Thus, the control effort metric is determined based on the velocity distribution obtained from the steady-state solution of Navier-Stokes equations for an uniform flow in the walled space (Munson et al., 2014). Fig. 1 shows an illustration of the scaled control effort metric in a 2D space (the result is comparable with the one in Folio and Ferreira, 2017).

Then, a performance measure based on the control effort metric can be developed as

$$J(\boldsymbol{p}) = \int_0^1 w(\boldsymbol{p}(l)) \cdot \|\dot{\boldsymbol{p}}(l)\| \, dl. \qquad (3)$$

Therefore, the optimum path with minimum effort can be found from

$$\boldsymbol{p}^* = \arg\min J(\boldsymbol{p}). \qquad (4)$$

Considering the image as a discrete domain, the first step for finding the solution of Eq. (4) by the dynamic programming (DP) approach (Kirk, 2012) is meshing this domain. After generating an adequately small mesh, each node has directly reachable neighbors. Thus, according to the

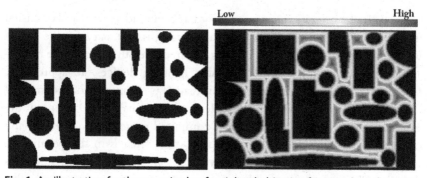

Fig. 1 An illustration for the magnitude of weighted objective function based on minimum effort, defined in Eq. (2), for a 2D image: The color bar demonstrates how this magnitude would be high or low. Note that the magnitude of this function is higher wherever the pressure is lower.

optimality principle (Kirk, 2012), for a path that contains the nodes G, H, and I, there is a total optimum path as $J^*_{GHI} = J^*_{GH} + J^*_{HI}$. For this purpose: (1) either destination or start point is considered as the initial point of DP (as the solution is reversible), (2) the minimum cost of a path from each node to its neighborhood nodes is calculated, and (3) different paths between start and destination points in the domain are analyzed and the optimum total path with the minimum total cost is obtained. In the perspective of time complexity, it is noteworthy that gradient-based methods are superior to the proposed method if the search space of problem (4) is smooth (Pourmand et al., 2019).

3 Dynamic modeling

The dynamic model of the monohelical swimming microrobot (shown in Fig. 2) is presented in this section. Study of the monohelical swimming microrobot is beneficial for generic acquaintance to microrobots dynamics, since large number of microrobots designed to have a bead (Belharet et al., 2011) and single (Dreyfus et al., 2005) or multiple flagella (Pourmand et al., 2018). The viscous force applied on each element of the helix (ds_h) has normal and tangential components obtained as

$$dF_\perp = -C_\perp V_\perp ds_h$$
$$dF_\| = -C_\| V_\| ds_h, \tag{5}$$

where from the geometry of the helix, normal and tangential velocities are formulated as

$$V_\perp = -V_a \sin\beta_p + R\dot{\Omega}\cos\beta_p$$
$$V_\| = V_a \cos\beta_p + R\dot{\Omega}\sin\beta_p \tag{6}$$

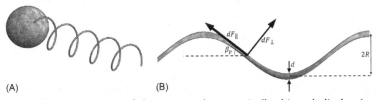

(A) (B)

Fig. 2 (A) Schematic picture of the proposed magnetically driven helical swimming microrobot: the spherical head can contain the operation essentials and the tail is just for exerting the propulsion, and (B) demonstration of the parameters and the tangential and normal forces on the helix.

and the normal and tangential viscous drag coefficients are given in Abbott et al. (2010) as

$$C_\perp = \frac{4\pi\eta}{\ln\dfrac{0.72\pi R}{d\sin\beta_p} + 0.5} > C_\| = \frac{2\pi\eta}{\ln\dfrac{0.72\pi R}{d\sin\beta_p}} \tag{7}$$

In Eqs. (6) and (7), V_a is magnitude of the robot velocity along its approach vector, Ω is magnitude of the robot angular velocity around its approach vector, R is the helix radius, β_p is the helix pitch angle. η and d are also the fluid viscosity and the microrobot filament diameter, respectively. Therefore, the propulsive force along the approach vector for each element along the helix is:

$$dF = dF_\| \cos\beta_p - dF_\perp \sin\beta_p \tag{8}$$

Substituting Eq. (6) in Eq. (5) leads to

$$dF_\perp = \left[C_\perp V_a \sin\beta_p - C_\perp R\dot\Omega \cos\beta_p \right] ds_h$$
$$dF_\| = \left[-C_\| V_a \cos\beta_p - C_\| R\dot\Omega \sin\beta_p \right] ds_h \tag{9}$$

, and substituting these equations in Eq. (8) leads to

$$dF = \Big\{ \left[-C_\| V_a \cos\beta_p - C_\| R\dot\Omega \sin\beta_p \right] \cos\beta_p$$
$$+ \left[C_\perp R\dot\Omega \cos\beta_p - C_\perp V_a \sin\beta_p \right] \sin\beta_p \Big\} ds_h \tag{10}$$

By integrating the resultant equation along the helix, we have:

$$F = 2\pi R N \left(R\left(C_\perp - C_\| \right)\dot\Omega \cos\beta_p - \left(C_\perp \sin\beta_p + C_\| \cos^2\beta_p \csc\beta_p \right) V_a \right), \tag{11}$$

where N denotes the number of helix turns.

The hydrodynamic drag force of the microrobot head is obtained using the drag coefficient of spherical beads as

$$\mathbf{F_d} = \frac{1}{2}\rho A \vec{v}_\infty^{\,2} C_d \vec{v}_U \tag{12}$$

in which ρ, A, C_d, $\vec{v}_\infty$, and $\vec{v}_U$ denote the fluid density, the reference area, the superficial velocity in an infinitely uniform flow, and the unit vector of the superficial velocity. Since the proposed microrobot operates inside the circulatory system, it is essential to consider the wall effect (Kehlenbeck and Felice, 1999) of the vessels as

$$\frac{\|\vec{v}\|}{\|\vec{v}_\infty\|} = \frac{1 - \lambda_p^{2.2}}{1 + \left(\dfrac{\lambda_p}{\lambda_0}\right)^{2.2}}, \tag{13}$$

where $\|\cdot\|$ is Euclidean norm, and λ_0 and λ_p are dimensionless parameters defined as

$$\lambda_0 = \frac{0.27 + 0.06 Re^{0.65}}{1 + 0.05 Re^{0.65}}; \quad \lambda_p = D/2r \tag{14}$$

in which D, r, and Re are the robot spherical head diameter, the vessel radius, and the Reynolds number, respectively. One of the most accurate hydrodynamic drag coefficients of spherical beads in laminar flow was proposed recently in Barati et al. (2014) as

$$\begin{aligned} C_d = C_1 \tanh\left(\frac{C_2 C}{\|\dot{x}\|}\right) + C_3 \tanh\left(\frac{C_4 C}{\|\dot{x}\|}\right) + C_5 \tanh\left(\frac{C_6 C}{\|\dot{x}\|}\right) \\ + C_7 \tanh\left(\frac{C_8 C}{\|\dot{x}\|}\right) + C_9 \tanh\left(\frac{C_{10} C}{\|\dot{x}\| + C_{11} C}\right) + C_{12} \end{aligned} \tag{15}$$

with the following coefficient values:

$$\begin{aligned} \left[C_1 \; C_2 \; C_3 \; C_4 \; C_5 \; C_6 \; C_7 \; C_8 \; C_9 \; C_{10} \; C_{11} \; C_{12} \right] \\ = [5.4856 \; 4.3774 \; 0.0709 \; 700.6574 \; 0.3894 \; 74.1539 \\ -0.1198 \; 7429.0843 \; 1.7174 \; 9.9851 \; 2.3384 \; 0.4744] \end{aligned} \tag{16}$$

and

$$C = \frac{\eta}{D\rho}.$$

Using Eqs. (11)–(16), Eq. (12) can be rewritten as

$$\begin{aligned} \mathbf{F_d} = \frac{\pi}{8}\rho \left(\frac{D}{1 - \left(\dfrac{D}{2r}\right)^{2.2}}\right)^2 \left(1 + \left(\frac{D\left(1 + 0.05\left(\dfrac{\|\dot{x}\|}{C}\right)^{0.65}\right)}{2r\left(0.27 + 0.06\left(\dfrac{\|\dot{x}\|}{C}\right)^{0.65}\right)}\right)^{2.2}\right)^2 \\ \left(C_1 \tanh\left(\frac{C_2 C}{\|\dot{x}\|}\right) + C_3 \tanh\left(\frac{C_4 C}{\|\dot{x}\|}\right) + C_5 \tanh\left(\frac{C_6 C}{\|\dot{x}\|}\right) \right. \\ \left. + C_7 \tanh\left(\frac{C_8 C}{\|\dot{x}\|}\right) + C_9 \tanh\left(\frac{C_{10} C}{\|\dot{x}\| + C_{11} C}\right) + C_{12} \right) \|\dot{x}\|\dot{x} \end{aligned} \tag{17}$$

where x is the robot position. A spherical magnetic bead attached to the helical tail produces a magnetic torque generating the corkscrew effect (Chwang and Wu, 1971) determined as $\vec{\Gamma}_m = V_m\,\vec{M} \times \vec{B}$, where V_m is the volume of magnetic material in the bead, and $\vec{M}$ and $\vec{B}$ are the bead magnetization vector and the oscillating magnetic field with the frequency of ω, respectively. Then, a feasible control input is defined as

$$u = \omega \frac{\vec{\Gamma}_m}{\left\|\vec{\Gamma}_m\right\|}. \tag{18}$$

Considering apparent weight as $\mathbf{W}_a = V_b(\rho_b - \rho)\,\vec{g}$ with V_b, ρ_b, and $\vec{g}$ as the bead volume, the bead density, and gravitational acceleration, respectively, knowing F and F_d from Eq. (11) and Eq. (17), and using Eq. (18), the dynamics (equation of motion, i.e., $m\ddot{x} = \mathbf{F} - \mathbf{F_d} + \mathbf{W}_a$) of the microrobot is formulated as

$$\ddot{x} = \frac{2\pi R^2 N\left(C_\perp - C_\parallel\right)\cos\beta_p}{m} u$$

$$- \left[\frac{2\pi RN}{m}\left(C_\perp \sin\beta_p + C_\parallel \cos^2\beta_p \csc\beta_p\right) - \frac{\pi\rho}{8m}\left(\frac{D}{1 - \left(\frac{D}{2r}\right)^{2.2}}\right)^2 \right.$$

$$\left(1 + \left(\frac{D\left(1 + 0.05\left(\frac{\|\dot{x}\|}{C}\right)^{0.65}\right)}{2r\left(0.27 + 0.06\left(\frac{\|\dot{x}\|}{C}\right)^{0.65}\right)}\right)^{2.2}\right)^2 \left(C_1 \tanh\left(\frac{C_2 C}{\|\dot{x}\|}\right)\right.$$

$$+ C_3 \tanh\left(\frac{C_4 C}{\|\dot{x}\|}\right) + C_5 \tanh\left(\frac{C_6 C}{\|\dot{x}\|}\right) + C_7 \tanh\left(\frac{C_8 C}{\|\dot{x}\|}\right)$$

$$\left. + C_9 \tanh\left(\frac{C_{10} C}{\|\dot{x}\| + C_{11} C}\right)\right) + C_{12}\,\|\dot{x}\| \bigg]\dot{x} + W_a \tag{19}$$

Based on Eq. (19), it can be written that

$$\ddot{x} = \beta u + f(\dot{x}), \tag{20}$$

where β and $f(\dot{x})$ can be obtained by comparing Eqs. (19), (20).

Other forces (e.g., van der Waals interaction) are not considered in this dynamic model (19) for the proposed microrobot because these forces are significantly smaller with lower order of magnitude compared to the hydrodynamic drag force.

4 Adaptive sliding mode control

The presence of structured uncertainties (such as inaccuracy of viscosity and density parameters) is possible in the modeled dynamics. Moreover, unstructured uncertainties due to existence of the pulsatile flow and negligence of the low-order forces will affect the system performance. Therefore, the application of robust controllers seems to be inevitable for these microrobotic systems. Although the sliding mode control scheme has been recommended as an appropriate robust controller (Slotine and Li, 1991) for uncertain dynamical systems, to exert this control method it is necessary to cope with two interconnected issues: the chattering phenomenon and the high magnitude of control action (Utkin and Poznyak, 2013). Accordingly, an adaptive version of the sliding mode strategy is employed in this work to control the microrobot inside the circulatory system, handle the chattering properly, and reduce the high activity of control action. Based on boundedness of f and β in Eq. (20), one can write:

$$\left| \hat{f} - f \right| < f_B; \quad \left| 1 - \frac{\hat{\beta}}{\beta} \right| < \beta_B, \tag{21}$$

where $\hat{\beta}$ and $\hat{f}$ are the estimated values of β and f in Eq. (20), respectively. The desired state (trajectory) of the microrobot is considered as $\begin{bmatrix} x_d & \dot{x}_d & \dots & x_d{}^{n-1} \end{bmatrix}$, where n is the order of system dynamics, and the position tracking error is defined as $\tilde{x} = x - x_d$. Since the microrobot has a second-order dynamics (19), i.e., $n = 2$, the sliding surface is defined as $s = \left(\dfrac{d}{dt} + \lambda \right)^{n-1} \tilde{x} = \dot{\tilde{x}} + \lambda \tilde{x}$, where λ is a positive constant. Differentiating this surface and using the system dynamics (20), it can be expressed that

$$\dot{s} = \beta u + f(\dot{x}) - \ddot{x}_d + \lambda \dot{\tilde{x}} \tag{22}$$

To make $\dot{s} = 0$, the best approximation for the control law is $\hat{u} = \frac{1}{\hat{\beta}}\left(-\hat{f}(\dot{x}) + \ddot{x}_d - \lambda\tilde{\dot{x}}\right)$. Now, by adding the robust term $\hat{k}\mathrm{sgn}(s)$ to the control input for converging to the sliding surface $s = 0$ in the presence of modeling uncertainties, we have:

$$u = \frac{1}{\hat{\beta}}\left(-\hat{f} + \ddot{x}_d - \lambda\tilde{\dot{x}} - \hat{k}\mathrm{sgn}(s)\right), \tag{23}$$

where $\hat{k}$ is a positive time-varying robust gain updating by the following adaptation law:

$$\dot{\hat{k}} = \varphi|s| \tag{24}$$

in which φ is a positive constant.

The closed-loop dynamics of the microrobot is obtained by substituting the control law (23) into Eq. (22) and doing some mathematical operations as follows:

$$\dot{s} = -\frac{\beta}{\hat{\beta}}\hat{k}\,\mathrm{sgn}(s) + f - \frac{\beta}{\hat{\beta}}\hat{f} + \left(\frac{\beta}{\hat{\beta}} - 1\right)\ddot{x}_d + \left(1 - \frac{\beta}{\hat{\beta}}\right)\lambda\tilde{\dot{x}} \tag{25}$$

To prove the stability and tracking convergence of the controlled micro-robot, the following Lyapunov function is considered:

$$V = \frac{1}{2}\left(s^2 + \frac{\beta}{\hat{\beta}\varphi}\left(\hat{k} - \gamma\right)^2\right) \geq 0, \tag{26}$$

where γ is a positive value that will be discussed later. The time derivative of the Lyapunov function is obtained:

$$\dot{V} = s\dot{s} + \frac{\beta}{\hat{\beta}\varphi}\dot{\hat{k}}\left(\hat{k} - \gamma\right). \tag{27}$$

Substituting the closed-loop dynamics (25) into Eq. (27), yields

$$\dot{V} = s\left(-\frac{\beta}{\hat{\beta}}\hat{k}\,\mathrm{sgn}(s) + f - \frac{\beta}{\hat{\beta}}\hat{f} + \left(\frac{\beta}{\hat{\beta}} - 1\right)\ddot{x}_d + \left(1 - \frac{\beta}{\hat{\beta}}\right)\lambda\tilde{\dot{x}}\right) + \frac{\beta}{\hat{\beta}\varphi}\dot{\hat{k}}\left(\hat{k} - \gamma\right) \tag{28}$$

Using the adaptation law (24), the time derivative of the Lyapunov control function is obtained as

$$\dot{V} = s\left(-\frac{\beta}{\hat{\beta}}\hat{k}\,\mathrm{sgn}(s) + f - \frac{\beta}{\hat{\beta}}\hat{f} + \left(\frac{\beta}{\hat{\beta}} - 1\right)\ddot{x}_d + \left(1 - \frac{\beta}{\hat{\beta}}\right)\lambda\tilde{\dot{x}}\right) + \frac{\beta}{\hat{\beta}}|s|\left(\hat{k} - \gamma\right) \tag{29}$$

After a few simplifications, Eq. (29) can be rewritten as

$$\dot{V} = |s| \frac{\beta}{\hat{\beta}} \left(\mathrm{sgn}(s) \left(\frac{\hat{\beta}}{\beta} f - \hat{f} + \left(1 - \frac{\hat{\beta}}{\beta} \right) \ddot{x}_d + \left(\frac{\hat{\beta}}{\beta} - 1 \right) \lambda \dot{\tilde{x}} \right) + \left(\hat{k} - \hat{k} \left(\mathrm{sgn}(s) \right)^2 \right) - \gamma \right)$$

(30)

The term $\left(\hat{k} - \hat{k} \left(\mathrm{sgn}(s) \right)^2 \right)$ is always zero unless $s=0$, in which case $\dot{\hat{k}}=0$ based on the update rule (24), i.e., $\hat{k}$ remains finite. Therefore, if $s=0$ (the microrobot trajectory converged to the sliding surface), γ is considered such large in the Lyapunov function (26) that

$$\hat{k} \leq \gamma$$

(31)

Thus, it is guaranteed that $\dot{V} \leq 0$. On the other hand, if $s \neq 0$, γ is considered such large that

$$\left| \frac{\hat{\beta}}{\beta} f - \hat{f} + \left(1 - \frac{\hat{\beta}}{\beta} \right) \ddot{x}_d + \left(\frac{\hat{\beta}}{\beta} - 1 \right) \lambda \dot{\tilde{x}} \right| + |s| \leq \gamma$$

(32)

Accordingly, $\dot{V} < 0$ in this case. Regarding the time derivatives of x that are assumed to be bounded, there exists a bounded (positive) value for γ, which is large enough to satisfy both inequalities (31) and (32). Therefore, by choosing a proper value for γ, the time derivative of the Lyapunov function is negative semidefinite. Consequently, Eqs. (30)–(32) result in

$$\dot{V} \leq -w_B(t) = -\frac{\beta}{\hat{\beta}} s^2 \leq 0$$

(33)

Since $w_B(t) = \left(\beta/\hat{\beta} \right) s^2$ is a uniformly continuous function, integrating Eq. (33) from zero to $t \to \infty$ gives

$$\lim_{t \to \infty} \int_0^t w_B(\tau) d\tau \leq V(0) - V(\infty)$$

(34)

Based on the negative semidefiniteness of $\dot{V} \leq 0$, $V(0) - V(\infty)$ is positive and finite in Eq. (34). Therefore, according to Eqs. (33), (34), $\lim_{t \to \infty} \int_0^t w_B(\tau) d\tau$ exists and it is positive and finite. As a result, based on Barbalat's lemma, it is obtained that

$$\lim_{t \to \infty} w_B(\tau) = \lim_{t \to \infty} \frac{\beta}{\hat{\beta}} s^2 = 0.$$

(35)

Thus, as $\left(\beta/\hat{\beta}\right)$ is nonzero and positive, Eq. (35) implies that $\lim_{t\to\infty} s = 0$. Regarding the definition of $s = \dot{\tilde{x}} + \lambda\tilde{x}$, the convergence to desired trajectory $\tilde{x} = (x - x_d) \to 0$ is concluded. Due to the positive definiteness of the Lyapunov function and the negative semidefiniteness of its time derivative, the proposed adaptive sliding mode control method guarantees the global stability, robustness against the bounded modeling uncertainties, and the racking convergence of the microrobot to the desired trajectory x_d.

5 Simulation results

The proposed control strategy for the endovascular swimming microrobot is evaluated via some simulations in this section. The human body circulatory system in arterioles with three different disturbances is included in these simulations. Parameters of the helix tail and head of microrobot are presented in Table 1, which are compatible with the data presented in Nourmohammadi et al. (2016) and Folio and Ferreira (2017).

One of the most challenging disturbances in arterioles is the pulsatile flow that is considered as (Munson et al., 2014; Belharet et al., 2011)

$$\sigma(t) = 0.001\left(1 - \left(\frac{r}{R_v}\right)^2\right)\left(1 + 1.15\cos\left(2\pi t + \frac{10\pi}{16}\right)\right)\vec{c}, \qquad (36)$$

where $\vec{c}$ is the unit vector along the vessel centerline. To more evaluate the robustness of the proposed control method, sinusoidal disturbances are considered in the fluid density and viscosity and applied in the simulations as

$$\eta(t) = 17.5 + 3\sin\left(\frac{2\pi t}{7}\right) \qquad (37)$$

$$\rho(t) = 1057 + 12\sin\left(\frac{2\pi t}{5}\right) \qquad (38)$$

Table 1 Parameters of the helix tail and head of the microrobot.

Parameters	Dimensions and values
β_p, helix angle	75 degrees
N, helix revolutions	5
d, filament diameter	20 (μm)
D, spherical head diameter	700 (μm)
R, helix amplitude	250 (μm)

The constants of the blood viscosity and density magnitude are extracted from Nelson et al. (2010). After finding the optimal path, corresponding to the minimum effort, inside a 3D complex phantom of endovascular system (shown in Fig. 3), the trajectory tracking is achieved in the presence of artificial pulsatile flow, density, and viscosity variation as illustrated in Figs. 3–5.

Fig. 4A demonstrates that the position error of the microrobot with respect to its desired trajectory that is controlled and converged to zero in the endovascular phantom with the mean diameter of 3 mm (Dong and Nelson, 2007; Nelson et al., 2010). In addition, Fig. 5A shows that after passing a bounded time, adaptive gains of the robust controller grow enough to overcome modeling uncertainties and disturbances and to control the robot position close to its desired trajectory (Fig. 3). Note that without any disturbance in the system, the robust gains are relatively small and autonomously reach small saturation levels as shown in Fig. 5B. The deviation of the microrobot velocity from the desired velocity in Fig. 4B remains small even in the frequency of artificial pulsatile flow and viscosity disturbance. The control inputs in 3 directions are also demonstrated in Fig. 4C. The magnitude of these control components is saturated to be <3 rad/s.

For the sake of comparison and demonstration of the proposed adaptive sliding mode control method's efficiency, the sliding mode gains (elements of $\hat{k}$) fixated on the constant values of 1 and 3 without employing the update rule (24). In these conditions, the control input, and trajectory tracking of the velocity and position in the presence of disturbances are illustrated in Fig. 6. Comparing Fig. 6A and 6D implies that the control effort is a little

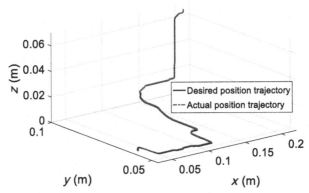

Fig. 3 Trajectory tracking of an optimal minimum-effort path in the presence of disturbances.

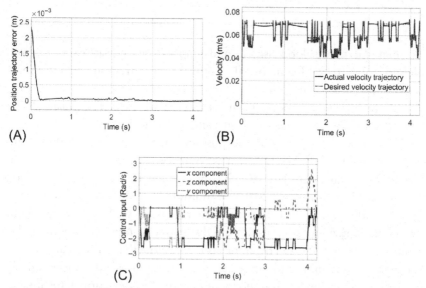

Fig. 4 (A) Convergence of the position error, (B) tracking of the desired velocity, and (C) control input of the swimming microrobot in the presence of artificial pulsatile flow, blood density, and viscosity disturbances.

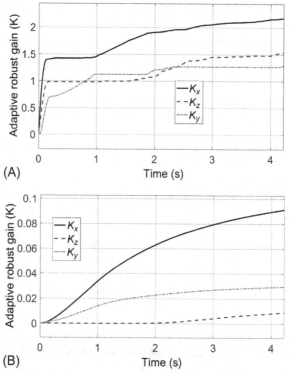

Fig. 5 Adaptation of robust gains during the microrobot motion (A) in the presence of disturbances and (B) without any disturbance.

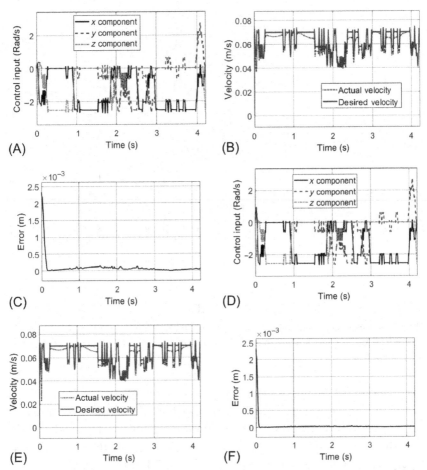

Fig. 6 (A) The control input, (B) the velocity tracking, and (C) the position error for the steady sliding mode gains with the value of 1 and (D) the control input, (E) the velocity tracking, and (F) the position error for the steady sliding mode gains with the value of 3.

higher at intervals when the robust gains are 3 in comparison with the case having the value of 1 for these gains. However, it could be mentioned that when the robust gains of the controller are larger, the position and velocity errors become smaller. The analogy of Figs. 4 and 6 clarifies how the online adjustment of the adaptive sliding mode gain $\hat{k}$ (Fig. 5) in the proposed control strategy is effective in decreasing the trajectory tracking errors by suggesting an efficient control effort.

Assuming the initial value of the gains is equal to 1 in the adaptive control scheme, the obtained results of the closed-loop system are demonstrated in

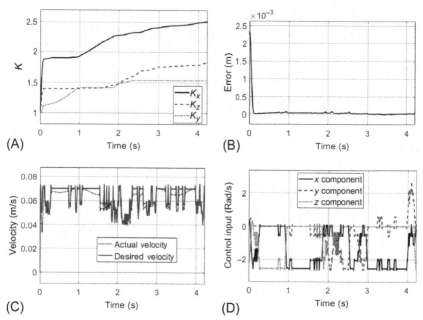

Fig. 7 The simulation results when the initial values of the robust controller gains is considered 1: (A) the gains adaptation, (B) the position error, (C) the velocity tracking, and (D) the control inputs.

Fig. 7. Note that the initial value of the robust gains was zero in Figs. 4 and 5. Based on Figs. 5A and 7A, it can be concluded that the adaptive performance of the controller in the presence of various uncertainties is achieved successfully that is in accordance with the control objectives and the Lyapunov stability analysis. Comparison of Fig. 4A with Fig. 7B and Fig. 4B with Fig. 7C intimates that the convergence of position and velocity trajectories is provided more appropriately when a larger initial value is assumed for the controller's time-varying robust gains. Specifically, a faster convergence rate of the transient response is seen in the initial part of the microrobot's motion. However, with reference to Figs. 4C and 7D, the control effort is lower in the case when the initial value of the robust gains is zero. From the control inputs of all four simulation studies presented in the presence of disturbances shown in Figs. 5–7, it is concluded that not only the control effort increases due to the rise of robust gains, but also the system chattering around the sliding surface occurs more often (the frequency and amplitude of this phenomenon are elevated). Therefore, it is recommended that the initial values of the controller's robust gains are considered small when the appropriate steady-state levels of these gains are unknown.

6 Conclusion

The main goals of the present chapter were (1) introducing an accurate and well-clarified dynamic model of the microrobot, (2) deployment of an optimal 3D path planning, and (3) mathematical formulation and designing a proper robust controller with adaptive gains. In this work, an adaptive sliding mode control method was presented for a monohelical-flagellated endovascular microrobot, for the first time. Since this type of microrobots deals with more challenging uncertainties in comparison with the magnetic resonance propulsion (MRP) systems due to the disturbances and inaccuracy of dynamic parameters, a robust adaptive control strategy was designed and adopted in this study. An optimality principle was employed for the path planning of the microrobot with a minimum effort criterion using dynamic programming (DP) in three dimensions to obtain the desired robot trajectory for the controller. The dynamic model of the microrobot was developed and its physical interaction inside the circulatory vascular system as the most dominant force (i.e., the hydrodynamic drag) was thoroughly taken into account. It was proven using the Lyapunov theorem that the controller guarantees the stability and tracking convergence in the presence of disturbances. Robustness and performance of the proposed control method is evaluated by some simulations considering the most important uncertainties in vascular system (the pulsatile flow, blood density, and viscosity fluctuations and inaccuracies).

References

Abbott, J.J., Peyer, K.E., Dong, L.X., Nelson, B.J., 2010. How should microrobots swim? In: Robotics Research. Springer.

Aghajanzadeh, O., Sharifi, M., Tashakori, S., Zohoor, H., 2018. Robust adaptive Lyapunov-based control of hepatitis B infection. IET Syst. Biol. 12, 62–67.

Arcese, L., Fruchard, M., Ferreira, A., 2012. Endovascular magnetically guided robots: navigation modeling and optimization. IEEE Trans. Biomed. Eng. 59, 977–987.

Arcese, L., Fruchard, M., Ferreira, A., 2013. Adaptive controller and observer for a magnetic microrobot. IEEE Trans. Robot. 29, 1060–1067.

Barati, R., Neyshabouri, S.A.A.S., Ahmadi, G., 2014. Development of empirical models with high accuracy for estimation of drag coefficient of flow around a smooth sphere: an evolutionary approach. Powder Technol. 257, 11–19.

Belharet, K., Folio, D., Ferreira, A., 2011. Three-dimensional controlled motion of a microrobot using magnetic gradients. Adv. Robot. 25, 1069–1083.

Chwang, A., Wu, T.Y., 1971. A note on the helical movement of micro-organisms. Proc. R. Soc. Lond. B: Biol. Sci. 178, 327–346.

De Cristofaro, S., Stefanini, C., Pak, N.N., Susilo, E., Carrozza, M., Dario, P., 2010. Electromagnetic wobble micromotor for microrobots actuation. Sens. Actuators A: Phys. 161, 234–244.

Dong, L., Nelson, B.J., 2007. Tutorial-robotics in the small part II: nanorobotics. IEEE Robot. Autom. Mag. 14, 111–121.

Dreyfus, R., Baudry, J., Roper, M.L., Fermigier, M., 2005. Microscopic artificial swimmers. Nature 437, 862.

Duffner, F., Schiffbauer, H., Glemser, D., Skalej, M., Freudenstein, D., 2003. Anatomy of the cerebral ventricular system for endoscopic neurosurgery: a magnetic resonance study. Acta Neurochir. 145, 359–368.

Evans, A.A., Lauga, E., 2010. Propulsion by passive filaments and active flagella near boundaries. Phys. Rev. E. 82, 041915.

Flake, A.W., 2003. Surgery in the human fetus: the future. J. Physiol. 547, 45–51.

Fluckiger, M., Nelson, B.J., 2007. Ultrasound emitter localization in heterogeneous media. In: 2007 29th Annual International Conference of the IEEE Engineering in Medicine and Biology Society (EMBS 2007). IEEE, pp. 2867–2870.

Folio, D., Ferreira, A., 2017. Two-dimensional robust magnetic resonance navigation of a ferromagnetic microrobot using Pareto optimality. IEEE Trans. Robot.

Hund, S.J., Kameneva, M.V., Antaki, J.F., 2017. A quasi-mechanistic mathematical representation for blood viscosity. Fluids 2, 10.

Jeong, S., Choi, H., Go, G., Lee, C., Lim, K.S., Sim, D.S., Jeong, M.H., Ko, S.Y., Park, J.O., Park, S., 2016. Penetration of an artificial arterial thromboembolism in a live animal using an intravascular therapeutic microrobot system. Med. Eng. Phys. 38, 403–410.

Kehlenbeck, R., Felice, R.D., 1999. Empirical relationships for the terminal settling velocity of spheres in cylindrical columns. Chem. Eng. Technol. 22, 303–308.

Kenner, T., 1989. The measurement of blood density and its meaning. Basic Res. Cardiol. 84, 111–124.

Khalil, I.S., Misra, S., 2017. Control of magnetotactic bacteria. In: Microbiorobotics, second ed. Elsevier.

Kim, H., Ali, J., Cheang, U.K., Jeong, J., Kim, J.S., Kim, M.J., 2016. Micro manipulation using magnetic microrobots. J. Bionic Eng. 13, 515–524.

Kirk, D.E., 2012. Optimal Control Theory: An Introduction. Courier Corporation.

Kristo, B., Liao, J.C., Neves, H.P., Churchill, B.M., Montemagno, C.D., Schulam, P.G., 2003. Microelectromechanical systems in urology. Urology 61, 883–887.

Kummer, M.P., Abbott, J.J., Kratochvil, B.E., Borer, R., Sengul, A., Nelson, B.J., 2010. OctoMag: an electromagnetic system for 5-DOF wireless micromanipulation. IEEE Trans. Robot. 26, 1006–1017.

Marino, H., Bergeles, C., Nelson, B.J., 2014. Robust electromagnetic control of microrobots under force and localization uncertainties. IEEE Trans. Autom. Sci. Eng. 11, 310–316.

Munson, B.R., Okiishi, T.H., Rothmayer, A.P., Huebsch, W.W., 2014. Fundamentals of Fluid Mechanics. John Wiley & Sons.

Nelson, B.J., Kaliakatsos, I.K., Abbott, J.J., 2010. Microrobots for minimally invasive medicine. Annu. Rev. Biomed. Eng. 12, 55–85.

Nourmohammadi, H., Keighobadi, J., Bahrami, M., 2016. Design, dynamic modelling and control of a bio-inspired helical swimming microrobot with three-dimensional manoeuvring. Trans. Inst. Meas. Control. https://doi.org/10.1177/0142331215627006.

Pourmand, M.J., Haghpanah, S.A., Taghvaei, S., 2019. Optimal robust navigation control of an endovascular microrobot. Iranian Journal of Science and Technology, Transactions of Mechanical Engineering. 2019 (in press).

Pourmand, M.J., Taghvaei, S., Vatankhah, R., Arefi, M.M., 2018. An underactuated bio-inspired helical swimming microrobot using fuzzy-PI controller with novel error detection method for 5-DOF micromanipulation. Designs 2, 18.

Sharifi, M., Moradi, H., 2017. Nonlinear robust adaptive sliding mode control of influenza epidemic in the presence of uncertainty. J. Process Control 56, 48–57.

Sharifi, M., Salarieh, H., Behzadipour, S., Tavakoli, M., 2017. Tele-echography of moving organs using an impedance-controlled telerobotic system. Mechatronics 45, 60–70.

Slotine, J.J.E., Li, W., 1991. Applied Nonlinear Control. Prentice Hall, Englewood Cliffs, NJ.

Torabi, M., Sharifi, M., Vossoughi, G., 2017. Robust adaptive sliding mode admittance control of exoskeleton rehabilitation robots. Sci. Iran.

Utkin, V.I., Poznyak, A.S., 2013. Adaptive sliding mode control. In: Advances in Sliding Mode Control. Springer.

Yesin, K.B., Vollmers, K., Nelson, B.J., 2006. Modeling and control of untethered biomicrorobots in a fluidic environment using electromagnetic fields. Int. J. Robot. Res. 25, 527–536.

Zaaroor, M., Kósa, G., Peri-Eran, A., Maharil, I., Shoham, M., Goldsher, D., 2006. Morphological study of the spinal canal content for subarachnoid endoscopy. Minim. Invasive Neurosurg. 49, 220–226.

CHAPTER EIGHT

Robotics in endoscopic transnasal skull base surgery: Literature review and personal experience

Alba Madoglio[a], Francesca Zappa[a], Davide Mattavelli[b], Vittorio Rampinelli[b], Marco Ferrari[b], Alberto Schreiber[b], Francesco Belotti[a], Andrea Bolzoni Villaret[b], Fabio Tampalini[c], Riccardo Cassinis[c], Lena Hirtler[d], Barbara Buffoli[e], Luigi Fabrizio Rodella[e], Piero Nicolai[b], Marco Maria Fontanella[a] and Francesco Doglietto[a]

[a]Unit of Neurosurgery, Department of Surgical Specialties, Radiological Sciences, and Public Health, University of Brescia, Brescia, Italy
[b]Unit of Othorinolaryngology, Department of Surgical Specialties, Radiological Sciences, and Public Health, University of Brescia, Brescia, Italy
[c]Department of Information Engineering, University of Brescia, Brescia, Italy
[d]Center for Anatomy and Cell Biology, Medical University of Vienna, Vienna, Austria
[e]Section of Anatomy and Physiopathology, Department of Clinical and Experimental Sciences, University of Brescia, Brescia, Italy

1 Introduction

Before 1970, open surgery was the most frequently used access for inflammatory diseases involving the frontal and maxillary sinus (Trevillot et al., 2013b). The Austrian surgeons Messerklinger and Stammberger pioneered a new, revolutionary procedure. They promoted a paradigm shift in the treatment of sinonasal inflammatory disease and developed the concept of functional endoscopic sinus surgery, taking profit of the magnification of anatomical structures provided by the endoscope (Trevillot et al., 2013a). This concept further evolved in endoscopic transsphenoidal surgery for the treatment of sellar tumors and, more recently, in Endoscopic transnasal Skull Base Surgery (ESBS) (Doglietto et al., 2005). The main advantage of transnasal endoscopic skull base approaches is providing a more direct and less invasive access to the skull base, avoiding craniofacial incisions and extensive bone removal, typical of open surgical approaches (Kupferman and Hanna, 2014; Doglietto et al., 2018).

With the increasing complexity of transnasal endoscopic surgery, the need for a bimanual dissection became evident (Cabuk et al., 2015) and the so-called four hands technique has been developed (Castelnuovo et al., 2006). The technique entails a combined work by at least two surgeons, of whom one holds the endoscope and the other is responsible for surgical dissection; the technique therefore requires a close collaboration between the two surgeons. Nonetheless, even among closely collaborating teams various issues can become evident, especially in long and complex operations (Trevillot et al., 2013a): the surgeon holding the endoscope has to coordinate with the other as to optimize visualization and surgical maneuverability of up to three instruments that might have to be introduced in a typically narrow and long surgical corridor; the need for a close-up but fixed image might be challenging in case of long procedures because of physiological tremor (Bolzoni Villaret et al., 2017).

The increasing complexity of transnasal endoscopic skull base surgery has therefore led to the demand for an endoscope holder (Chan et al., 2016). Until recently, though, only a few centers have used endoscope holders systematically (Paraskevopoulos et al., 2016, 2018), mainly because a change of view usually requires the use of one hand by the surgeon and the maneuverability of the endoscope is suboptimal. A robotic endoscope holder would theoretically have the same advantages of nonrobotic holders, including tremor filtration (Blanco and Boahene, 2013), but also a higher accuracy in endoscope movements, which might be performed without the need of a surgeon's hand (Bolzoni Villaret et al., 2017).

Robotics, in the last two decades, has been applied to multiple surgical specialties, including cardiac surgery, urology, gynecology, and, most recently, minimally invasive head and neck surgery (O'malley and Weinstein, 2007). Da Vinci® Robotic System (Intuitive Surgical, Sunnyvale, CA, USA) and Flex® Robotic System (Medrobotics, Raynham, MA, USA) are commercially available robots approved for head and neck surgery, though mainly limited to thyroid or transoral surgery. These systems have not been designed for ESBS and consequently they show several limitations when applied to it. These limitations include poor ergonomics and large dimensions, which require a working space that is typically not available in ESBS (Bolzoni Villaret et al., 2017).

Significant advances in surgical robotics have been made, but a central role for robot-based applications in transnasal skull base surgery has not been defined (Blanco and Boahene, 2013). Recently, several prototypes for endoscopic transnasal skull base surgery have been developed, but with some disadvantages, including ergonomics and prolonged set-up time (Bolzoni Villaret et al., 2017).

To overcome the limits of previous robotic prototypes, novel hybrid solutions, conceptually dissimilar from a purely robotic surgery, have been also developed: the robotic system has the task of holding the endoscope, remaining entirely dependent on the surgeon (Bolzoni Villaret et al., 2017). This solution is defined hybrid as the robotic system is not "pure," but the robotic holder works with the surgeon at the patient's side; for this reason, it is also defined "collaborative."

Recently, a hybrid robotic solution for ESBS has become available (EndoscopeRobot®, Medineering, Munich, Germany): it is made of a robotic arm together with a smaller robot that acts as an endoscope holder that can be controlled with a foot pedal. As the system has been introduced only recently, there are no preclinical or clinical reports on it.

The aim of this chapter is to provide a comprehensive critical overview of nonrobotic endoscope holders and of the different robotic prototypes that have been developed for ESBS, together with a brief presentation of preclinical and clinical data on the use, at the University of Brescia—Italy, of this commercially available hybrid solution for ESBS.

2 Nonrobotic endoscope holders

The increasing complexity of transnasal endoscopic skull base surgery has led to the augmentation of the time dedicated to tumor removal, leading to an increasing demand for an endoscope holder (Chan et al., 2016), as it can provide a steady vision, eliminating the human physiological tremor, which increases with fatigue. Until recently, though, only few centers have used endoscope holders systematically (Paraskevopoulos et al., 2016, 2018), mainly because a change of view usually requires the use of one hand by the surgeon and the maneuverability of the endoscope is suboptimal.

Over the years, several solutions have been reported. To understand the evolution of endoscope holders (EH), Paraskevopoulos et al. (2016, 2018) conducted an overview of available devices, identifying their main features, advantages, and limitations.

Endoscope holders can be divided into mechanical and pneumatic-based systems (Table 1).

2.1 Mechanical fixation type

Leyla arm is the cheapest holding device and two microsurgical arms attached to a plate that holds the endoscope comprise it (Jimenez, 2019). The position of the flexible holding arm and the Leyla retractor can be

Table 1 Summary of the analyzed features for each endoscope holder.

Endoscope holder	Fixation type	Advantages	Limitations
Point Setter® (Mitaka Kohki, Co., Ltd., Tokyo)	Pneumatic	Nitrogen-powered endoscope stabilizer Quick release mechanism Single-hand control 6 Joints (including one manual joint)	Bulky It requires a gas supply (N_2 or compressed air) Quite expensive
Unitrac® (Aesculap AG, Tuttlingen, Germany)	Pneumatic	Pneumatically powered retraction and scope holding system Connect with NeuroPilot for fine adjustment	It requires compressed air source (or CO_2 cartridge)
Leyla arrm	Mechanical	Wide availability Inexpensiveness Familiarity	Cumbersome It requires of 2 arms for safe endoscope fixation
"Snake"-type of friction holder (Aesculap AG, Tuttlingen, Germany)	Mechanical	Similar to Leyla arm	Bulky, crude, and outdated Some occasional downward drift
Neuroarm® (Adeor Medical AG, Pullach, Germany)	Mechanical	It holds different tools Simple	Cumbersome It requires of 2 arms for safe endoscope fixation
M-TRAC® (Aesculap AG, Tuttlingen, Germany)	Mechanical	Flexible holding device Small, supple joints for fine positioning Autoclavable	Minimal downward drift
Endoarm® (Eskandari et al., 2008) (Olympus Co., Tokyo, Japan)	Hybrid endoscope-holder system	Solid fixation and smooth relaxation. Easily maneuverable Bayonet-shaped endoscope Integrated with a video system to facilitate endoscopic microneurosurger	Bulky
Endocrane® (Karl Storz GmbH & Co. KG, Tuttlingen, Germany)	Piezoelectric locking mechanism	Faster positioning compared to manual positioning systems Rapid locking Single-hand use Fail-safe function	NA

adjusted in three dimensions. The major advantages of the system are its availability, cost, and the familiarity with its use by most neurosurgeons. As for every mechanical holder, its main limit is the requirement of two arms for safe endoscope fixation.

Another similar mechanical device is the **"snake"-type** friction holder, available by Aesculap AG (Tuttlingen, Germany). It includes several endoscope adapter inserts and a tip for fine adjustments. As for other systems, it has some occasional downward drift upon fixation.

Fine corrections or adjustments are necessary after positioning the neuroendoscope to receive the optimal endoscopic image. With traditional holding devices only a rough positioning is possible; the **NeuroPilot Micro Steering System** is a mechanical system that provides a precise and fine steering of the neuroendoscope to achieve clearer visualization.

Neuroarm® (Neuroarm, 2019) (Adeor® Medical AG, Pullach, Germany) has been created in order to hold different tools, such as retractor blades, endoscopes, or other instrument in a fixed position. It is a simple system, similar to Leyla arms, and with the same limitations.

M-TRAC® (M-TRAC®, 2019) is another mechanical holding arm by Aesculap (Aesculap AG, Tuttlingen, Germany). It is a flexible holding device with mechanical fixation by clamping handle. The device has small, supple joints for fine positioning and it is autoclavable. A minimal downward drift can be compensated by Neuropilot micromanipulator.

2.2 Pneumatic fixation holders

The minimal downward shift that characterizes mechanical systems is usually not present in pneumatic-based endoscope holders.

One of the most useful holders is probably the **Point Setter** (Point Setter, 2015), developed originally by Mitaka (Mitaka Kohki, Co., Ltd., Tokyo, Japan) and distributed outside Japan by Karl Storz (Karl Storz GmbH & Co. KG, Tuttlingen, Germany and KSEA, CA, USA). The Point Setter works as a third arm that can hold various endoscopes of many manufacturers and it can be attached on either rail of the operating table through an adapter. This endoscope holder is a nitrogen-powered endoscope stabilizer and it provides precise positioning (Arnholt and Mair, 2002). The holder arm has a total of six joints, including one manual joint and three ball joints. For optimal positioning, the surgeon can adjust the angle with the manual ball joint adjust knob. The Point Setter has a reliable safety mechanism, which causes the arm to be locked when the power or nitrogen gas pressure is down. The system offers comfortable use, low friction, good balance and a rigid, safe design. The quick release mechanism is quite useful and single-hand control is possible.

The **Unitrac®** (Unitrac®, 2019) holding arm from Aesculap (Aesculap AG, Tuttlingen, Germany) is used in procedures that require limited endoscope movements. It is a pneumatically powered retraction and scope-holding system that allows the surgeon to perform a more efficient procedure.

EndoArm (Olympus Co., Tokyo, Japan) was developed by the collaboration of neurosurgeons from three institutions (University of Tokyo, Tokyo Women's Medical University, Kinki University) with a commercial manufacturer (Olympus Co., Tokyo, Japan). A pneumatic arm with several ball-and-socket style joints allows for movement of the endoscope. A single button locks and releases the pneumatic clutch device and allows for movement in all planes with one hand. For a solid fixation and smooth relaxation mechanism, a negatively actuated air-locking system was developed, which locks the scope automatically and releases it by activating the air pressure (Eskandari et al., 2008).

In daily surgical practice, most teams do not routinely use pneumatic arms to secure the endoscope (they require compressed air and are quite expensive), but have found them useful in selected lengthy cases (Chowdhry and Cohen, 2013).

2.3 Piezoelectric fixation holders

A recent technological alternative is represented by **Endocrane** from Karl Storz that features a special piezoelectric locking joint mechanism. It can achieve positioning without misalignment thanks to a rapid locking mechanism. The compact holding arm features a fail-safe function, which prevents a loss of retention force in the case of malfunction, i.e., power failure.

2.4 Survey on endoscope holders

Paraskevopoulos et al. (2016, 2018) recently performed a survey among neurosurgeons regarding the use of endoscopic holders.

The results of the survey underlined that most neurosurgeons prefer not to use an endoscope holder and those who responded that they use an EH, mainly use a holder for ventricular rather than skull base procedures. Furthermore, most surgeons regularly work with a cosurgeon.

The responders stressed the positive and negative features of endoscope holders:

– Positive features: stability, number of joints, maneuverability, safety;
– Negative features: raw movements, limited degrees of freedom as compared to free hand, downward drift, lack of flexibility, iatrogenic injury, expensive, too bulky.

The same group conducted a second, wider, and more specific survey (Paraskevopoulos et al., 2018) after the pilot one; in particular, they tried to differentiate between different uses of EHs, such as intraventricular versus skull base, specific procedures, and pitfalls. In this second study, the authors reported pitfalls of endoscope holders in different applications.

Pitfalls that were reported and were specific to skull base surgery included:

– Cleaning difficulties: maintaining a clear vision is of paramount importance in surgery; when the endoscope is introduced in the nose, even minor bleeding might impair visualization and lead to the need of lens cleaning. This is usually achieved with commercially available cleaning systems that flush saline solution on the lens; when this is not enough, the endoscope might have to be removed from the nose to be cleaned;
– Need to incorporate the limitations of the holder into the operative planning.

Regarding pitfalls in both intraventricular and skull base surgery:

– Not dynamic with limited maneuverability: limited degrees of freedom as compared to free hand;
– Time consuming and cumbersome: the main issue is usually related to the space that is required by the holder, as this might occupy significant space, usually used by the assistant surgeon; set-up time might be time consuming;
– Difficult to get in the appropriate position: with some systems positioning is not optimal due to the delay in the locking system or due to the absence of minor movements that are needed to obtain an optimal position in confined, small areas;
– Restricted range of motion because of geometry of the holder;
– Shifting of holder or accidental loosening (esp. mechanical arm): scope may slip and may be inadvertently displaced by assistant or surgeon, so scope holder is one more tool to pay attention to during surgery;
– Failure to hold the scope;
– Downward drift: articulated arms can help to prevent fatigue in lengthy cases, but are often prone to drifting downwards during initial positioning,

thus making minor adjustments necessary, either in anticipation of this downward drift or after initial positioning.

Although endoscopic holders have a role in cranial neuroendoscopy, their use seems limited and their features are regarded as suboptimal by most neurosurgeons. To overcome the limitations of nonrobotic holders, in the last two decades, modern technology allowed robotic applications to be considered (Paraskevopoulos et al., 2018).

3 Prototypes for endoscopic transnasal skull base surgery: Literature review and personal experience

A recent systematic literature review on the current status of robotics for endoscopic transnasal skull base surgery (ESBS) (Bolzoni Villaret et al., 2017) was updated, searching Pubmed and Google Scholar with the following key words: "Robotic skull base surgery" and "Robotic transnasal."

The results of the updated review are summarized in Table 2.

Different features have been analyzed for each prototype, when available (Table 2):
– interface;
– tools under robotic control;
– safety features;
– set-up time and operative time (as compared to standard, nonrobotic).

3.1 Robotic interface

There are two types of interface: telemanipulation mode or "cooperative mode."

Telemanipulation mode is used to control endoscope positioners by, for example, voice, head motion, foot pedal, joystick (Trevillot et al., 2013b). Cooperative mode is an interesting alternative, in which surgeon and robotic arm cooperate side by side to hold the endoscope. In a cooperative mode, the surgeon manually moves the endoscope as in conventional surgery (Bolzoni Villaret et al., 2017), but the robot follows the motion of the endoscope and maintains it in position when the surgeon no longer holds it (Trevillot et al., 2013b).

All described robots in this review are controlled in telemanipulation mode, except two of them (Table 2).

The first one is an image-guided robot system described by Xia et al. (2008) developed to improve safety by preventing the surgeon from accidentally damaging critical neurovascular structures during drilling

Table 2 Summary of the analyzed features for each prototype that has been described for endoscopic transnasal skull base surgery.

First author (year of publication)	Robot name	Type of control interface	Robotic task	Safety	Set up time/operating time
Nimsky et al. (2004)	**EVO1**	Joystick (TM)	Endoscope holder	NR	Long (30 min)/prolonged
Wurm et al. (2005)	**A73**	Automatic robot Joystick (TM)	Neuroptik T30 (Endoscope holder)	3D navigation system and "Loss of control" mode	Long/prolonged
Nathan et al. (2006)	**AESOP**	Voice control (TM)	Endoscope holder	Three saved positions	Long (several minutes)/normal
Strauss et al. (2007)	–	Joystick (TM)	Endoscope holder	Integrated navigation system Easy switch to manual endoscopy	Short (2 min)/prolonged
Xia et al. (2008)	–	CM	Endoscope or drill holder	Integrated navigation system	Long/unknown
Yoon et al. (2011)	–	Double joystick (TM)	Active bending spring backbone endoscope	NR	Long/prolonged
Eichhorn and Bootz (2011)	**Tx40**	Joystick (TM) and autonomous tracking movements	Endoscope holder	Navigation system	Unknown/prolonged
Swaney et al. (2012)	–	Surgeon's console (TM)	Quadramanual robot: 2 grippers, suction, and flexible endoscope with light source	NR	NR

Continued

Table 2 Summary of the analyzed features for each prototype that has been described for endoscopic transnasal skull base surgery.—cont'd

First author (year of publication)	Robot name	Type of control interface	Robotic task	Safety	Set up time/operating time
Trevillot et al. (2013a),b	**HYBRID**	CM	Endoscope holder	Force threshold	Long/unknown
Schneider et al. (2013)	**TENTACLE-LIKE**	Joystick (TM)	Instruments manipulator	NR	Unknown/normal
Cabuk et al. (2015)	**SP ROBOTIC SYSTEM**	Joystick (TM)	Endoscope holder	Resistance felt on haptic arm in case of contact or friction with adjacent tissues	Normal/prolonged
Swaney et al. (2015)	–	The surgeon manipulates a user interface that controls the position and orientation of the robot manipulator (TM)	One arm outfitted with a gripper and the other with an angled ring curette	Ability to change the axial orientation of the angled ring curette without changing the tip position or orientation of the robot	Unknown/unknown
Chan et al. (2016)	**FREE**	Inertial measurement unit and vocal control (TM)	Endoscope holder	Force threshold, vocal command	Short (<3 min)/normal (<7 min for maxillary antrostomy)
Bolzoni Villaret et al. (2017)	**BEAR**	Head control (marked glasses)	Endoscope holder	Force threshold	Potentially limited set up time/unknown

Abbreviations: CM, cooperation mode; *NR*, not reported; *TM*, telemanipulation.

procedures. This system includes a modified NeuroMate® robot (Integrated Surgical Systems, Sacramento, CA) in cooperative control mode, StealthStation® navigation system, 3D Slicer software for intraoperative visualization and workstation for application logic and robot control. The NeuroMate® robot was modified converted into a cooperative robot by attaching a six-DoF force sensor (JR3 Inc., Woodland, CA, USA) at the end effector, between the final axis and the surgical instrument (Anspach eMax drill, Palm Beach Gardens, FL, USA). The disadvantages were the inaccuracy of about 1 mm, due to an initial placement error, calibration issues, and robot kinematic inaccuracy (Trevillot et al., 2013a).

The second one is a hybrid solution (HYBRID—compact, ergonomic, and safe endoscope positioner), which was developed as a combination of EVOLAP (a prototype built for minimally invasive laparoscopic surgery) and VIPER (a conventional six-active DOF industrial robot). The advantages of the hybrid prototype are active control of the penetration, intuitive, stable and safe motion, compactness, no interface between the right hand of the surgeon and the local manipulator. The drawback is that the rotation about the main endoscope axis is not controlled (Trevillot et al., 2013b).

Most of the reported prototypes were joystick-controlled:
- **Evolution 1** (Nimsky et al., 2004) was initially designed as an endoscope-holding device for use in endoscopic ventriculostomy and then modified for transsphenoidal endoscopic surgery. It is a robotic teleoperation system controlled directly by the surgeon via a joystick and is based on a Stewart platform (hexapod design) with a seven axis (z-axis). The kinematic structure of the hexapod allows movements in all six degrees of freedom and the workspace is increased by the additional z-axis. It was certified for actual patient use, but the presented system remains a prototype due to its relatively large dimensions;
- **A-73** (Wurm et al., 2005) is an automatic robot that allows the definition of the ideal trajectory on preoperative imaging, so during the procedure the remote-control unit allows the surgeon to correct a trajectory or complete an insufficient resection. Its major advantages are precision and multifunction, but the disadvantage is that it was not tested for any operation except anterior sphenoidotomy and its large dimensions limit its potential clinical use (Trevillot et al., 2013a; Bolzoni Villaret et al., 2017);
- **Tx40** (Eichhorn and Bootz, 2011) is a robotic holder that performs autonomous tracking movements, adjusted by a surgeon via a joystick. Moreover, it is equipped with an automatic lens cleaning system. Even in this

case, the robot dimensions represent a major limit, as they limit the surgeon's range of movement;
- the telemanipulated endoscopic holder dedicated to functional endoscopic endonasal surgery described by Strauss et al. (2007) presented a high accuracy, was quicker and more accurate and precise than the human in positioning the endoscope inside the nose, but was still judged inferior to the classic, nonrobotic set-up in most of the preclinical tests; the authors indeed stressed the importance of developing a man–machine interface to optimize endoscope maneuverability;
- **Stewart Platform (SP)-based robotic system** (Cabuk et al., 2015) was developed as an endoscope positioner and holder that is able to change the position with the help of an assistant. The transsphenoidal sellar approach was possible with the endoscope holder attached to the Stewart platform. Stewart platform has a special structure with mobile and fixed plates, six linear motors and joints. The robotic endoscope holder allows the main surgeon to use both hands, optimizing also ergonomics during surgery. The robot is though controlled by an assistant with a joystick, with the difficulty of controlling at the same time both the operating area (i.e., the robot in its entry point) and the screen (where the endoscopic picture is projected);
- a double joystick is needed to control the active bending of the endoscope prototype coupled with a spring backbone developed by Yoon and coworkers (Yoon et al., 2011), which was designed to be bent up to 180 degrees.

Most Authors indeed agree on the significant limitation of joystick-controlled telemanipulation: the technique implies the need for an added surgeon, who has to be extremely well coordinated with the primary surgeon. None of these systems have been applied to transnasal surgery, possibly due to robots dimensions and the requirement of a separate hand, by an assistant surgeon, to manipulate the joystick.

Other types of telemanipulation have been suggested to overcome the limits of the joystick based one.

Automated Endoscopic System for Optimal Positioning (**AESOP**) robot (Nathan et al., 2006) represents a voice-controlled prototype; its most interesting feature is the possibility to memorize certain positions, which could be returned to with a single voice command. It has seven degrees of freedom and supports continuous or incremental voice commands. However, it only responds to easy commands thus not supporting complex combinations of movement that are available in a free-hand technique

(Paraskevopoulos et al., 2016). The large working space occupied by the robot and its high cost prevents its use in routine surgery (Trevillot et al., 2013a).

To further improve previous designs, the use of a foot-control was investigated (Chan et al., 2016) (Foot-Controlled Robotic-Enabled Endoscope—**FREE**) and its feasibility tested for functional endoscopic sinus surgery. FREE is controlled by an IMU (inertial measurement unit), the foot control interface, which is attached to the surgeon's foot and communicates with the control unit via Bluetooth. This setup allows the measurement of the foot's relative orientation in real time. Eversion or inversion of the foot changes the active joint in an increasing or decreasing manner. To provide the desired movement command to the active joint, the heel of the foot is raised to 15 degrees from the ground and moved left or right.

3.2 Continuum robotics

Other described prototypes for ESBS include continuum robotics, which use a new concentric tube technology. The use of these robots was a challenge, because many degrees of freedom must be controlled simultaneously in a compact and unobtrusive package in the operating room. While significant progress in designing these actuation units has been made for both larger, nonmedical continuum robots, and for certain types of surgical continuum robots, the design of clinically relevant actuation units for concentric tube robots remains a comparatively understudied area of research, according to different Authors (Swaney et al., 2012).

Some researchers (Swaney et al., 2012) designed a novel 24 degrees of freedom quadramanual slave robot, teleoperated for use in single-nostril transnasal skull base surgery. The robot is one component of a larger system that shall include a surgeon-controlled console conceptually similar to that of the da Vinci robot. In this system, a camera and three instruments has been included, to allow simultaneous use of two surgical tools with suction/irrigation and a camera for visual guidance. The camera and instruments are mounted at the tip of concentric tube manipulators and they enable the surgeon to visualize areas not otherwise seen with traditional endoscope while minimizing damage to healthy tissues. The same group (Swaney et al., 2015) recently added to the robotic system the ability to rotate the end effector while leaving the robot fixed in space. In the actuation unit, used to translate and rotate the tubes, each tube is grasped at its base and may be translated and rotated independent of the others, creating a tentacle-like motion.

Moreover, another author (Schneider et al., 2013) reported the functional requirements for an ideal robotic system for PSSB (paranasal sinuses and skull base) surgery, in which one instrument can be forwarded transnasally through a concentric tube, while being telemanipulated by the surgeon (Friedrich et al., 2017). The many existing instruments designed for PSSB surgery can be adapted to the new technology by mounting them at the tip of this tube. This robot achieved "tentacle-like" motion and it was composed of several flexible, percurved tubes that are nested within each other.

3.3 Hybrid robotics

Bolzoni Villaret et al. (2017) recently described an additional hybrid prototype, BEAR (Brescia Endoscope Assistant Robotic holder), developed and tested preclinically at the University of Brescia. To overcome the limits of commercially available systems, our group developed this hybrid solution, which is conceptually dissimilar from purely robotic surgery solution: the robotic system holds the endoscope, remaining entirely dependent on the surgeon, with a high safety standard, thanks to force feedback. This prototype uses the surgeon's head position to achieve an intuitive control of the movement of the robot-held endoscope. The system gathers the surgeon's head position using an off-the-shelf optical sensor (Microsoft Kinect), which actually measures the position of two especially adapted glasses that must be worn by the surgeon (Figs. 1 and 2). A second version of the system has been designed more recently and uses Microsoft Kinect 2.0 sensor, which has higher precision and identifies the main surgeon's shape even in operating room drapes, therefore eliminating the need for special glasses. BEAR shares with other systems the limits typical of a prototype, which was developed adapting an industrial robot, which had nonoptimal dimensions; furthermore, during the preclinical tests other limitations became evident, such as suboptimal joint movements and excessive inertia (Bolzoni Villaret et al., 2017).

4 Clinical applications of robotics in transnasal endoscopic skull base surgery: Literature review

Different prototypes have been described for endoscopic transnasal skull base surgery, with clinical applications limited to a robotic arm rest (Ogiwara et al., 2017) or to Trans-Oral Robotic Surgery (TORS) combined with Extended Endonasal Approach (EEA-TORS) (Carrau et al., 2013).

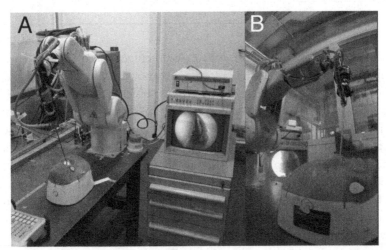

Fig. 1 BEAR at the Robotics laboratory of the University of Brescia. An industrial manipulator (Kawasaki RS03N) was used: it is an anthropomorphic robotic arm with six rotational joints and six degrees of freedom. It was coupled with a force sensor: its transducer was positioned between the wrist and the endoscope holder (Bolzoni Villaret et al., 2017). (A, B) The robot holds the endoscope that has been positioned in a phantom (S.I.M.O.N.T®—SInus Model Othorino Neuro Trainer by ProDelphus, Olinda, Brazil); the endoscopic screen shows the picture of the simulated left nasal cavity during the intranasal robotic navigation.

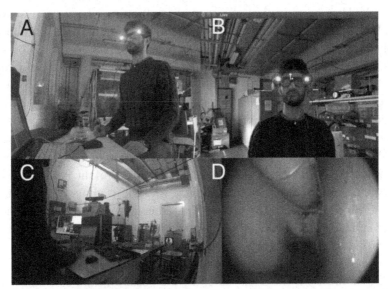

Fig. 2 BEAR and its telemanipulation system. An optical system recognized the position of markers attached to glasses and therefore allowed the surgeon to move the endoscope with intuitive head movements. (A–C) The surgeon is positioned away from the robot and the glasses are recognized by the system; (B) The movements of the surgeon's head are recognized and the simulated endoscope [*green tube (light gray in the print version)* in D] is thus navigated inside the simulated nasal cavity.

Ogiwara et al. (2017) described an intelligent arm-support system, i-ArmS, developed as an operation support robot. It is controlled by force sensors, brakes, and encoders. There are two types of iArms: one for right-handed use and one for left-handed use. It follows the movement of surgeon's nondominant arm and fixes it at an adequate position. The system has three modes: transfer (Free), arm holding (Hold), and arm free (Wait). When the surgeon's arm is placed on the arm holder, the mode changes from Wait to Hold. When the surgeon's arm moves to the desired position and holds still, the mode changes from Free to Hold. The mode changes from Hold to Free with a click action by the surgeon's arm. It has been designed to prevent hand tremor and to alleviate fatigue during surgery. It is therefore used as a robotic armrest for the surgeon. The authors reported on the application of this robotic device to endoscopic endonasal transsphenoidal surgery (ETSS) and evaluated their initial clinical experience in 43 patients in different surgical procedures. During the nasal and sphenoid phases, iArmS was used to support the surgeon's nondominant arm, which held the endoscope. In these phases, the simultaneous use of two instruments was not essential, so two-hand technique with iArms was better than the use of a scope holder (reducing operation time and achieving excellent operability). Then, during the sellar and reconstruction phases, its role was finished and a UniArm endoscope holder (Mitaka Kohki) was introduced and the scope was fixed to it. The intelligent armrest seemed to be safe and effective during ETSS. The main limit of the system is that it does not substitute the surgeon's arm but is indeed an armrest.

Carrau et al. (2013) described the EEA-TORS approach, which was studied and practiced on cadavers, and then applied clinically to two patients. The EEA-TORS technique was used on human subjects to resect an adenoid cystic carcinoma of the nasopharynx with extension both laterally into the infratemporal fossa and inferiorly below the level of the hard palate and resect a clival chordoma with extension into the craniocervical junction inferiorly to C2. Both patients had total resection with no complications and minimal postoperative morbidity. The EEA-TORS approach provided excellent exposure to the posterior skull base, nasopharynx, and infratemporal fossa. The main advantage of using this technique to manage skull base tumors is the ability to reach the posterior skull base below the level of the Eustachian tube, which is the inferior limit of the EEA, with TORS. This study, by an extremely experienced group, confirms the present limits of robotics, as the EEA phase was not performed with the robot.

5 A novel, commercially available hybrid system: Initial preclinical and clinical experience

Endoscope Robot® by Medineering Surgical Robots® (Munich, Germany) has recently become available for clinical practice in endoscopic skull base surgery. It is a compact robot specifically developed to work as an endoscope holder during transnasal operations. As the product follows the same philosophy and fully develops the concept of BEAR, scientific collaboration was developed in 2016 with the company to test the system preclinically and possibly validate its use for endoscopic transnasal skull base surgery.

The system is made of four components: a small, robotic, endoscope holder; the Positioning Arm; the control unit (foot pedal) and the power source (Fig. 3). The Positioning Arm is fixed directly to the OR table; it moves in every direction of space thanks to seven joints. Endoscope Robot® is attached to Positioning Arm and holds and moves the endoscope.

Endoscope Robot® was tested for the first time in November 2016 on a cadaveric specimen in the Center for Anatomy and Cell Biology of the Medical University of Vienna and then in the Laboratory of Endoscopic Anatomy of the University of Brescia (Figs. 4 and 5). The first version was optimized for speed and type of movement. A commercially available surgical foot pedal was adapted to work as a robot manipulator (Figs. 3–5): the main surgeon can then control the robot with the foot pedal joystick (which moves the endoscope in a coronal plane), moving the endoscope in or out with the corresponding anterior pad. Foot pedal switches can be designed to return the robot to a saved position ("home position") and to switch the joystick–controlled movement (i.e., rotation or translation of the optic).

A first preclinical study has been recently performed to investigate the acceptance rate and possible benefits of Endoscope Robot® for transnasal robotic-assisted endoscopic surgery (Zappa et al., 2019a). All (i.e., attendees and faculty) skull base surgeons, Neurosurgeons, or ENT surgeons, attending the Joint European Diploma of Endoscopic Skull Base Surgery 2018–2019 in Paris, were enrolled in the study in December 2018.

Before the test phase, a questionnaire was administered to all participants, principally concerning personal surgical experience and habits. The test phase was structured in a 2-day session. A modified NeuroEndoTrainer (Singh et al., 2016), which is a low-cost training model that has been validated for transsphenoidal surgery, was modified to allow for a bimanual task.

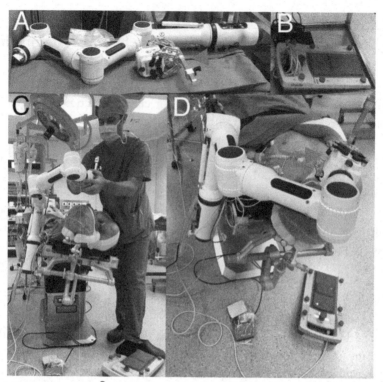

Fig. 3 Endoscope Robot® in the operating room at the University of Brescia. Once the patient is asleep, the robot holder is attached to the surgical bed and the surgeon can check the final position of the Endoscope Robot® attached to it. (A) The robotic arm and EndoscopeRobot® [*red arrow (light gray in the print version)*] before their positioning. (B) The foot pedal. (C) The robotic holder has been attached to the surgical bed; the surgeon checks the positioning of Endoscope Robot®. (D) overview of Endoscope Robot® in a simulated surgical position before draping.

Surgeons had to perform two different tasks with a 0–degree endoscope: the first one was a simple grasping task, while the second one was a relatively complex task, with the need of positioning a simulated nasoseptal flap over two pins. A tutorial video was shown to all participants before the tasks. Surgeons had 3 min to get accustomed to the robot before performing the task with it.

To control the effects of surgical learning, the participants were randomized in two groups so that one group performed the task initially robotically and then manually, vice versa for the other.

Participants will be evaluated with a modified GEARS-E score (Takeshita et al., 2018), which is an objective assessment tool in minimally

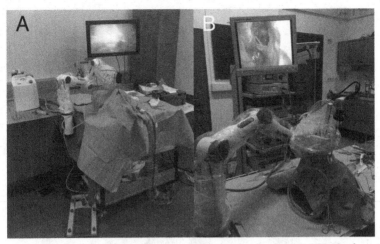

Fig. 4 Laboratory set-up at the University of Brescia (A) and of Vienna (B). The foot pedal is on the floor; the endoscope is held by Endoscope Robot and positioned inside the right nasal cavity of the specimen; the surgical field is displayed on a high definition (HD—3D and 2D, respectively) video screen.

invasive surgery that comprises different domains representing various skill-related variables:
– Endoscope control and depth perception;
– Bimanual dexterity;
– Tissue handling;
– Efficiency.
A 5-point Likert scale (i.e., 1 denotes the lowest proficiency, while 5 is the highest) was used to assess each domain.

At the end of the test phase, a modified NASA-Task Load Index Test (Dixon et al., 2014) was conducted via "Survey Monkey," in order to evaluate the task load with a highly standardized tool.

At the moment data analysis, including task execution time, GEARS-E score and the answers to the questionnaires, is still ongoing. Nonetheless, preliminary data analysis shows that most surgeons believe that robotic might provide a significant benefit in endoscopic skull base surgery, possibly proving that many surgeons feel the need for a solution to the present limits of endoscope maneuverability.

At the same time, the clinical study ROBOTHOLDER (RHOLD—Brescia Ethics Committee approval number: NP3080) has been initiated at the University of Brescia with the aim of prospectively evaluating efficacy and efficiency of this novel hybrid system (Figs. 3 and 6) (Zappa et al., 2019b).

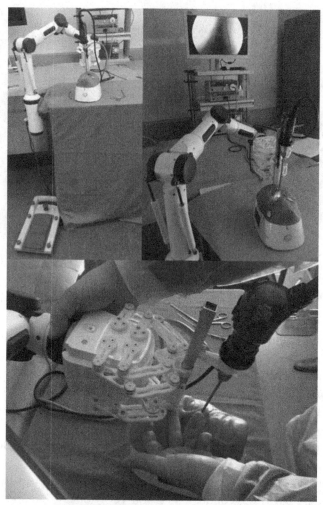

Fig. 5 Pictures from the first preclinical tests performed at the University of Brescia with a prototype by Medineering® (Munich, Germany). Surgical set-up was optimized and feasibility tests were performed in a model for transsphenoidal surgery (S.I.M.O.N. T®—SInus Model Othorino Neuro Trainer by ProDelphus, Olinda, Brazil).

RHOLD has a period of 20 months (from September 2018 to April 2020) (Doglietto et al., unpublished data).

The main objective is to evaluate the utility of robotic holder. The primary outcome variable is the duration of the surgery (by comparison between patients operated with and without robotic holder, using historical comparison), in particular:

- duration of the robotic phase of surgery, comparing the recording with historical data;
- duration of the surgical set-up phase, including the time required for the robot set-up, with comparison with data for surgery of the same type performed without robotic holder (historical comparison).

The secondary objective is to evaluate the efficiency and utility of the robotic system in different endoscopic transnasal approaches.

Study inclusion criteria are:
- Patients undergoing endoscopic skull base surgery during the study period

Study exclusion criteria are:
- Minor patients and patients unable to provide consent.

The study is ongoing and 21 robotic endoscopic transnasal procedures have been performed. Preliminary data document that surgical set-up is always <10 min and can be performed during other surgical set-up maneuvers. The robot is compact and does not hinder the classic access to the patient; it can be prepared at the beginning of surgery and can be "stored" in a safe position before the robotic phase is commenced (Fig. 6). Subjective data include a significant advantage while holding angled endoscopes and when holding either endoscopes close to the target and in confined spaces (e.g., removing a craniopharyngioma during an extended, transtuberculum approach).

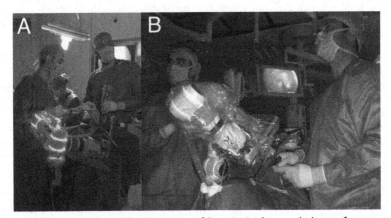

Fig. 6 Clinical evaluation at the University of Brescia. In the nasal phase of surgery, the surgeon holds the endoscope and the robot is positioned at the patient's head, away from the surgical field (A). In the neurosurgical phase, the endoscope is held by Endoscope Robot® and positioned inside the patient's right nasal cavity. The primary surgeon performs a bimanual procedure, while controlling the robot with the foot pedal (B).

6 Conclusions

Many robotic prototypes dedicated to endoscopic skull base surgery have been described, but their application has been limited by large dimensions, inefficient control, and limited precision. Only recently, a hybrid robotic system has become available for clinical practice in endoscopic transnasal skull base surgery. Preclinical and clinical data are extremely promising, but will need to be confirmed by further studies.

References

Arnholt, J.L., Mair, E.A., 2002. A 'third hand' for endoscopic skull base surgery. Laryngoscope 112, 2244–2249.

Blanco, R.G., Boahene, K., 2013. Robotic-assisted skull base surgery: preclinical study. J. Laparoendosc. Adv. Surg. Tech. A 23, 776–782.

Bolzoni Villaret, A., Doglietto, F., Carobbio, A., Schreiber, A., Panni, C., Piantoni, E., Guida, G., Fontanella, M.M., Nicolai, P., Cassinis, R., 2017. Robotic transnasal endoscopic skull base surgery: systematic review of the literature and report of a novel prototype for a hybrid system (Brescia Endoscope Assistant Robotic holder). World Neurosurg. 105, 875–883.

Cabuk, B., Ceylan, S., Anik, I., Tugasaygi, M., Kizir, S., 2015. A haptic guided robotic system for endoscope positioning and holding. Turk. Neurosurg. 25, 601–607.

Carrau, R.L., Prevedello, D.M., De Lara, D., Durmus, K., Ozer, E., 2013. Combined transoral robotic surgery and endoscopic endonasal approach for the resection of extensive malignancies of the skull base. Head Neck 35, E351–E358.

Castelnuovo, P., Pistochini, A., Locatelli, D., 2006. Different surgical approaches to the sellar region: focusing on the "two nostrils four hands technique" Rhinology 44, 2–7.

Chan, J.Y., Leung, I., Navarro-Alarcon, D., Lin, W., Li, P., Lee, D.L., Liu, Y.H., Tong, M.C., 2016. Foot-controlled robotic-enabled endoscope holder for endoscopic sinus surgery: a cadaveric feasibility study. Laryngoscope 126, 566–569.

Chowdhry, S.A., Cohen, A.R., 2013. Intraventricular neuroendoscopy: complication avoidance and management. World Neurosurg. 79, S15 e1–10.

Dixon, B.J., Daly, M.J., Chan, H., Vescan, A., Witterick, I.J., Irish, J.C., 2014. Augmented real-time navigation with critical structure proximity alerts for endoscopic skull base surgery. Laryngoscope 124, 853–859.

Doglietto, F., Prevedello, D.M., Jane Jr., J.A., Han, J., Laws Jr., E.R., 2005. Brief history of endoscopic transsphenoidal surgery—from Philipp Bozzini to the First World Congress of Endoscopic Skull Base Surgery. Neurosurg. Focus 19, E3.

Doglietto, F., Ferrari, M., Mattavelli, D., Belotti, F., Rampinelli, V., Kheshaifati, H., Lancini, D., Schreiber, A., Sorrentino, T., Ravanelli, M., Buffoli, B., Hirtler, L., Maroldi, R., Nicolai, P., Rodella, L.F., Fontanella, M.M., 2018. Transnasal endoscopic and lateral approaches to the clivus: a quantitative anatomic study. World Neurosurg. 113, e659–e671.

Eichhorn, K.W., Bootz, F., 2011. Clinical requirements and possible applications of robot assisted endoscopy in skull base and sinus surgery. Acta Neurochir. Suppl. (109), 237–240.

Eskandari, R., Amini, A., Yonemura, K.S., Couldwell, W.T., 2008. The use of the Olympus EndoArm for spinal and skull-based transsphenoidal neurosurgery. Minim. Invasive Neurosurg. 51, 370–372.

Friedrich, D.T., Scheithauer, M.O., Greve, J., Hoffmann, T.K., Schuler, P.J., 2017. Recent advances in robot-assisted head and neck surgery. Int. J. Med. Robot.. 13.

Jimenez, D.F., 2019. Endoscopes and instrumentation. In: Jimenez, D.F. (Ed.), Endoscopic Neurological Surgery. The Health Sciences Publisher.

Kupferman, M.E., Hanna, E., 2014. Robotic surgery of the skull base. Otolaryngol. Clin. North Am. 47, 415–423.

M-TRAC®, 2019. (Online). Available from: https://www.bbraun.com/en/products/b/m-trac.html (Accessed 25.04.2019).

Nathan, C.O., Chakradeo, V., Malhotra, K., D'agostino, H., Patwardhan, R., 2006. The voice-controlled robotic assist scope holder AESOP for the endoscopic approach to the sella. Skull Base 16, 123–131.

Neuroarm, 2019. (Online). Available from: https://www.rumed.ch/wp-content/uploads/2016/09/adeor_Micro_Instruments.pdf (Accessed 25.04.2019).

Nimsky, C., Rachinger, J., Iro, H., Fahlbusch, R., 2004. Adaptation of a hexapod-based robotic system for extended endoscope-assisted transsphenoidal skull base surgery. Minim. Invasive Neurosurg. 47, 41–46.

Ogiwara, T., Goto, T., Nagm, A., Hongo, K., 2017. Endoscopic endonasal transsphenoidal surgery using the iArmS operation support robot: initial experience in 43 patients. Neurosurg. Focus 42, E10.

O'malley Jr., B.W., Weinstein, G.S., 2007. Robotic anterior and midline skull base surgery: PRECLINICAL investigations. Int. J. Radiat. Oncol. Biol. Phys. 69, S125–S128.

Paraskevopoulos, D., Roth, J., Constantini, S., 2016. Endoscope holders in cranial neurosurgery: part I—technology, trends, and implications. World Neurosurg. 89, 343–354.

Paraskevopoulos, D., Constantini, S., Bal, J., Roth, J., 2018. Endoscope holders in cranial neurosurgery: part 2—an international survey. World Neurosurg. 111, e632–e643.

Point Setter, 2015. User Manual (Online). https://mitakausa.com/wp-content/uploads/2018/08/PS-manual-150115-final-ver.pdf (web archive link, 25 April 2019). Available from: https://mitakausa.com/wp-content/uploads/2018/08/PS-manual-150115-final-ver.pdf (Accessed 25.04.2019).

Schneider, J.S., Burgner, J., Webster 3rd, R.J., Russell 3rd, P.T., 2013. Robotic surgery for the sinuses and skull base: what are the possibilities and what are the obstacles? Curr. Opin. Otolaryngol. Head Neck Surg. 21, 11–16.

Singh, R., Baby, B., Damodaran, N., Srivastav, V., Suri, A., Banerjee, S., Kumar, S., Kalra, P., Prasad, S., Paul, K., Anand, S., Kumar, S., Dhiman, V., Ben-Israel, D., Kapoor, K.S., 2016. Design and validation of an open-source, partial task trainer for endonasal neuro-endoscopic skills development: Indian experience. World Neurosurg. 86, 259–269.

Strauss, G., Hofer, M., Kehrt, S., Grunert, R., Korb, W., Trantakis, C., Winkler, D., Meixensberger, J., Bootz, F., Dietz, A., Wahrburg, J., 2007. HNO 55, 177–184. Manipulator assisted endoscope guidance in functional endoscopic sinus surgery: proof of concept.

Swaney, P.J., Croom, J.M., Burgner, J., Gilbert, H.B., Rucker, D.C., Webster, R.J., Weaver, K.D., Russell, P.T., 2012. Design of a quadramanual robot for single-nostril skull base surgery. In: 387–393. ASME 2012 5th Annual Dynamic Systems and Control Conference Joint With the JSME 2012 11th Motion and Vibration Conference. American Society of Mechanical Engineers.

Swaney, P.J., Gilbert, H.B., Webster 3rd, R.J., Russell 3rd, P.T., Weaver, K.D., 2015. Endonasal skull base tumor removal using concentric tube continuum robots: a phantom study. J. Neurol. Surg. B: Skull Base 76, 145–149.

Takeshita, N., Phee, S.J., Chiu, P.W., Ho, K.Y., 2018. Global evaluative assessment of robotic skills in endoscopy (GEARS-E): objective assessment tool for master and slave transluminal endoscopic robot. Endosc. Int. Open 6, E1065–E1069.

Trevillot, V., Garrel, R., Dombre, E., Poignet, P., Sobral, R., Crampette, L., 2013a. Robotic endoscopic sinus and skull base surgery: review of the literature and future prospects. Eur. Ann. Otorhinolaryngol. Head Neck Dis. 130, 201–207.

Trevillot, V., Sobral, R., Dombre, E., Poignet, P., Herman, B., Crampette, L., 2013b. Innovative endoscopic sino-nasal and anterior skull base robotics. Int. J. Comput. Assist. Radiol. Surg. 8, 977–987.

Unitrac®, 2019 (Online). Available from: https://www.aesculapusa.com/assets/base/doc/DOC905_Unitrac_Retraction_and_Holding_Brochure.pdf (Accessed 25.04.2019).

Wurm, J., Dannenmann, T., Bohr, C., Iro, H., Bumm, K., 2005. Increased safety in robotic paranasal sinus and skull base surgery with redundant navigation and automated registration. Int. J. Med. Robot. 1, 42–48.

Xia, T., Baird, C., Jallo, G., Hayes, K., Nakajima, N., Hata, N., Kazanzides, P., 2008. An integrated system for planning, navigation and robotic assistance for skull base surgery. Int. J. Med. Robot. 4, 321–330.

Yoon, H.-S., Oh, S.M., Jeong, J.H., Lee, S.H., Tae, K., Koh, K.-C., Yi, B.-J., 2011. Active bending endoscope robot system for navigation through sinus area. In: 2011 IEEE/RSJ International Conference on Intelligent Robots and Systems. IEEE, pp. 967–972.

Zappa, F., Mattavelli, D., Madoglio, A., Ferrari, M., Rampinelli, V., Fontanella, M., Nicolai, P., Doglietto, F., PR Group, 2019a. Hybrid robotics for endoscopic skull base surgery: PEER preclinical evaluation of surgeons' first impression. (Submitted for publication).

Zappa, F., Mattavelli, D., Schreiber, A., Taboni, S., Ferrari, E., Madoglio, A., Belotti, F., Fontanella, M., Nicolai, P., Doglietto, F., 2019b. Hybrid robotics for endoscopic skull base surgery: single-centre case series. (Submitted for publication).

Further reading

Telescopes, 2019. Visualization and Documnetation Systems (Online). Available from: https://www.karlstorz.com/cps/rde/xbcr/karlstorz_assets/ASSETS/2142150.pdf (Accessed 25.04.2019).

CHAPTER NINE

Strategies for mimicking the movements of an upper extremity using superficial electromyographic signals

J. Antonio Ruvalcaba[a], R. Muñoz[a], A. Altamirano[b], A. Vera[a] and L. Leija[a]

[a]Centro de Investigación y de Estudios Avanzados del Instituto Politécnico Nacional, Mexico, Mexico
[b]Departement de Recherche, Institution Nationale des Invalides CERAH—Ministere des Armees, Woippy, France

1 Introduction

The term prosthesis is a generic name for any device that replaces a part of the body (Ramírez García, 2011). With the Second World War came a technological upturn and with current technology, we have the possibility of making compact systems with physical characteristics close to the part of the body that needs to be replaced, and of using myoelectric signals for the voluntary control of prosthetic joints (Escudero Uribe, 2002). The first prostheses had a purely aesthetic use. The new prostheses were designed and implemented to restore a percentage of the functionality of the lost limb, as well as "proprioception." This was achieved using control techniques such as the use of EMG (electromyographic) signals (Moreno Pérez, 2010).

Transhumeral prostheses currently on the market have two to three degrees of freedom (DOF), flexion-extension, pronation-supination, opening-hand closure, however, their human equivalent has between seven and eight DOF (Feng et al., 2005). The electromechanical actuators available for commercial prostheses are assembled to work sequentially. This simplifies the control methods applied to the mechanisms. However, it limits the load capacity of the system. To solve this problem and try to increase the DOF of the prosthesis to improve its performance, the actuators are implemented in parallel (Grant, 2007).

In the 1960s, the German company Otto Bock developed a myoelectric control for forearm prosthesis. In the same years, the Massachusetts

Control Systems Design of Bio-Robotics and Bio-mechatronics with advanced applications
https://doi.org/10.1016/B978-0-12-817463-0.00009-5

245

technology developed the *Boston Elbow*, a one DOF myoelectric elbow, later marketed by Liberty Mutual (Moreno Pérez, 2010). In the 1970s, the University of Chicago developed the *Utah Arm*, later marketed by Motion Control. In 2005, the American government organization Defense Advances Orthotics and Prosthetics Association (AOPA) announced the prosthetic research program "Revolutionizing prosthetics" (Moreno Pérez, 2010).

Prostheses are classified as passive, those that only simulate in appearance the lost limb, but without any functionality other than its aesthetic appearance, and active prostheses, which provide functionality to the joints. This type of prosthesis is propelled by the body or batteries (Ramírez García, 2011). The higher the level of amputation, the more complex the prosthetic systems become, as well as the control method (Scott and Parker, 1988).

A major challenge is related to the naturalness of the movements developed by the prosthetic devices. In an active prosthesis, the more joints you have, the greater the number of inactive motors you will have. This represents a load for the system because the motors that are not contributing to the movement generate an excess load to the system (Escudero Uribe, 2002). Commercial prostheses such as the Utah Arm and the Edinburgh Modular Prosthesis have a motor for pronation; however, that motor becomes a load for the bending motor (Jacobsen et al., 1982) (Gow et al., 2001). It is for this reason that prostheses that use parallel mechanisms allow the generated forces to converge in the same point or mechanical link (Escudero Uribe, 2002). The DOF are determined by the amount of control signals that can be generated and coordinated voluntarily. The current prostheses on the market for amputations above the elbow have only up to three DOF (Escudero Uribe, 2002). The prostheses for amputations above the elbow work sequentially, that is, they cannot perform more than one simultaneous joint movement. An important challenge in the development of prosthetics is to achieve naturalness of movement (Ramírez García, 2011).

There have been several projects developed at the Bioelectronic Department at CINVESTAV-IPN in Mexico City related to the subject of EMG signal analysis for prosthetic control purposes.

In 2002, the design and construction of a new prosthetic mechanic device was proposed to replace an up the elbow limb with parallel topology of linear actuators in which at least two actuators will work simultaneously in each one of the three active movements of the elbow, while the actuator that does not participate in the movement gives structural support (Escudero Uribe, 2002).

In 2003, a system for identifying seven hand movements from superficial electromyographic (sEMG) signals registered over the forearm was developed (León Ponce, 2003). They use artificial neural networks (ANN) as classifier to identify patterns that related the EMG signal with the final position of the hand. In 2012, a system was proposed for identifying 27 movements of the upper limb for the simultaneous control of three DOF of a virtual robotic arm (León Ponce, 2012). Using offline analysis of wrist movements using eight channels for acquisition, features from both time and frequency domain were extracted and three classifiers (linear discriminant analysis (LDA), ANN, and support vector machines (SVM)) were used.

In 2010, a virtual training system was developed for the amputee to become familiar with a myoelectric control prosthesis (Barraza Madrigal, 2010). This computer interface simulates an upper limb prosthesis for an above elbow amputee. It was developed in MATLAB® and the control algorithm was generated from a single differential EMG channel from which three levels of force were defined. The user was able to practice and identify each of the force levels generated by the contraction of the remaining muscle in his or her stump by receiving a visual feedback from the interface. In 2016, an ambulatory system was developed for the analysis of upper arm movements (Barraza Madrigal, 2010). This system provides information referring to the position and orientation of the upper limb, due to the shoulder rotation. This system makes it possible to monitor, reproduce, and follow the movement of the shoulder articulation; this is done by measuring the range of the movement described by the upper limb during its active elevation.

In 2010, the design and construction of a movement control system was proposed for a trans-humeral prosthesis with four active DOF (Moreno Pérez, 2010). The control hardware consists of four "cycloconvertors circuits," a "power board," and a "main board," with an embedded 16 bits DSP microcontroller.

In 2011, an automatic system to decompose myoelectric signals was developed using wavelets analysis and support vectors machines for classification processes (Márquez Lázaro, 2011). This system was able to classify the signals when a simulated noise was added.

In 2011, an upper limb prosthesis was proposed, the movement of which would use actuators with a set of angles measured during different activities, such as drinking water, waving the hand, answering the phone, opening a door, and serving a glass of water from a jar (Ramírez García, 2011).

In 2012, a prototype of an artificial hand was proposed with five DOF, with anthropomorphic and anthropometric dimensions able for grasping and moving a handle (Altamirano Altamirano, 2012). The author proposed a modular prosthesis that can be attached to different types of prostheses and connectors. The work is based on the anatomical and biomechanical features and functions of the hand to perform grasping movements performed by the fingers. In 2017, the author proposed a process to analyze multichannel sEMG signals using wavelet transform, Hilbert-Huang transform and methods like Kalman and Göertzel filters as techniques to detect, measure, filter, and decompose these sEMG signals to identify patterns in time, frequency, space, or a combination, for flexion–extension movements of the fingers of the hand using link-fingers superficial muscles in order to predict hand movements and with this, reduce computing time (Altamirano Altamirano, 2017).

In 2014, a novel classification of upper limb movements was proposed by analyzing the transient state of the muscle contraction using Hjorth's parameters (Pla Mobarak, 2014). With this proposal they manage to identify movements with classification accuracy higher than 95%, suggesting the existence of highly relevant info in the dynamic part of the muscle contraction as to be able to propose myoelectric control schemes from its analysis.

In 2015, developed an active electrode was developed capable of acquiring superficial EMG signal in a different way using a double reference proposal for noise suppressing (Ruvalcaba Granados, 2015).

In 2017, an ambulatory system was developed for analyzing the movement of an upper limb (Contreras Rodríguez, 2017). The author designed and fabricated sensing cards involving inertial sensors and a DsPic for data acquisition and processing. They also implemented a radio frequency (RF) communication to send the data between each one of the sensing cards—this in order to establish it as an ambulatory system.

In this chapter, a review of several works developed in the bioelectronic department at CINVESTAV-IPN in Mexico City is described. These works aim to develop a novel ambulatory active electrode for superficial EMG signal acquisition systems in order to obtain a robust database of the forearm signals while doing wrist and fingers movements in order to produce cleaner and faster signal processing. A work related to feature extraction using Hilbert-Huang transform and Kalman and Göertzel filters as classifiers in order to obtain prediction of several fingers' movements is reported. These researches have the aim of reducing the time consumed since the signal acquisition until the signal interpretation, and thus, reduce the delay that a prosthesis takes to respond to a command.

2 Process of prostheses control

In this section there are described many different types of prosthetic devices developed for the replacement of an upper limb. In addition, there are described some control strategies that regulate the movement of these prostheses that use sEMG.

2.1 Type of prostheses

The prostheses are classified into two types: active prostheses (with independent energetic capacity); and passive prostheses (they require muscle energy to function) (Moreno Pérez, 2010).

They can also be classified based on their functions and characteristics:
- Cosmetic prosthetics: This is a type of passive prosthesis. Its main function is "cosmetic restoration," and is usually constructed of PVC, latex and/or silicone. The functionality of this type of prosthesis is limited and has a short lifespan (Moreno Pérez, 2010).
- Body powered: This type of prosthesis uses harnesses and cables and the device is controlled through the movement of the body.
- Electrical: In this type of prosthesis the joints are activated by small electric motors controlled by myoelectric signals generated during a muscular contraction of the remaining limbs.
- Hybrid: This type of prosthesis combines the use of electric motors with the harnesses and straps used in body powered.

2.2 Types of prostheses control

There are several forms of control for active prostheses:
- Body power: They are the simplest control options, controlled by cables. This type of control requires a great effort for action, but in turn provides "proprioceptive" feedback. Through cable control, terminal devices and active elbows can be controlled (Doeringer and Hogan, 1995).
- When using electrically driven prostheses with myoelectric control, the user is free of straps and harnesses.

2.3 Protocol for prosthesis control using sEMG signals

The effectiveness of classification algorithms based on EMG signals and pattern recognition is based on an effective implementation of three modules

that should be considered when selecting an optimal vector of characteristics (Ruiz–Olaya et al., 2015):
- preprocessing;
- feature extraction; and
- pattern classification.

The implementation of an algorithm for pattern recognition based on the EMG signal has the following difficulties (Ruiz–Olaya et al., 2015):
- The EMG signal is variable in time and highly nonlinear.
- The EMG signal is highly affected by noise, such as ECG cross–talk, electromagnetic induction of the line, and movement of the cables and the arm (movement artifacts).
- There are factors such as interference of surrounding muscles, physiological conditions (fatigue), and impedance of the skin, among others.
- The level of activity of each muscle for a certain movement is different for each individual.

The sEMG signal can be used to activate or continuously control some external assistance devices including prosthesis and electric orthoses (Farina and Negro, 2012). The control of a device through EMG can be achieved through two main types of sEMG signal processing (Song et al., 2013):
- Data–driven machine learning: The idea is to associate a desired control command or some observed action with the underlying EMG signal patterns (Farina et al., 2014; Matrone et al., 2012; Muceli et al., 2014).
- Model driven approach: Uses the superficial EMG signal as input for specific, physically correct models of the skeletal muscle system.

These are used to reconstruct all neurophysiological and musculoskeletal transformations from the beginning of the EMG signal to the body function.

3 Electrodes for sEMG signals acquisition

In this section, different types of electrodes (wet and dry) used in the sEMG signal recordings are presented. A method is proposed for the design and testing of an active electrode sensor for the acquisition of sEMG signals using dry electrodes.

3.1 Types of electrodes

There are two main types of electrodes used to detect the electromyographic signal: surface and insertable (needle).

3.1.1 Surface electrodes

There are two types of surface electrodes: passive and active. Passive electrodes consist of a conductive surface that detects the current in the skin through the electrode/skin interface. The main characteristic of active electrodes is that they have an electronic amplifier with high input impedance in the same housing as the electrodes. A basic example of a passive electrode consists of a silver disk attached to the skin. By using conductive gel, the electrical conductivity at the electrode/skin interface is improved.

Passive electrodes

There are two types of electrodes: perfectly polarizable and perfectly nonpolarizable. This refers to what happens to an electrode when a current flows between it and the electrolyte.
- Perfectly polarizable: those in which no load crosses the electrode/electrolyte interface when a current is applied. The electrode behaves like a capacitor.
- Perfectly nonpolarizable: those where the current passes freely through the electrode-electrolyte interface, without requiring energy to carry out the transmission.

The silver/silver chloride electrode (Ag/AgCl) is a perfectly nonpolarizable electrode, consisting of a metal covered with a layer of ionic compound poorly soluble in this metal with a suitable anion. The solution is immersed in an electrolyte containing the anion in a relatively high concentration (Webster, 2009).

Active electrodes

For many years, systems for biopotentials based on dry electrodes have been developed and marketed (Burke and Gleeson, 2000). However, these can be polarized and have a high impedance at the electrode/skin interface, causing variation in contact electrode resistance and polarization potential (Zipp and Ahrens, 1979). The wet Ag/AgCl electrodes have a variable impedance between 1 and 100 kΩ (Huigen et al., 2002; Grimnes, 1983). This imbalance is produced when the common mode voltage present in the body interferes with the differential measurement, and an interfering differential voltage occurs due to currents coupled by electromagnetic interference (EMI) from the cables (Huhta and Webster, 1973). This problem can be solved using active buffers of unitary gain connected directly to the dry electrodes; this produces a high input impedance in the electrodes and low output impedance to drive the coupled currents to ground, avoiding the use of meshed cables (Fernandez and Pallas-Areny, 1997; MettingVanRijn et al., 1996).

3.1.2 Electrodes configuration

The electrical activity on the surface of the skin is acquired by placing an electrode with a single detection surface with respect to a reference electrode located in an electrically neutral zone. This type of electrodes is known as monopolar, used in the clinical area due to its simplicity. This configuration has the disadvantage of detecting any signal adjoining the electrode, including unwanted signals.

The bipolar configuration compensates for this limitation; in this case, two electrodes are used to detect two potentials in the muscle of interest, each one with respect to the reference electrode. Subsequently, the two signals are fed to a differential amplifier, which will amplify the difference of both signals thus eliminating any common mode component. In the clinical setting, the skin is often degreased using cotton wool with alcohol before the placement of the electrode, which helps to remove some of the dead, loose cells external to the stratum corneum and lipid substances of low conductivity on the skin surface (Chi et al., 2010). When a wet gel for electrodes is applied to the skin, previously cleaned with alcohol, the gel penetrates the degreased skin quickly, once the electrode has been in contact with the skin for several minutes, causing a rapid decrease in the impedance of the skin and a decrease of movement artifacts (Xu et al., 2017).

3.2 Active electrodes

3.2.1 Introduction

The failure of a chemical equilibrium in the metal/electrolyte junction causes a polarization potential, which affects the amount of current flowing to the electrode. The active electrodes were developed to eliminate the need to preprepare the skin and the conductive medium. These electrodes are also known as dry or paste-less electrodes. These electrodes are not reliable for long-term acquisition because their dielectric properties are susceptible to change due to the effects of perspiration and to erosion of the substance of the dielectric. To improve the prediction of activation and muscle strength using sEMG signals, it is important to reduce unwanted variabilities, so that real muscle activity can be best represented by the acquired signals. Previous research has contributed to the optimization of technical aspects, instrumentation (Clancy et al., 2002; Godin et al., 1991; Klijn and Kloprogge, 1973;

Nishimura et al., 1992; Webster, 1984), and of appropriate filtering techniques (Webster, 2006; Clancy and Hogan, 1995).

Like the myoelectric prostheses, a good signal acquisition, even cleaning it at the moment of processing it, depends on the electrodes and the type of electrode used and a system that delivers EMG signals in a clean way and with good amplitude. When designing an electrode, the electronic processing device is usually the first to be developed, whereas the electrode design for the sensing is left to the end, this is a design error (Coughlin and Driscoll, 1987). It is important that the acquired signals are clear, without distortion, and free from artifacts. When monitoring the electrodes, if they are not carefully chosen, this will cause significant problems, which make the signal obtained difficult to analyze (Coughlin and Driscoll, 1987).

A serious problem occurs due to the change of the contact impedance due to the frequency. The dependence of the frequency on the contact impedance is a consequence of the parasitic capacitance at the electrode/electrolyte interface (Coughlin and Driscoll, 1987). The electrode/skin impedance is highly affected by the preparation of the skin and is inversely related to the surface of the electrodes. Rubbing the skin with abrasive surgical soap is, without a doubt, the best treatment to reduce the electrode/skin impedance; rubbing with alcohol is not very effective. Electrode noise is generally the most important source of noise in the EMG signal register and is, therefore, the limiting factor for detecting very small potentials (De Luca et al., 2006). Noise is affected by the treatment of the skin. A small abrasion of the skin or epilation with adhesive cloth reduces electrode/skin impedance, noise, DC voltages, and movement artifacts (De Luca et al., 2006).

Some wearables systems for the acquisition of superficial EMG signals use conductive gel to have a better conduction at the electrode/skin interface. Some other authors develop their own dry electrodes, to avoid the constant use of gel; however, they still use a commercial system for data acquisition (DAQ). Other authors have recently developed novel flexible electrodes, coupled to a flexible acquisition system. However, the manufacture of these devices is still very complex and expensive.

3.2.2 Dry electrodes

In 2015, a dry brass electrode was designed (Ruvalcaba Granados et al., 2015); due to its low level of oxidation (Perry et al., 1997) this is a good conductor and, because it is a base material, to be plated in gold for future use. A dry electrode is proposed due to the limitations and disadvantages that

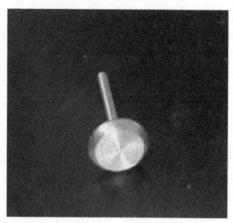

Fig. 1 Machined brass electrode.

derive from the use of wet electrodes, such as signal distortion in sEMG signal acquisitions for long periods of time; this is because the electrolytic gel begins to dry, causing the quality of the acquired signal to become deficient over time.

The proposed dry brass electrode is shown in Fig. 1. The electrode is machined from a brass bar of 9.6 mm in diameter; the head has a thickness of 1.5 mm and a 2 cm long by 0.5 mm in diameter stem (this stem is used to connect it to the acquisition board).

The electrode/skin impedance of the brass electrodes was measured by the 3-electrode method (Salazar Muñoz, 2004). This measurement was also made with the Ag/AgCl electrodes in order to compare the result with a commercial wet electrode. The measurement frequency range varies from 5 to 2 kHz. The results are shown in the graph of Fig. 2.

The highest impedance in the Ag/AgCl electrodes is observed in the lowest frequency and its value decreases as the frequency increases. Its impedance range is from 134 kΩ–5 Hz to 3.33 kΩ–2 kHz, having an impedance of 10.54 kΩ at 450 Hz, which is the cutoff frequency of the system. As for the brass electrode, it has an impedance range of 220 kΩ–5 Hz at 4.32 kΩ–2 kHz and having an impedance of 14.81 kΩ at the cutoff frequency of 450 Hz. The values obtained for the brass electrodes, which are very close to those obtained in the Ag/AgCl electrodes, showing good electrode/skin impedance.

In 2017, conductive cloth electrodes were used as dry acquisition sensors, with dimensions of 20 mm of diameter and distribution of three

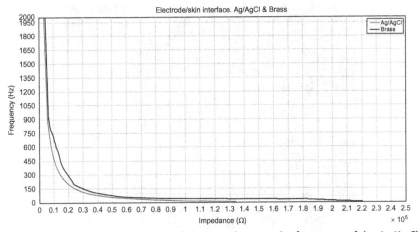

Fig. 2 Relationship of the response of the impedance to the frequency of the Ag/AgCl and brass electrodes.

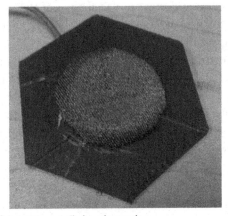

Fig. 3 Conductive fabric 20 mm Ø dry electrode.

electrodes along the long palmar muscle in a row, with an IED (inter-electrode distance) of 25 mm (Fig. 3) (Ruvalcaba Granados et al., 2017a, 2017b).

3.2.3 Signal acquisition system
Signal acquisition systems are used in sports medicine to measure and monitor the performance of athletes in order to improve their physical performance. It is also used to measure the evolution in the increase of the

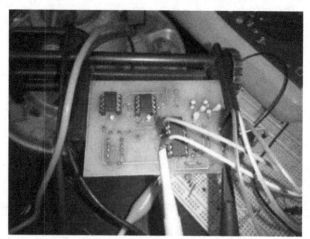

Fig. 4 Circuit board of the EMG acquisition system.

strength of contraction and resistance to fatigue of the muscles and, in this way, to corroborate the progress, success, or failure of a specific sports training plan, or for a plan of rehabilitation after an injury.

In 2014, the electronic design of a system for acquiring sEMG signals was proposed (Fig. 4) (Ruvalcaba Granados et al., 2014). Two INA128UA instrumentation amplifiers were used in cascade due to their high common mode rejection rate (CMRR)—120 dB—and cascaded to not saturate the amplification, of 100 and 10, respectively. An active high-pass filter in integrator configuration is used, whose feedback is connected to the INA128 reference terminal to reduce the offset voltage. The cut-off frequency of the filter is calculated at 20 Hz. The next circuit is a low-pass, second-order filter, Sallen-Key type of unit gain with a cut-off frequency calculated at 400 Hz. After the buffer follows a passive filter of first-order passes high, with a cutoff frequency at 0.05 Hz for electrode coupling with the skin. The acquisition system is powered with ±9 Vcd by two rechargeable 9 V batteries of 1300 mAh. The first electrodes used to test the system and acquire the electromyographic signal were disposable wet electrodes for the 3M brand electrocardiogram. A differential acquisition is used to obtain the corresponding muscle signal between two electrodes. For the configuration of the reference electrodes, a proposal is used to place them near the acquisition electrodes, in order to suppress the noise to the acquisition electrodes and suppress the noise that surrounds the muscle section that is to be sampled; this configuration is shown in Fig. 5.

Fig. 5 Electrodes array proposal.

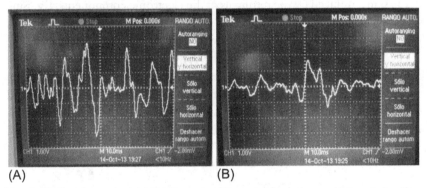

(A)   (B)

Fig. 6 (A) Signal acquired at the longus palmar muscle during a strong contraction.
(B) Signal acquired at the longus palmar muscle during a light contraction.

Fig. 6A and B shows the recording of two types of signals to verify the correct operation of the sEMG signal acquisition system. Two differential electrodes of the long palmar muscle and two reference electrodes were placed at 90° to the acquisition electrodes. Using an oscilloscope, two types of signals were recorded: a light contraction (slight flexion of the wrist) and a strong contraction of the muscle (strong flexion of the wrist).

Using a signal generator and an oscilloscope, the CMRR of the complete sEMG signal acquisition system was measured. Fig. 7 shows a final version of the first circuit board developed, changing all electronic components for surface components. This is done with the purpose of reducing the noise

Fig. 7 Final design of the EMG acquisition system using surface elements.

of the signal induced by the terminals of the electronic components. By using surface-mounted devices, it helps to reduce the size of the electronic circuit, in this case, going from an 8 cm x 6 cm board to a 3 cm diameter board.

In 2017, the authors added to the acquisition system a noninverting amplifier adder to send the negative part of the EMG signal to positive and thus be able to transmit the complete signal (both the negative and the positive part) to the A/D converter (Fig. 8) (Ruvalcaba Granados et al., 2017a). They added a 2.2 Vdc voltage to the input of this system.

Fig. 8 sEMG acquisition system with miniaturized electronics.

The output of this circuit is the input of an amplification stage with a gain of 2, before being sent to the A/D converter to compensate for the loss of amplitude due to the adder.

The manufacture of a circuit for the transmission of data is also described. An ATMEGA328-P microcontroller is used for A/D conversion and serial transmission with the computer. A USB–serial driver FT232-RL is also used to make the serial connection through the USB port. The system is powered by two 9 V batteries (± 9 V) and the microcontroller by a +5 V battery.

Figs. 9 and 10 show two signals corresponding to a wrist flexion, Ag/AgCl electrodes and conductive fabric electrodes, placed in the long palmar muscle. This to compare the signals obtained with the new dry electrode and a commercial wet electrode. The CMRR of the signal acquisition system was also measured, obtaining a CMRR of 94 dB.

In A. Ruvalcaba et al., using the circuits and electrodes described in the previous work (Ruvalcaba Granados et al., 2017a), the authors recorded on the computer—using the acquisition system described in (Ruvalcaba Granados et al., 2017b)—and compared the signal obtained when performing an elbow flexion, with the electrodes connected to the biceps (Ruvalcaba Granados et al., 2017b). The signals obtained between the dry electrodes of conductive fabric and the commercial wet Ag/AgCl

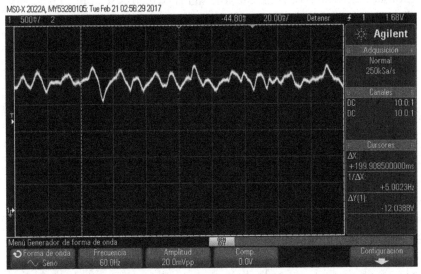

Fig. 9 Long palmar muscle response to a wrist flexion using the dry electrodes wearable system. The signal has a 2.2 Vdc offset in order to send it by the analog input to the microcontroller.

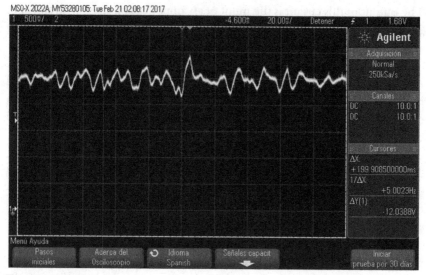

Fig. 10 Long palmar muscle response to a wrist flexion using Ag/AgCl wet electrodes wearable system. The signal has a 2.2 Vdc offset in order to send it by the analog input to the microcontroller.

electrodes are compared, observing that it is possible to use the dry electrodes of conductive fabric to acquire EMG surface signal. However, they produce an offset at the time of recording the signal, which is seen in Fig. 11; this offset is corrected with postprocessing techniques.

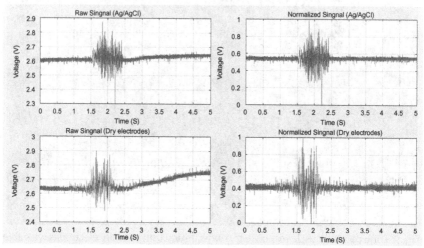

Fig. 11 Elbow flexion registered at the bicep muscle using (A) wet electrodes and (B) dry electrodes. Raw signal and normalized signal.

3.2.4 Active electrodes developed

By using dry electrodes on unprepared skin, the impedance of the electrode/skin interface becomes relatively high and variable, making them susceptible to EMI. This problem can be solved by an electronic design. The base signal obtained using dry electrodes takes more time to stabilize. This problem is solved by using active electrodes.

In 2015, a new acquisition system was designed, where the high passive filter of 0.05 Hz is eliminated because this filter does not contribute substantially to the cleaning of the signal, and removing it from the circuit saves space and electronic noise by decreasing components (Ruvalcaba Granados et al., 2015). The circuit board developed with the electronics previously described uses only surface mounted devices (Fig. 12). The circuit board measures 3.5 cm in diameter and 2 cm in height. A plastic housing was designed and built to carry and protect the system. A threaded housing was designed to reduce the size of the housing and ensure a good seal with the acquisition system.

To establish that the acquisition system is working correctly and that the signal delivered to the output of the circuit is a valid EMG signal without noise, the CMRR of the system is measured. The CMRR obtained from the system is 116 dB; way above what is established by the International

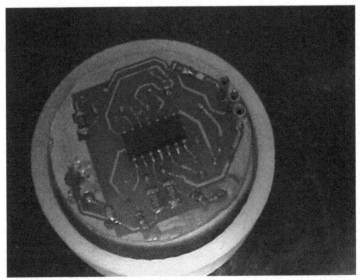

Fig. 12 Final stage of the electrodes integrated to the sEMG signal acquisition system.

Society of Electrophysiology and Kinesiology (ISEK) standards. The actual cutoff frequency is obtained, calculated according to the commercial resistance value implemented, using Eq. (1).

$$f_c = \frac{0.707}{2\pi(2.7k)(0.1\mu)} = 416.7\,Hz \tag{1}$$

A frequency sweep is performed to obtain the response of the system, graphically observing the real cutoff frequency, when the signal drops to −3 dB (Fig. 13), obtaining a real cutoff frequency of 450 Hz.

The system is powered by two 11 Vdc batteries to have a bipolar voltage. The current consumed by the device is 17.2 mA. The batteries have a load supply of 1000 mAh, therefore, calculating the discharge time, the device can operate for 58 h continuously.

An EMG signal of the muscular response of a wrist flexion was acquired with the system connected in the long palmar muscle; the signal quality and its amplitude can be observed in Fig. 14.

Finally, the characteristics of the developed system are compared with a similar commercial system (Delsys, 2018) in Table 1, obtaining similar characteristics in CMRR, type of acquisition, and input impedance, the difference between both systems is the material used in the sensing electrodes.

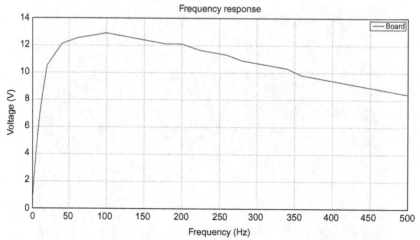

Fig. 13 Frequency response of the low-pas filter.

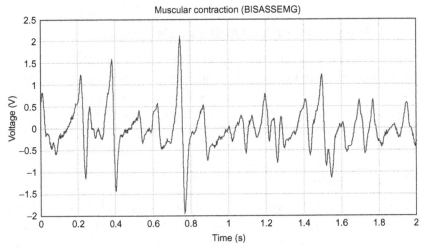

Fig. 14 Signal obtained from a high contraction from the palmar muscle using the brass electrodes integrated to the sEMG acquisition system.

Table 1 Comparison of specifications between the system proposed by A. Ruvalcaba et al. and the Bagnoli™ commercial system.

Specifications	Proposed system	Bagnoli®
Mechanical		
Type	Single differential	Single differential
Number of contacts	2	2
Contact dimensions	2 mm × 9 mm diameter	10.0 × 1.0 mm diameter
Contact spacing	20 mm	10 mm
Contact material	Brass	99.9% Ag
Detection area	29 mm^2	10 mm^2
Case dimensions	35 mm diameter × 20 mm height	41 × 20 × 5 mm
Case material	ABS	Polycarbonate
Electrical		
Bandwidth	0–450 Hz ± 30	Open
CMRR	116 dB	100 dB
Power consumption	17.2 mA	20 mW
Input impedance	>10^{10} Ω/0.2 pF	>10^{15} Ω/0.2 pF
Sensor contacts	4 brass round electrodes 2 mm × 9 mm diameter	2 silver bars 10 mm × 1 mm diameter
Sensor dimensions	30 mm diameter	19.8 mm × 5.4 × 35 mm

4 Superficial EMG (sEMG) signals

This section talks about the basics of the superficial EMG signals such as frequency range and amplitude of the signals. Further, the state of the art regarding feature extraction and classifiers meant to recognize hand movements is reported. In addition, a novel method for feature extraction (Hilbert-Huang transform) and a hand movement classifier is described (Göertzel and Kalman filters).

4.1 Introduction

Although a muscle can contain hundreds or thousands of motor units, the muscle fibers of a motor unit are not adjacent to each other but intermingled with the muscle fibers of other motor units; however, they contract and relax in parallel, thanks to the action potential generated in the myoneural junction by the impulse of the nerve fiber (Ramírez García, 2011). The sEMG signal is a complicated signal, which is affected by the anatomical and physiological properties of the muscles, the control scheme of the peripheral nervous system, and by the characteristics of the instrumentation used to detect and observe it (Basmajian and De Luca, 1985).

The electromyographic signal has become a fundamental tool to achieve control of artificial limb movement, functional electrical stimulation and rehabilitation (Castellanos Abrego, 1997). The interpretation of the instructions received by the muscles from the nervous system to induce movements are useful for the development of artificial elements such as prostheses and orthoses (Altamirano Altamirano, 2017). sEMG signals provide ample information about muscle activity in a noninvasive way; this information is really useful as a source of control. sEMG signals are very noisy. This noise may be due to the electronic system, or due to interference from electromyographic sources (Altamirano Altamirano, 2017).

The raw EMG signal is processed mainly to extract information regarding the amplitude of the signal [root mean square (RMS), average rectified value (ARV), and linear envelope (LE)] and its spectral power density (Fourier or autoregressive approximations). The mean and median frequency (MNF and MDF) are obtained from the spectral power density (Merletti and Farina, 2016). These parameters of the sEMG, provide information on the muscular contraction force and the frequency content of the signal. Processing the signal means applying algorithms to extract parameters or characteristics, which will be used for some purpose such as signal classification or quantification of changes. The estimation of the amplitude of the sEMG signal is used as the input in the control of myoelectric prostheses.

In the frequency domain, the dominant change in the EMG signal, during a high effort and sustained contraction, is a compression of the signal spectrum towards lower frequencies. The measurement of this compression is related to muscle fatigue. In the time domain, the dominant changes in the single-channel EMG signal are: (a) the modulation of the standard deviation of the signal (RMS) and (b) spectral changes due to muscular effort and/or fatigue. The spectral changes can be evaluated in the time domain with simple techniques, based on counting the zero crossings of the signal or in the analysis of peaks (Merletti and Farina, 2016).

4.1.1 EMG signal bases

The electromyographic signal is the electrical manifestation of the neuromuscular activation associated with an order of muscular contraction. The signal represents the current generated by the ionic flux through the membrane of the muscle fibers, which propagates to the surface of the electrode, which is the sensor of this EMG signal. The amplitude of the EMG signal ranges from 50 µV to 20–30 mV. The value of the amplitude of this biopotential is proportional to the intensity of the contraction (Merletti et al., 2003). The frequency range in which the EMG signal is located ranges from 20 to 450 Hz (Paul et al., 2014).

4.2 sEMG signal processing

4.2.1 Introduction

One of the areas where pattern recognition has recently been applied is in biomechanics, mainly in the detection of EMG signals that produce natural muscle movements in a person; these signals are used in the control of artificial devices that mimic natural movement (León Ponce, 2003). This involves pattern recognition based on the acquisition, analysis, and processing of the signal generated by the muscles in contraction (Miller et al., 2008; Huang et al., 2008; Hargrove et al., 2009; Chu and Lee, 2009; Simon et al., 2011; Fougner et al., 2011). The raw EMG signal has a lot of noise. This type of signal can be very useful when it is correctly analyzed, quantified, and classified (Basmajian and De Luca, 1985).

In biomedical applications, the acquisition of the signal is not enough. It is required to process the acquired signal to obtain the relevant information buried in it. This is because the information of interest is not visible inside the signal (Bronzino, 1999). A large number of signal processing methods are available, and to correctly select an algorithm, one must know the objective of the processing, the test conditions, and the characteristics of the signal (Bronzino, 1999). Recent advances in technology for signal processing and mathematical models have allowed the development of advanced techniques for the detection of EMG signals and their analysis.

Several mathematical techniques and artificial intelligence (A.I.) have received great attention. Some mathematical models include the wavelet transform (WT), time-frequency approximations, Fourier transform (FT), Wigner-Ville distribution (WVD), statistical measurements, high order statistics, and Hilbert-Huang transform (HHT). Some artificial intelligence approaches for signal recognition include ANN, dynamic recurrent neural networks (DRNN), and fuzzy logic systems (FLS). Genetic algorithms (GA) have also been used for the mapping of EMG entries for desired hand movements. Some methods, such as mean frequency (MNF), median frequency (MDF), peak frequency (PKF), median power (MNP), spectral moments or spectral frequency variance, are not good for classifying EMG signals (Phinyomark et al., 2011).

4.2.2 Signal processing techniques

In 2017, a novel proposal for the analysis of sEMG signals was reported, obtained from the forearm muscles for the prediction of flexion and extension movements of the fingers, in order to reduce the computation time (Altamirano Altamirano, 2017). The proposal consists of the use of wavelets and Hilbert-Huang transformations, as well as the use of signal analysis methods such as Kalman and Göertzel filters, used to detect, measure, filter, and decompose these signals and thus achieve the identification of movements.

The generation of a database is started here, placing Ag/AgCl surface electrodes on five muscles of the forearm for registering the movements of the subject's fingers. Using a four channel electrode array, these signals are sampled and recorded. Subsequently, the signals are normalized and previously visualized when analyzed. A database of six finger movements was obtained. The BIOPAC MP35 acquisition system with four channels was used to acquire the EMG surface signals from five healthy volunteers. The signals for the database were recorded using the software BLS PRO-321111.7 with a sampling frequency of 2 kHz and a gain of 1000. The BIOPAC system has an integrated filter IIR Chebyshev 2 of sixth order selected for a frequency range of 10–500 Hz. Each channel has a differential acquisition based on instrumentation amplifiers with an external reference. The signals of six finger movements were obtained: (1) flexion of the index finger, (2) flexion of the middle finger, (3) flexion of the ring finger and little finger, (4) flexion of the thumb, (5) closing the hand, and (6) opening the hand.

Normalization of the processed signals

Normalization is fundamental in every method of signal processing. The normalization is done by obtaining the α factor, which is reciprocal to the maximum absolute value of the myoelectric signal in each channel, in a sampling of 1000 ms. See Eq. (2).

$$V_{max_i} = \max[V_i(t)]; \quad i = 1, ..., 4\alpha_i = \frac{1}{V_{max_i}}; \quad i = 1, ..., 4i$$
$$= No.\, of\ channels \qquad (2)$$

The muscle intensity level was recorded and compared with the six movements. Three intensity levels were established: high intensity (0.7–1 V); average intensity (0.35–0.65 V); and null intensity (0–0.3 V). Fig. 15 shows the intensity map for user 1.

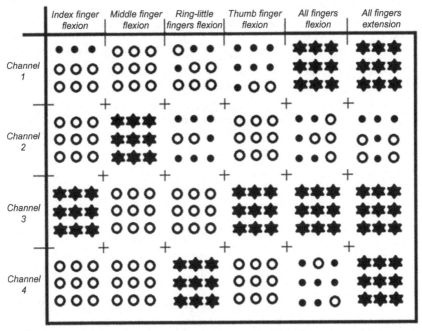

Fig. 15 Muscular contraction intensity of subject 1 for four channels versus six movements of the fingers. The map shows the intensity present in the four channels when a movement was performed: the star represents the 0.7–1 V level, the dot represents the 0.35–0.65 V level, and the circle represents 0–0.3 V. Each icon represents one of nine repetitions.

4.2.3 Feature extraction methods

With respect to stationary analysis, this consists of the methods and processes proposed to analyze the data recorded using transformed wavelets and Hilbert-Huang (Altamirano Altamirano, 2017).

Wavelet transform analysis

For the processing of EMG signals, the wavelet transform is an alternative for another time-frequency representation. While wavelet transformation provides a flexible time-frequency resolution, it suffers from a relatively low resolution in the high frequency range. The comparison was made with scalograms obtained with a Daubechies 44 wavelet (Fig. 16) and a Meyer wavelet (Fig. 17). This comparison shows that the Meyer wavelet has a better definition in time and frequency than the Daubechies 44.

Hilbert-Huang transform analysis

The purpose of the EMD (empirical mode decomposition) is to decompose a multicomponent signal (as the myoelectric signal) into a number of mono-components called IMFs (intrinsic mode functions). The EMD

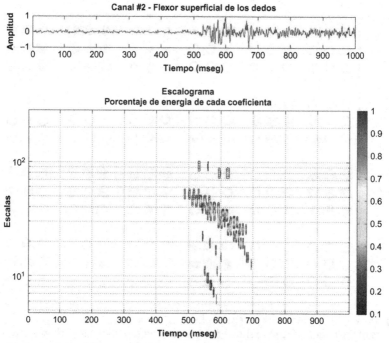

Fig. 16 Scalogram of CBA4_234 obtained with Daubechies 44 Wavelet.

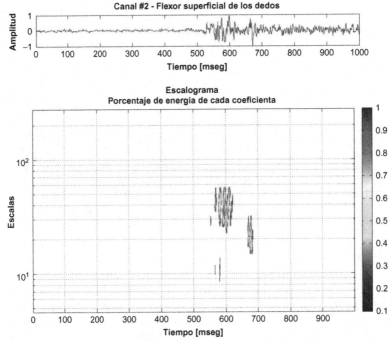

Fig. 17 Scalogram of CBA4_234 obtained with Meyer Wavelet.

signal processing technique is suitable for filtering the EMG signal. The great disadvantage of the EMD method is that it is more sensitive to the presence of noise. Fig. 18 shows the signal sEMG and its IMFs of a single channel related to the flexion and extension of the fingers (Altamirano Altamirano, 2017).

For EMD, using the first normalization, the number of IMFs obtained was between 8 and 19 per channel. For the second standardization, between 5 and 11 IMFs were obtained. Subsequently, the Hilbert-Huang transform and the instantaneous frequency were applied to each IMF obtained from each channel. The resulting data are shown in Fig. 19. The instantaneous envelopes are used to obtain the instantaneous frequencies of each IMF. The ranges of these frequencies result in 105–310 Hz. The instantaneous frequency is useful for locating significant changes in signal energy, for example, voluntary contractions and relaxation.

The main frequencies of the IMFs are in the order of 200 ± 20 Hz, detected in the first IMFs. The low frequencies detected in the first IMFs were 12, 8, and 6 Hz. The detected high energy frequencies are a group

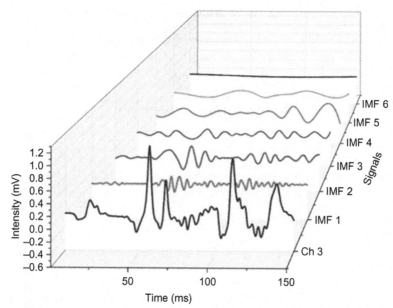

Fig. 18 sEMG signal and its IMFs of Channel 3, related to the flexion and extension of all fingers. Six levels of decomposition are shown. IMF 2 shows two motor unit action potentials (MUAPs) in 50 ms and approximately 100 ms.

of AM/FM signals with average frequencies of 83.3 Hz (73.57–85.9 Hz), 96.7 Hz (94.35–99.82 Hz), 59 Hz (58.5–61.3 Hz), and 113.3 Hz (111–117.04 Hz). By analyzing the resulting instantaneous frequencies, it is possible to identify changes in the first four IMFs, which are related to the activation of the flexion movement.

4.2.4 Classification methods
During the nonstationary analysis of the signal it is expected to find the conditions, methods, techniques, and processes to perform in short-time, or less than 100 ms of real-time processing, using Kalman and Göertzel filters to identify the characteristics or patterns of the signal myoelectric. The Kalman filter improves noise cleaning and reconstruction to predict the input signal. The output of the Kalman filter goes to the Göertzel filter, which detects specific signals using the DFT in the patterns of the model. The results of these filters could be applied directly to a prosthetic system as a control signal or be applied to a classification system, depending on the complexity of the acquisition system, channels or movements.

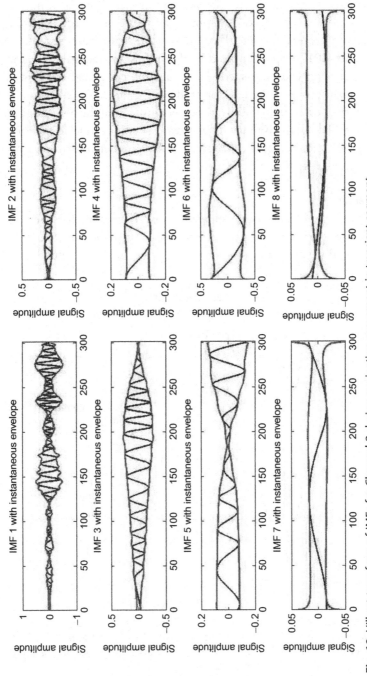

Fig. 19 Hilbert transform of IMFs for Channel 2 during supination movement in transient segment.

Kalman filter

In Altamirano's proposal, after the acquisition stage, for real-time applications, it is necessary to decode the information present in the sEMG signal. The author proposed filtering the signal to achieve this purpose (Altamirano Altamirano, 2017). Specific filtering algorithms are required to clean the noise of the signal to be treated. To reduce the computational cost of the identification process, a system for predicting states in the filtering process is necessary in order to make advanced decisions about the behavior of muscle activity. To use the Kalman filter, a representative model in the form of a state space is required (Fig. 20A and B). This model consists of a process equation and a measurement equation.

Göertzel filter

If the specific frequencies of the standard signal are known, then the Göertzel filter is able to recognize these frequencies instantaneously. In the same way, a rapid myoelectric identification can be made to detect a specific

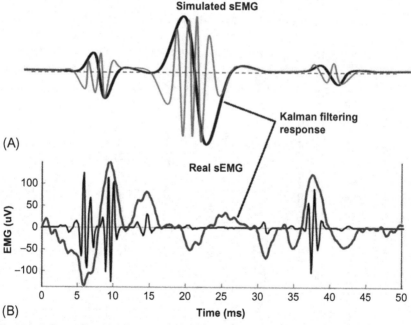

Fig. 20 Kalman filter responses in simulated and real sEMG signals. (A) Simulated sEMG signal (red line—gray line in print version) is fitted with the math sEMG model established in the filtering parameters. (B) Real sEMG signal (black line) is fitted with the math sEMG model settled in the parameters.

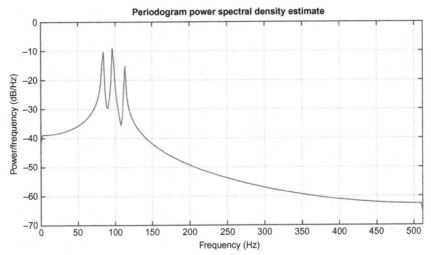

Fig. 21 Periodogram of the PSD estimated with FFT for $x_{pattern}$ signal without noise.

movement. The Göertzel filter uses the known frequency values of the standard signal to identify the frequencies present in an input signal within a time window. If the known frequencies match one or more of these, then the filter responds. Using $f1$, $f2$, and $f3$, of $x_{pattern}$, the Göertzel filter was applied to an input x [n].

Fig. 21 shows a periodogram of $x_{pattern}[n]$ input signal without noise for the spectral density power (SDP) present in the signal. This periodogram was calculated using the fast Fourier transform (FFT). The three frequency components, 83.3, 96.7, and 113.3 Hz are detected. Fig. 22 shows the discrete Fourier transform of the signal $x_{pattern}$ without noise. The density of each frequency component is different, related to the weight of the sinusoidal component.

A new methodology proposal was described for signal processing for the extraction of characteristics of the raw sEMG signal and to obtain an identification and prediction of the movements of the fingers of the hand (Altamirano Altamirano, 2017). The wavelet transform is useful for the detection of motor unit action potentials (MUAPs) in the presence of white noise. The novel proposal for the extraction of characteristics through the use of the Hilber-Huang transform helps to find intrinsic characteristics of the sEMG signal in real-time applications, by identifying the main frequency components involved, in this case, in the movements of the fingers.

By using the Kalman filter the noise is reduced and it provides a fast reconstruction of the desired shape of the reported pattern. Finally, the

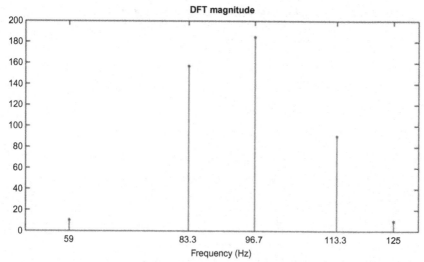

Fig. 22 Discrete Fourier transform of the $x_{pattern}$ signal without noise obtained by the Göertzel algorithm.

Göertzel filter provides a simple method of identifying the frequency patterns in less than 5 ms, only to locate the desired frequencies. The consumption time for the identification, detection, prediction, and correction of the myoelectric signal can be completed in less than 100 ms. The compendium of this proposal developed is summarized in Fig. 23.

5 Discussions

In the content of this chapter, A. Ruvalcaba proposes an electronic system that acquires the sEMG signal using dry brass electrodes with an in-line arrangement in parallel with the muscle fibers of the forearm. Finally, the acquisition device is validated by assuring its operating values are keeping with the ones that the SENIAM (Surface ElectroMyoGraphy for the Non-Invasive Assessment of Muscles) proposed for sEMG signal acquisition devices, measuring a CMRR of 96, a signal amplitude of 0.5–2 Vp-p, a frequency bandwidth of 20–450 Hz, and a high impedance input. The electrode/skin impedance ranges from 220 kΩ at 5 Hz to 4.32 kΩ at 2 kHz. These values are similar between each other, establishing that our electrodes can be used to acquire sEMG signals according to the standards described by the AAMY. These signals will be used to generate a database that is processed

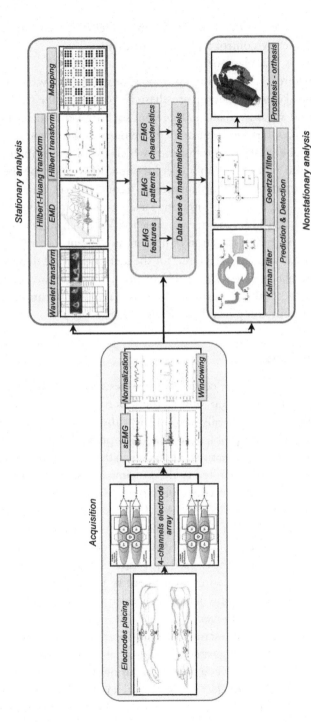

Fig. 23 EMG prediction system block diagram proposal.

to obtain characteristics that can identify movements of the wrist in a short time to subsequently control a physical prosthesis.

This work is complemented with the research carried out by Alvaro Altamirano where he describes and proposes the next phase, which is to perform the control of a myoelectric prosthesis by processing the sEMG signals in order to find a feature that can represent each movement. In this novel proposal he makes the identification of movements of the fingers of the hand by using the Hilbert-Huang transform, which decomposes multicomponents signals into a number of mono-components (IMFs). Using the instantaneous envelopment, the instantaneous frequencies of each IMF are obtained. By analyzing the resulting instantaneous frequencies (83.3, 96.7, 99.82, 59, and 113.3 Hz) it is possible to identify changes in the IMFs, which are related to the activation of flexion movement. Later he proposes the use of Kalman and Göertzel filters for the identification of these frequencies and thus, obtain a system capable of predicting the movements that, computationally speaking, manages to make the identification of these movements in something less than 80 ms.

6 Conclusion

The purpose of a myoelectric limb prosthesis, or a part of it, is to return lost function. A prosthesis is not only aesthetic, it must be functional, the user must be comfortable, and finally, it must allow the user to integrate into the labor market. One of the problems with robotic prostheses is the slowness in their action, producing frustration in users when they cannot be used in daily life. The investigations described in the previous work attack this problem from two fronts. The work carried out in the LAREMUS laboratory in the bioelectronics group of CINVESTAV-IPN, is focused on solving the response speed of a superior limb prosthesis. The regulation of the movement is intended with the voluntary EMG signal of the user, with the identification and prediction of the desired movement in less than 80 ms, sending the activation code of the desired movement to the biomechanical system (prosthesis). Each movement will have an activation code and an established sequence to produce this movement.

The first investigation described tries to solve this by improving hardware and sensors for the acquisition of superficial EMG signals. To acquire the muscular signal that moves a prosthesis during the day, it is necessary to stop using wet electrodes of electrolytic gel and it is necessary to use dry electrodes for the acquisition. Now, to be able to use the dry electrodes as sensors for the acquisition of sEMG signals it is necessary to use another type of hardware for the amplification and interpretation of the signals, before they are transmitted to store and process them, or to be already implemented in a myoelectric prosthesis.

In the research described by A. Ruvalcaba et al., a novel system proposed for the acquisition of sEMG signals is shown. This system improves the quality of the signal since its acquisition stage, avoiding costly postprocessing time that worsens the prediction time of the movement that the prosthesis should identify and perform.

As a work in the future, this signal identification and prediction system needs to be implemented, both in a virtual prosthesis and later in a real prosthesis to verify and demonstrate its true effectiveness in the use of everyday prostheses. It is expected to improve computational cost time and thus the prediction and anticipation of movements that are expected to be performed using a myoelectric prosthesis with voluntary movements that mimic the characteristics of natural movements.

References

Altamirano Altamirano, A., 2012. Propuesta de un Prototipo de Mano Antropomórfica para ser usada como Prótesis. Center for Research and Advanced Studies of the National Polytechnic Institute.

Altamirano Altamirano, A., 2017. Predicción de patrones EMG para movimientos de miembro superior con base en la transformación wavelet y Hilbert-Huang. Center for Research and Advanced Studies of the National Polytechnic Institute.

Barraza Madrigal, J.A., 2010. Desarrollo de una prótesis virtual para extremidad superior con amputación por arriba de la articulación del codo. Center for Research and Advanced Studies of the National Polytechnic Institute.

Basmajian, J.V., De Luca, C.J., 1985. Muscles Alive: Their Functions Revealed by Electromyography, fifth ed. Lippincott Williams and Wilkins.

Bronzino, J.D., 1999. Biomedical Engineering Handbook, Volume II, first ed. CRC Press.

Burke, M.J., Gleeson, D.T., 2000. A micropower dry-electrode ECG preamplifier. IEEE Trans. Biomed. Eng. 47 (2), 155–162. https://doi.org/10.1109/10.821734.

Castellanos Abrego, P., 1997. Electrofisiología humana: un enfoque para ingenieros, first ed. UAM, Unidad Iztapalapa.

Chi, Y.M., Jung, T., Cauwenberghs, G., 2010. Dry-contact and noncontact biopotential electrodes. IEEE Rev. Biomed. Eng. 3, 106–119.

Chu, J.U., Lee, Y.J., 2009. Conjugate-prior-penalized learning of Gaussian mixture models for multifunction myoelectric hand control. IEEE Trans. Neural Syst. Rehabil. Eng. 17 (3), 287–297.

Clancy, E.A., Hogan, N., 1995. Multiple site electromyograph amplitude estimation. IEEE. Trans. Biomed. Eng. 42 (2), 203–211.

Clancy, E.A., Morin, E.L., Merletti, R., 2002. Sampling, noise-reduction and amplitude estimation issues in surface electromyography. J. Electromyogr. Kinesiol. 12 (1), 1–16.

Contreras Rodríguez, L.A., 2017. Desarrollo de un sistema ambulatorio para el análisis de movimiento de la extremidad superior. Center for Research and Advanced Studies of the National Polytechnic Institute.

Coughlin, R.F., Driscoll, F.F., 1987. Circuitos Integrados Lineales y Amplificadores Operacionales, second ed. Prentice-Hall Hispanoamericana.

Delsys, 2018. Bagnoli™ EMG System, User's Guide. Delsys Incorporated.

Doeringer, J.A., Hogan, N., 1995. Performance of above elbow body-powered prostheses in visually guided unconstrained motion tasks. IEEE Trans. Biomed. Eng. 42 (6), 621–631. https://doi.org/10.1109/10.387202.

Escudero Uribe, A.Z., 2002. Desarrollo de una Prótesis de 4 Grados de Libertad Activos para Reemplazo por Arriba del Codo. Center for Research and Advanced Studies of the National Polytechnic Institute.

Farina, D., Negro, F., 2012. Accessing the neural drive to muscle and translation to neuro-rehabilitation technologies. IEEE Rev. Biomed. Eng. 5, 3–14. https://doi.org/10.1109/RBME.2012.2183586.

Farina, D., Jiang, N., Rehbaum, H., Holobar, A., Graimann, B., Dietl, H., Aszmann, O.C., 2014. The extraction of neural information from the surface EMG for the control of upper-limb prostheses, emerging avenues and challenges. IEEE Trans. Neural Syst. Rehabil. Eng. 22, 797–809.

Feng, Y., et al., 2005. An algorithm for simulating human arm movement considering the comfort level. Simul. Model. Pract. Theory 13 (5), 437–449.

Fernandez, M., Pallas-Areny, R., 1997. A simple active electrode for power line interference reduction in high resolution biopotential measurements'. In: Proceedings of 18th Annual International Conference of the IEEE Engineering in Medicine and Biology Society. IEEEpp. 97–98. https://doi.org/10.1109/IEMBS.1996.656864.

Fougner, A., Scheme, E., Chan, A.D.C., Englehart, K., Stavdahl, Ø., 2011. Resolving the limb position effect in myoelectric pattern recognition. IEEE Trans. Neural Syst. Rehabil. Eng. 19 (6), 644–651.

Gow, D.J., et al., 2001. The development of the Edinburgh modular arm system. Proc. Inst. Mech. Eng. H 215 (3), 291–298.

Godin, D.T., Parker, P.A., Scott, R.N., 1991. Noise characteristics of stainless-steel surface electrodes. Med. Biol. Eng. Comput. 29 (6), 585–590.

Grant, M., 2007. Introduction to the Stepper Stall Detector Module, Application Note 3330, Rev. 0. Freescale Semiconductor Inc., USA, pp. 1–24

Grimnes, S., 1983. Impedance measurement of individual skin surface electrodes. Med. Biol. Eng. Comput. 21 (6), 750–755. https://doi.org/10.1007/BF02464038.

Hargrove, L.J., Li, G., Englehart, K.B., Hudgins, B.S., 2009. Principal components analysis preprocessing for improved classification accuracies in pattern-recognition-based myo-electric control. IEEE Trans. Biomed. Eng. 56 (5), 1407–1414.

Huang, H., Zhou, P., Li, G., Kuiken, T.A., 2008. An analysis of EMG electrode configuration for targeted muscle reinnervation based neural machine interface. IEEE Trans. Neural Syst. Rehabil. Eng. 16 (1), 37–45.

Huhta, J.C., Webster, J.G., 1973. 60-Hz interference in electrocardiography. IEEE Trans. Biomed. Eng. BME-20 (2), 91–101. https://doi.org/10.1109/TBME.1973.324169.

Huigen, E., Peper, A., Grimbergen, C.A., 2002. Investigation into the origin of the noise of surface electrodes. Med. Biol. Eng. Comput. 40 (3), 332–338.

Jacobsen, S.C., et al., 1982. Development of the Utah artificial arm. IEEE Trans. Biomed. Eng.. https://doi.org/10.1109/TBME.1982.325033.

Klijn, J.A., Kloprogge, M.J., 1973. Cable artefact suppressor for electrophysiological recording. Electromyogr. Clin. Neurophysiol. 13 (1), 87–92.

León Ponce, M., 2003. Desarrollo de un sistema para la identificación de 7 movimientos de la mano basado en la señal mioeléctrica del antebrazo. Center for Research and Advanced Studies of the National Polytechnic Institute.

León Ponce, M., 2012. Clasificación de patrones mioeléctricos para la operación de un dispositivo antropomórfico. tesis doctorado, diciembre. CINVESTAV – IPN, México, D.F., p. 2012.

De Luca, C.J., et al., 2006. Decomposition of surface EMG signals. J. Neurophysiol. 96 (3), 1646–1657. https://doi.org/10.1152/jn.00009.2006.

Márquez Lázaro, A.P., 2011. Estrategia para la descomposición de señales mioeléctricas basadas en las técnicas de transformada wavelet y vector soporte. Center for Research and Advanced Studies of the National Polytechnic Institute.

Matrone, G.C., Cipriani, C., Carrozza, M.C., Magenes, G., 2012. Real-time myoelectric control of a multi fingered hand prosthesis using principal components analysis. J. Neuroeng. Rehabil. 9, 40.

Merletti, R., Farina, D., Gazzoni, M., 2003. The linear electrode array: a useful tool with many applications. J. Electromyogr. Kinesiol. 13 (1), 37–47.

Merletti, R., Farina, D., 2016. Merletti, R., Farina, D. (Eds.), Surface Electromyography: Physiology, Engineering, and Applications. John Wiley & Sons, Inc., Hoboken, New Jersey, https://doi.org/10.1002/9781119082934

MettingVanRijn, A.C., Kuiper, A.C., Dankers, T.E., Grimbergen, C.A., 1996. Low-cost active electrode improves the resolution in biopotential recordings. In: Proc. 18th Int. IEEE EMBS Conf. pp. 101–102.

Miller, L.A., et al., 2008. Improved myoelectric prosthesis control using targeted reinnervation surgery: a case series. IEEE Trans. Neural Syst. Rehabil. Eng. 16 (1), 46–50. https://doi.org/10.1109/TNSRE.2007.911817.

Moreno Pérez, A.D., 2010. Desarrollo del control para una prótesis de brazo con cuatro grados de libertad activos y actuadores paralelos. Center for Research and Advanced Studies of the National Polytechnic Institute.

Muceli, S., Jiang, N., Farina, D., 2014. Extracting signals robust to electrode number and shift for online simultaneous and proportional myoelectric control by factorization algorithms. IEEE Trans. Neural Syst. Rehabil. Eng. 22 (3), 623–633. https://doi.org/10.1109/TNSRE.2013.2282898.

Nishimura, S., Tomita, Y., Horiuchi, T., 1992. Clinical application of an active electrode using an operational amplifier. IEEE Trans. Biomed. Eng. 39 (10), 1096–1099.

Paul, G., et al., 2014. The development of screen printed conductive networks on textiles for biopotential monitoring applications. Sens. Actuators A Phys. 206, 35–41. https://doi.org/10.1016/j.sna.2013.11.026.

Perry, R.H., Green, D.W., Maloney, J.O., 1997. Perry's Chemical Engineers' Handbook, seventh ed. McGraw-Hill.

Phinyomark, A., Limsakul, C., Phukpattaranont, P., 2011. Application of wavelet analysis in EMG feature extraction for pattern classification. Meas. Sci. Rev. https://doi.org/10.2478/v10048-011-0009-y.

Pla Mobarak, M., 2014. Análisis del estado transitorio en la señal EMG multicanal con fines de identificación de movimientos de la mano. Center for Research and Advanced Studies of the National Polytechnic Institute.

Ramírez García, A., 2011. Implementación funcional de una prótesis para un amputado por encima del codo. Center for Research and Advanced Studies of the National Polytechnic Institute.

Ruiz-Olaya, A.F., Callejas-Cuervo, M., Perez, A.M., 2015. EMG-based pattern recognition with kinematics information for hand gesture recognition. In: 2015 20th Symposium on Signal Processing, Images and Computer Vision (STSIVA). IEEE, pp. 1–4. https://doi.org/10.1109/STSIVA.2015.7330409.

Ruvalcaba Granados, J.A., et al., 2014. Multichannel EMG acquisition system for arm and forearm signal detection. In: 2014 IEEE International Instrumentation and Measurement

Technology Conference (I2MTC) Proceedings. IEEE, pp. 1075–1078. https://doi.org/10.1109/I2MTC.2014.6860907.

Ruvalcaba Granados, J.A., et al., 2015. Design and measurement of the standards of a miniaturized sEMG acquisition system with dry electrodes integrated. In: 2015 International Conference on Mechatronics, Electronics and Automotive Engineering (ICMEAE). IEEE, pp. 99–104. https://doi.org/10.1109/ICMEAE.2015.34.

Ruvalcaba Granados, J.A., 2015. Diseño y desarrollo de un electrodo integrado a un sistema de adquisición de señales superficiales EMG de músculo. (BISASSEMG). Center for Research and Advanced Studies of the National Polytechnic Institute.

Ruvalcaba Granados, J.A., et al., 2017a. Desarrollo de una tarjeta miniatura de adquisición de señales para el monitoreo continuo de señales superficiales electromiográficas. In: VIII Congreso Internacional de Investigación en Rehabilitación.

Ruvalcaba Granados, J.A., et al., 2017b. Design and test of a dry electrode array implemented on wearable sEMG acquisition sleeve for long term monitoring. In: 2017 Global Medical Engineering Physics Exchanges/Pan American Health Care Exchanges (GMEPE/PAHCE). IEEE, pp. 1–5. https://doi.org/10.1109/GMEPE-PAHCE.2017.7972111.

Salazar Muñoz, Y., 2004. Caracterización de tejidos cardíacos mediante métodos mínimamente invasivos y no invasivos basados en espectroscopia de impedancia eléctrica. Polytechnic University of Catalonia.

Scott, R.N., Parker, P.A., 1988. Myoelectric prostheses: state of art. J. Med. Eng. Technol. 12 (4), 143–151.

Simon, A.M., Hargrove, L.J., Lock, B.A., Kuiken, T.A., 2011. A decision-based velocity ramp for minimizing the effect of misclassifications during real-time pattern recognition control. IEEE Trans. Biomed. Eng. 58 (8), 2360–2368.

Song, R., et al., 2013. Myoelectrically controlled wrist robot for stroke rehabilitation. J. Neuroeng. Rehabil. 10 (1), 52. https://doi.org/10.1186/1743-0003-10-52.

Webster, J.G., 1984. Reducing motion artifacts and interference in biopotential recording. IEEE Trans. Biomed. Eng. 31 (12), 823–826.

Webster, J.G., 2006. In: Webster, J.G. (Ed.), Encyclopedia of Medical Devices and Instrumentation. John Wiley & Sons, Inc, Hoboken, NJ, USA, https://doi.org/10.1002/0471732877.

Webster, J.G., 2009. Medical Instrumentation Application and Design, fourth ed. Wiley.

Xu, J., et al., 2017. Active electrodes for wearable EEG acquisition: review and electronics design methodology. IEEE Rev. Biomed. Eng. 10, 187–198. https://doi.org/10.1109/RBME.2017.2656388.

Zipp, P., Ahrens, H., 1979. A model of bioelectrode motion artefact and reduction of artefact by amplifier input stage design. J. Biomed. Eng. 1 (4), 273–276.

CHAPTER TEN

Automated transportation of microparticles in vivo

Xiaojian Li[a,b] and Dong Sun[b]
[a]School of Management, Hefei University of Technology, Hefei, China
[b]The Department of Biomedical Engineering, City University of Hong Kong, Hong Kong, China

1 Introduction

Biological research mainly aims to discover, understand, and utilize the law of life activity. With the evolution of life, biological structures evolve from simple single-cell organisms into a wide variety of complex life individuals. To protect life activity from external damage, organisms evolve out of a relatively closed and independent environment, that is, the in vivo environment. This environment provides a stable condition for life operation, resulting the in increasingly complex form of life. A closed in vivo environment provides protection for our lives but has hindered our research on life activities. The in vivo environment is difficult to observe and control because of technological restrictions. To address this limitation, scholars opt to conduct in vitro experiments. However, the in vivo condition is difficult to completely simulate in vitro, regardless of improvements in in vitro experiments, because the organism is a complex whole.

With the development of science and technology, advance tools and techniques have emerged to help us observe and control in vivo environments. Such technologies are vital for in vivo explorations. Optical tweezers (OTs) are a new but effective tool used to control in vivo environments. OTs, also known as optical trap, use a highly focused laser beam to trap and manipulate microscopic, neutral objects with noncontact force. This tool can be used to noninvasively manipulate cells in living animals. The first in vivo manipulation experiment with OTs was reported in 2013. Zhong et al. (2013a,b) used infrared OTs to trap and manipulate red blood cells (RBCs) within subdermal capillaries in living mice. In this experiment, RBCs in the capillary at the depth of ~40 µm beneath the surface of mouse ear skin was manipulated to block and clear the capillary to simulate the formation and removal of thrombus. OTs can not only be used as an operating tool but also as a mechanical measurement tool. In Johansen et al. (2016), OTs were used to analyze adhesion properties and membrane deformation in endothelium and macrophages in a living zebrafish. Overall, OTs exhibit potential for various applications, including studies on cancer metastasis mechanism (Gupta and Massagué, 2006; Clevers, 2011; Oskarsson et al., 2014; Reya et al., 2001), eliminating thrombus (Zhong et al., 2013b), targeted drug delivery (Forbes et al., 2003), screening of nanoparticle–cell interactions for cancer therapy (Tan et al., 2012), and research on tissue invasion in cancer and infection biology (Wu et al., 2011). Although OTs are a promising tool, their applications are hindered by many factors. Manual operation limits the handling of biological particles in complex in vivo environments, which are subject to various disturbances and uncertainties. With manual control, human operators encounter difficulty in retrieving cells once they escape from the OTs. The cells rapidly move away from the blood flow, and human operators cannot quickly find the lost cells. Manual control is also difficult considering the presence of disturbances, uncertainties, and collision avoidance problem in vivo.

The limitations of manual control can be addressed by robotics technology. An automated robot-aided OTs system integrates robotics with OTs manipulation to achieve fast, stable, and efficient manipulation of microparticles. Robot-tweezer manipulation is widely used in in vitro manipulation tasks, including cell transportation (Thakur et al., 2014; Chowdhury et al., 2014; Hu and Sun, 2011b; Banerjee et al., 2012; Ju et al., 2014; Li et al., 2015b), cell pairing (Xie et al., 2015), cell sorting (Chapin et al., 2006), and cell assembly (Tanaka et al., 2013). However, OTs with robotics technology have been rarely used in vivo. Hence, studies of in vivo robot-aided OTs system can develop the potential of OTs in in vivo research and exhibits important implications for biomedicine.

In this chapter, an automated cell transportation system that integrates robotics and OTs technologies is introduced. This system can automatically identify, capture, and transfer single cells in in vivo environments. In order to increase the success rate of transportation, the disturbance of blood flow in blood vessels and collision with free cells are considered. Disturbance compensation and obstacle avoidance functions of the transportation system are developed. The effectiveness of the proposed system is verified by transporting red blood cells in living zebrafish.

The remainder of this chapter is as follows. In Section 1, the background of the in vivo manipulation is introduced, including the dynamic model of microparticles in blood vessels and the mechanical properties of OTs.

Section 2 shows the identification and tracking methods for in vivo particles used to construct a visual feedback control system. Section 3 presents the position control system of in vivo microparticles that use OTs to drive microparticles from one location to another in blood vessels of living zebrafish. Two kinds of position controllers are introduced in this section. To solve the collision problem between the controlled particles and the free cells in the in vivo transportation process, a collision avoidance method called collision-avoidance vector methods is showed in Section 4. Section 5 summarizes the content of this chapter and looks forward to future work.

1.1 In vivo environment

The blood vessels are a part of the circulatory system that transports blood throughout the body. Blood delivers necessary substances, such as nutrients and oxygen, to the cells and transports metabolic waste products away from the same cells. As an important system of the body, the blood vessels have been extensively studied (Cohen and Taylor, 2002; Pries et al., 2000; Davies et al., 1997). Therefore, the internal environment of blood vessels has been studied in this chapter as the main environment for cell transport in vivo.

In this chapter, zebrafish was selected as experimental model to conduct cell transport experiments in its blood vessels. Zebrafish is an important vertebrate model organism in scientific research. Zebrafish has become a popular model system because its embryos and larvae are small, transparent, and undergo rapid development ex utero, allowing in vivo analysis of embryogenesis and organogenesis. In cardiovascular research, zebrafish is used to model blood clotting, blood vessel development, heart failure, and congenital heart and kidney diseases (Drummond, 2005). Zebrafish also represents a common cancer model system used in cancer research (Evensen et al., 2016; Teng et al., 2013; Stoletov and Klemke, 2008). Fig. 1 shows a 30-hour-postfertilization (hpf) zebrafish.

To control RBCs in the blood vessels, the mechanical properties of these cells must be study first. When blood flows in microcirculation, several features are observed: low Womersley number, low Reynolds number (Re), pulsating fluid, and interaction between RBCs and blood vessels. These features differentiate in vivo and in vitro cell manipulation.

Low Re is similar to that in in vitro manipulation. Re is a dimensionless quantity used to predict similar flow patterns in different fluid flow situations. This parameter is defined as the ratio of inertial forces to viscous forces

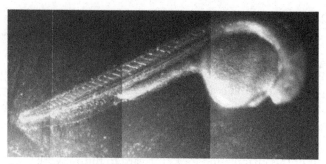

Fig. 1 The image of 30-hpf zebrafish.

and quantifies the relative importance of the two types of forces for given flow conditions. Re is defined as

$$Re = \frac{\rho v_f D}{\mu} = \frac{v_f D}{\nu},\tag{1}$$

where D is the hydraulic diameter of the pipe, v_f is the mean velocity of the fluid (SI units: m/s), μ is the dynamic viscosity of the fluid [Pa s or N s/m^2 or kg/(m s)], ρ is the density of the fluid (kg/m^3), and ν is the kinematic viscosity ($\nu = \mu/\rho$) (m^2/s). The Re of microcirculation systems is very low. For example, the Re in dogs are as follows:

Arterioles	$Re = 0.09$,
Capillaries	$Re = 0.001$,
Venules	$Re = 0.035$.

When $Re \ll 1$, the viscous forces are larger than the inertial forces. The relationship between force and speed of motion is provided by Stokes' law (Dusenbery, 2009).

Stokes' law is illustrated as follows:

$$\boldsymbol{F}_d = -6\pi\mu R \boldsymbol{v}_p,\tag{2}$$

where $\boldsymbol{F}_d$ is the frictional force, also known as Stokes' drag, which acts on the interface between the fluid and the particle (N); R is the radius of the spherical object (m); and $\boldsymbol{v}_p$ is the velocity of the particle (m/s).

Womersley number (α) is a dimensionless number in biofluid mechanics and expresses pulsatile flow frequency in relation to viscous effects. This parameter is defined as (Womersley, 1955):

Table 1 Womersley numbers in different human blood vessels.

Vessel	Aorta	Artery	Arteriole	Capillary	Venule	Veins	Vena cava
Diameter (m)	0.025	0.004	3×10^{-5}	8×10^{-6}	2×10^{-5}	0.005	0.03
α		13.83	2.21	0.0166	4.43×10^{-3}	0.011	2.77	16.6

$$\alpha = r_v \sqrt{\frac{\omega}{\nu}} = \sqrt{\frac{\text{Local inertial force}}{\text{Viscous force}}}, \tag{3}$$

where r_v is the radius of the blood vessels, and ω is the angular frequency of the oscillations. When α is less than 1 and the frequency of pulsations is sufficiently low, the velocity distribution along the radial direction of the blood vessel is parabolic. The blood flow velocity near the blood vessel wall is slow, and the blood flow velocity in the central region is fast. In addition, the flow will be very near in phase with the pressure gradient.

A list of estimated Womersley numbers in different human blood vessels is shown in Table 1.

When $\alpha \ll 1$, the heartbeat-induced oscillations and the local inertial force can be negligible. Therefore, the velocity and pressure of blood can be regarded as constant in microcirculation systems for a period of time.

Because of the low Womersley number and Re, the force balance equation of RBCs can be written as follows:

$$\Delta p S + F_d + F_t = 0 \tag{4}$$

Substituting Eq. (2) into Eq. (4) provides

$$v_p = \frac{\Delta p S + F_t}{6\pi\mu R} \tag{5}$$

where Δp is the average pressure difference across the RBC, and S is the sectional area of the RBC.

The pressure difference Δp is usually zero in vitro and can be negligible. However, Δp plays an important role in promoting blood flow. Although this parameter is constant within a short period of time, the pressure significantly changes under prolonged period. Pressure difference is affected by many factors, such as positional relationship between RBCs and blood pressure changes. Therefore, Δp must be considered when cells are controlled in vivo.

When the cells flow freely with blood, the force of the OTs $F_t = 0$. According to Eq. (5), the following equation can be obtained:

$$\Delta p S = 6\pi\mu R v_l, \tag{6}$$

where v_l is the blood flow velocity. Substituting Eqs. (2), (6) into Eq. (4), the dynamic equation can be written as follows:

$$F_t - 6\pi\mu R \left(v_p - v_l \right) = 0 \qquad (7)$$

1.2 Optical tweezers

Optical tweezers (OTs), also known as optical trap, use a highly focused laser beam to trap and manipulate microscopic, neutral objects with noncontact force. Since their development four decades ago, OTs have been widely employed for actively manipulating and positioning biological objects at the nano/microscale. Arthur Ashkin performed the first optical trapping experiment over 40 years ago (Ashkin, 1970); in this study, particles align themselves and move along the axis of the beam through a phenomenon commonly known as optical guiding. This year marks the 25th year since the inception of the popular version, namely, "optical tweezers" (Ashkin et al., 1986). Currently, OTs are commercially available and applied particularly in biological sciences (Molloy and Padgett, 2002). For example, tweezers are used to measure the compliance of bacterial tails (Block et al., 1989), evaluate forces exerted by motor proteins (Finer et al., 1994), stretch DNA molecules (Wang et al., 1997), and manipulate biological cells (Li et al., 2015a; Chowdhury et al., 2014; Hu and Sun, 2011a; Zhang and Liu, 2008).

Scattering and gradient forces exerted on the particles depend on the wavelength of the laser beam (λ) and particle size (r). Particles trapped by the OTs can be divided into three regimes, namely, Mie regime ($r \gg \lambda$); Rayleigh regime ($r \ll \lambda$); and regime in between them ($r \approx \lambda$) (Ashkin and Dziedzic, 1987). In different regimes, the principle of trap force varies. In this study, the transportation of cells, with size ranging from 5 to 40 μm, is focused. Therefore, only the Mie regime should be considered. In this regime, the trap force of the OTs can be due to changes in the optical momentum. The object often exhibits higher refractive index than that of water. The laser beam is refracted from the water into the object, and the direction of the laser beam changes, leading to momentum change in the laser beam. According to momentum conservation, the amount of change in momentum produces force acting on the object. In this principle, force caused by the OTs will pull the target to the area of high-light intensity for spherical objects. According to Hu and Sun (2011b), the trap force on a 3-μm yeast cell was measured. Fig. 2 shows the relationship between the lateral trapping force and the offset when the OTs is forced on a yeast cell.

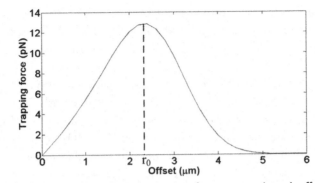

Fig. 2 Example of yeast cell: lateral trapping force versus lateral offset (Hu and Sun, 2011b).

The offset is the distance from the center of the yeast cell to the center of the laser beam. The trapping force increases as the offset increases, in an approximately linear relationship, until the offset is larger than r_0, and then decreases to zero as the offset increases.

Trap force on spherical objects can be approximated by the following equation:

$$F_t = k_{OT} d_o, \quad \|d_o\| \leq r_0, \tag{8}$$

where F_t is the trap force, d_o denotes the offset from the center of the yeast cell to the center of the laser beam, k_{OT} is the stiffness of OTs, and r_0 is the escape radius.

With the development of technology, HOTs have been proposed to flexibly generate multiple OTs and control their position with a computer (Curtis et al., 2002; Wang et al., 2006; Wulff et al., 2006; Fournier et al., 2008). HOTs use a computer-designed diffractive optical element (DOE) to split a single collimated laser beam into several separate beams; each beam is focused into an OT by strongly converging lens (Dufresne et al., 2001; Dufresne and Grier, 1998). As originally demonstrated with microfabricated DOEs (Grier, 1998), HOTs are implemented with computer-addressed liquid crystal spatial light modulators (SLM) (Reicherter et al., 1999; Liesener et al., 2000). Projecting a sequence of computer-designed holograms reconfigures the resulting pattern of traps.

In general, OTs exhibit the following advantages: noncontact force, active or passive force clamp, well-defined geometries, and few material requirements and this can be used to control one or more objects flexibly.

Few scholars have confirmed that OTs can be used to manipulate cells in vivo. Zhong et al. (2013a,b) and Johansen et al. (2016) reported the use of OTs to manually manipulate in vivo cells in mice and zebrafish, respectively.

Zhong et al. (2013a,b) reported the first experiment of trapping cells in living animals in 2013; in this in vivo cell manipulation with OTs, RBCs in living mice were transported to produce or eliminate thrombosis. In general, RBCs in capillaries flow at a speed of 0.1–2 mm/s. RBCs are trapped in a capillary at $\sim$5 μm (and $\sim$15 μm) diameter observed at depth of $\sim$40 μm (and $\sim$45 μm) beneath the surface of mouse ear skin. At deep locations in the mouse ear ($\sim$60 μm), OTs cannot trap cells because of two main reasons. First, the trapping force is reduced significantly by the loss of laser power at a deep location. Second, an appropriate gradient force might not exist to hold the cells because of severe aberration in focusing through thick tissues. The experiment confirmed that cells can be manipulated in vivo with OTs.

Johansen et al. (2016) reported the optical micromanipulation of nanoparticles and cells in living zebrafish in 2016. In this study, different cells and the injected nanoparticles and bacteria can be trapped; as such, adhesion properties and membrane deformation of endothelium and macrophages can be analyzed. This noninvasive micromanipulation inside an entire organism provides direct insights into cell interactions that are inaccessible using existing approaches. This technique can be applied in screening nanoparticle-cell interactions for cancer therapy or tissue invasion studies in cancer and infection biology. This study shows that the manipulation ability of OTs in zebrafish is more powerful than that in mice because the transparent zebrafish tissues minimally affect laser projection.

2 Identification and tracking of microparticles in vivo

The movement of cells in vivo is difficult to predict because of disturbances and uncertainty in in vivo environments. Automatic in vivo cell transport is impossible to realize with an open-loop controller. The cell position must be identified in real time to build a close-loop control system for in vivo cell transport. The image signal is the main output signal of most in vivo imaging tools, such as charge-coupled device (CCD), optical coherence tomography, ultrasound, MRI, functional near-infrared spectroscopy, and radiography. Information on cell position should be extracted from the in vivo image with image processing. In this section, an image processing module is designed to identify and track the target cell based on the in vivo image captured by the CCD camera.

2.1 The identifying of fluorescently labeled microparticles

Precise measurement of cell position is necessary to an in vivo control system. In contrast to in vitro application, the image showing an in vivo environment is chaotic. Fluorescently labeled cells are easier to identify than nonfluorescently labeled cells. The labeled cells show a different color from the background, which makes the cell easy to be segmented. Here, cells labeled by green fluorescent protein were injected into a 30-hpf zebrafish and observed by the OTs system.

As shown in Fig. 3A, an image of the fluorescently labeled cells obtained through a CCD camera can be denoted as a matrix $O(k) \in \mathbb{R}^{W \times H \times 3}$

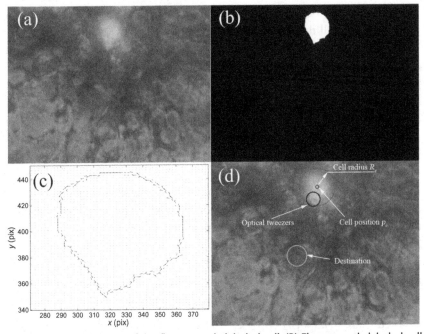

Fig. 3 (A) Original image of the fluorescently labeled cell. (B) Fluorescently labeled cell area after binarization. (C) Vector graph of the direction $[-Mx, My]^T$. (D) Detection of the fluorescently labeled cell.

constituted by $W \times H$ vector elements $\boldsymbol{o}_{x,y}(k) = [o_{x,y,r}(k), o_{x,y,g}(k), o_{x,y,b}(k)]^T$, where $o_{x,y,r}(k)$, $o_{x,y,g}(k)$, and $o_{x,y,b}(k)$ denote the red–green–blue (RGB) intensity value of the pixel located at (x, y). $k \in \mathbb{N}$ is the frame number. W and H are the width and height of the image, respectively.

Fluorescently labeled cell image can be obtained through threshold segmentation. The image of the fluorescently labeled cells in Fig. 3B can be denoted as $\boldsymbol{N}(k) \in \mathbb{R}^{W \times H \times 1}$, with vector elements that can be obtained as follows:

$$n_{x,y}(k) = \begin{cases} 1, & o_{x,y,r}(k) \geq T_r \text{ or } o_{x,y,b}(k) \geq T_b \text{ or } o_{x,y,g}(k) \geq T_g \\ 0, & \text{else} \end{cases}, \quad (9)$$

where T_r, T_b, and T_g are positive thresholds for red, blue, and green, respectively; and $n_{x,y}(k)$ denotes the intensity value of the pixel located at (x, y) in new image. In Eq. (9), the original image can be binarized. $n_{x,y}(k) = 1$ is the white area or the fluorescently labeled cell area, and $n_{x,y}(k) = 0$ is the black area or the tissue area.

The area of fluorescently labeled cells has been separated. Based on this phenomenon, cell detection will become easy.

First, the contour of the area should be extracted by using edge detection operators (Nixon and Aguado, 2012). A vector graph of the direction $[-Mx, My]^T$ as shown in Fig. 3C can be obtained by the convolution of the binarized image using the Sobel operator. The contour of the fluorescently labeled cell can be extracted by selecting any point of the edge on the vector graph (that is, the nonzero point) and searching along the direction $[-Mx, My]^T$. The process can be achieved using the cvFindContours function in OpenCV.

Second, the external rectangle of the cell contour should be determined. After contour extraction, a set of contour pixels can be obtained. Based on these pixels, we will calculate the center and radius of the cell. We will determine the smallest x, the smallest y, the largest x, and the largest y in the contour pixels, which are expressed as $x_{\min}$, $y_{\min}$, $x_{\max}$, and $y_{\max}$, respectively. The external rectangle can be formed by four lines, which are $x = x_{\min}$, $y = y_{\min}$, $x = x_{\max}$, and $y = y_{\max}$. The measured cell position is denoted as $\tilde{\boldsymbol{p}}_c = \left[\tilde{p}_{cx}, \tilde{p}_{cy}\right]^T$, where $\tilde{p}_{cx} = (x_{\min} + x_{\max})/2$ and $\tilde{p}_{cy} = (y_{\min} + y_{\max})/2$. The measured cell radius can be obtained by $r_c = \sqrt{(x_{\max} - x_{\min})(y_{\max} - y_{\min})}/2$. In this way, the fluorescently labeled cell is detected as shown in Fig. 3D.

2.2 The identifying of the nonfluorescently labeled microparticles

Images of nonfluorescently labeled cells are difficult to distinguish from tissue images compared with fluorescently labeled cells. Therefore, the nonfluorescently labeled cell is difficult to extract from a single frame. Fortunately, the tissue image is always stable and can be regarded as a background. The moving nonfluorescently labeled cells can be separated from the stable background and identified.

An image of the nonfluorescently labeled cells is obtained through a CCD camera (Fig. 4A). After comparison between the two adjacent frames (frames $k-1$ and k), the stable area image, which is the same in the both frames, can be extracted by the following equation:

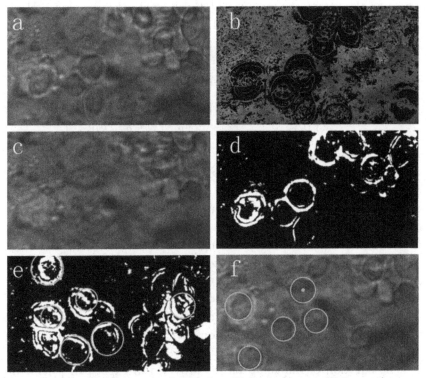

Fig. 4 The step of background segmentation. (A) Original image. (B) Extract the same area between adjacent frames. The black area is the different part. (C) The extracted background. (D) The image after background segmentation. The white area is the moving cells. (E) The detected moving cells with a blue circle (gray circle in print version). (F) The tracked cell with a green center (gray dot in print version).

$$s_{x,y}(k) = \begin{cases} o_{x,y}(k), & \left\| o_{x,y}(k) - o_{x,y}(k-1) \right\| \leq T \\ 0, & \left\| o_{x,y}(k) - o_{x,y}(k-1) \right\| > T \end{cases} \tag{10}$$

where $s_{x,y}(k)$ denotes the RGB intensity value of the stable area image, and T is the positive threshold. Fig. 4B shows the stable area image, in which the intensity of the black areas is zero, representing the moving area. The intensity of the remaining areas is the same as the frame k, representing the stable area.

The image background in Fig. 4C can be denoted as $B(k) \in \mathbb{R}^{W \times H \times 3}$ with vector elements that can be expressed as follows:

$$b_{x,y}(k) = \begin{cases} \tau b_{x,y}(k-1) + (1-\tau)s_{x,y}(k), & s_{x,y}(k) \neq 0 \\ b_{x,y}(k-1), & s_{x,y}(k) = 0 \end{cases} \tag{11}$$

$$b_{x,y}(0) = 1, \tag{12}$$

where $b_{x,y}(k)$ denotes the intensity RGB value of the background, and τ is the forgetting factor between 0 and 1.

Comparison between the original image $O(k)$ and the background $B(k)$ shows that the image of the moving cells in Fig. 4D can be denoted as $N(k)$, and vector elements can be obtained as follows:

$$n_{x,y}(k) = \begin{cases} \left\| b_{x,y}(k) - o_{x,y}(k) \right\|, & \left\| b_{x,y}(k) - o_{x,y}(k) \right\| \geq T_c \\ 0, & \left\| b_{x,y}(k) - o_{x,y}(k) \right\| < T_c \end{cases} \tag{13}$$

where T_c is a positive threshold.

In this manner, the image of the moving cells can be obtained. However, the moving cell image is more messy and trivial than the image of fluorescently labeled cells. The image of the moving cells is thus difficult to be identified by the introduced method in Section 2.1. Therefore, a different method is necessary to identify moving cells.

As shown in Fig. 4, the RBCs of zebrafish before 48 hpf show a regular spherical shape. HT is a commonly used algorithm for identifying arbitrary shapes and can be used to identify moving cells (Ballar, 1981).

First, the edge of the moving cells can be obtained by the convolution of the binarized image with edge detection operators. The edge of the moving cells should be formed in a circle.

The equation of the curve can be provided in explicit or parametric form. In explicit form, HT can be defined by considering the equation for a circle given by

$$(x - p_{cx})^2 + (x - p_{cy})^2 = r_c^2 \tag{14}$$

This equation defines a locus of points (x, y) centered on an origin (p_{cx}, p_{cy}) and with radius r_c. This equation can be visualized in two ways as follow: as a locus of points (x, y) in an image and as a locus of points (p_{cx}, p_{cy}) centered on (x, y) with radius r_c.

According to Eq. (14), HT mapping is defined by

$$\begin{cases} p_{cx} = x - r_c \cos(\beta) \\ p_{cy} = y - r_c \sin(\beta) \end{cases}, \tag{15}$$

where $\psi \in (-\pi, \pi]$ is the angle of the polar coordinate system. This equation defines the points in the accumulator space dependent on radius r_c. All moving cells can be identified using HT to process the moving image $N(k)$ (Fig. 5).

2.3 In vivo tracking microparticle

To date, in vivo cells are identified with image processing techniques. However, the video is discontinuous in time. We need to judge which of the identified cells is the target cell that we are tracking. In this chapter, positional correlation method is used to track in vivo cells.

In our experiments, the blood flow velocity was less than 6×10^{-5} m/s, and the capture rate was set to 20 Hz. The cell movement distance between the two adjacent frames was less than 3 μm, which is slightly smaller than the radius of the tracked cells ($\sim$4–5 μm). Therefore, the cell can be easily tracked by finding the closest cell in the next frame. As the velocity of blood flow increases, the method can still be used by increasing the capture rate.

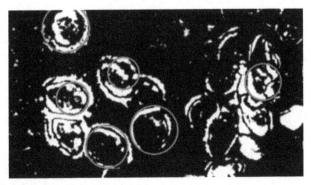

Fig. 5 Moving cell detection.

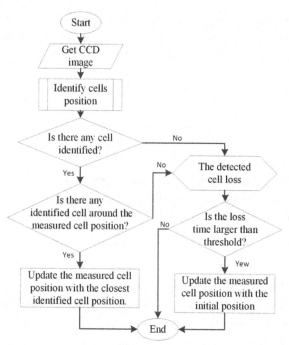

Fig. 6 Flowchart of the cell detection scheme.

Fig. 6 illustrates a flowchart of detection of moving cells. The measured cell position is denoted as $\tilde{\boldsymbol{p}}_c \in \mathbb{R}^{2\times1}$. Given the measurement noise during imaging, $\tilde{\boldsymbol{p}}_c$ includes the true value and error and expressed as

$$\tilde{\boldsymbol{p}}_c = \boldsymbol{p}_c + \boldsymbol{\omega}, \tag{16}$$

where $\boldsymbol{p}_c \in \mathbb{R}^{2\times1}$ is the actual position of the cell, and $\boldsymbol{\omega} \in \mathbb{R}^{2\times1}$ is the measurement error.

3 Transportation of microparticles in vivo

Cell position information can be obtained in real time by using the image processing technique introduced in Section 2. The closed-loop OTs system for in vivo cell transportation can also be established. The essence of cell transportation is position control; the system will automatically move the cells from the initial point to a given destination. In contrast to in vitro transport, in vivo transport is not only affected by the disturbance of blood flow but also showed a possibility of collision between the target cell and the other flowing cells. To achieve high success rate, in vivo cell transport requires a transport strategy. In this section, automatic in vivo cell transport is studied. Two kinds of controller are designed for in vivo cell transport with the OTs system. The performance of the controllers is analyzed and compared by simulation and experiment.

3.1 In vivo cell transport with P-type controller

As introduced in Section 1.2, the trapping force increases as the offset increases, in an approximately linear relationship, until the offset is larger than r_0, and then decreases to zero as the offset increases. Therefore, the manipulation is efficient when the offset is less than r_0. A nonlinear saturation function is defined as follows:

$$f_\sigma(x) = \frac{\|x\| + \sigma - |\|x\| - \sigma|}{2\|x\|} x, \tag{17}$$

where $x \in \mathbb{R}^{2\times1}$ is the argument of the function $f_\sigma(x)$, and σ is a positive number representing the upper bound of the function. The radius would be different for different cell types, and so is r_0. In this chapter, we assume:

$$r_0 = 0.7r_c \tag{18}$$

With this saturation function, the optical trap position will be limited in an efficient area. Then, a saturation P-type controller can be designed as follows (Li et al., 2015a):

$$u_{OT} = \tilde{p}_c + f_{r_0}\left(k_p\, \tilde{e}\right) \tag{19}$$

$$\tilde{e} = p_d - \tilde{p}_c = e - \omega, \tag{20}$$

where $u_{OT} \in \mathbb{R}^{2\times1}$ denotes the coordinate of the optical trap, $p_d \in \mathbb{R}^{2\times1}$ denotes the coordinate of the destination, $e = p_d - p_c$ denotes the error between the desired and actual positions of the cell, $\tilde{e}$ denotes the estimate of the error e, and k_p is a positive real number as gain factor.

In low Reynolds number environment, the inertia force of cells can be ignorable (Arai et al., 2004; Wu et al., 2011). According to $v_p = \dot{p}_c$, the relative speed of the trapped cell with the blood flow can be defined as follows:

$$v = \dot{p}_c - v_l \tag{21}$$

and Eq. (7) can be rewritten as:

$$F_t - 6\pi\mu R v = 0 \tag{22}$$

The offset of optical trap can be calculated as:

$$d_o = u_{OT} - p_c \tag{23}$$

According to Eq. (8), the trapping force can be described as:

$$F_t = k_{OT}(u_{OT} - p_c) \tag{24}$$

when

$$\|u_{OT} - p_c\| \le r_0 \tag{25}$$

Assuming the measurement error $\omega = 0$ and substituting Eq. (19) into Eq. (24), we obtain

$$F_t = k_{OT} f_{r_0}(k_p e) \tag{26}$$

Because of the nonlinear saturation function (17), format (25) always holds.

Substituting Eqs. (21), (26) into Eq. (22), we obtain the following closed-loop dynamics equation:

$$\dot{p}_c = v_l + \frac{k_{OT}}{6\pi\mu r} f_{r_0}(k_p e) \tag{27}$$

When $\dot{p}_c = 0$, the system reaches stable state. Substituting $\dot{p}_c = 0$ into Eq. (27), we obtain

$$v_l = -\frac{k_{OT}}{6\pi\mu r} f_{r_0}(k_p e) \tag{28}$$

To ensure the equation has a solution, the necessary and sufficient condition is

$$\|v_l\| \le \frac{k_{OT} r_0}{6\pi\mu r} \tag{29}$$

When Eq. (29) is satisfied, the following equation can be obtained from Eq. (28):

$$p_c = p_d + \frac{6\pi\mu r}{k_p k_{OT}} v_l \tag{30}$$

If the blood flow velocity v_l is constant, (30) will have a unique solution. Apparently, the position control system has steady-state error $6\pi\mu r v_l / k_p k_{OT}$. According to Eq. (29), the length of the steady-state error is less than r_0. The integral term, which is used to eliminate the steady-state error, will cause overshoot and have a great impact on system real-time performance. In contrast, a steady-state error is acceptable.

3.2 In vivo cell transport with the disturbance compensation controller

As the analysis in Section 3.1, the P-type controller fails to overcome the influence of the drag force caused by blood flow, which results in a variation of the cell transportation trajectory and relatively large steady-state error. To eliminate this effect, a disturbance compensation controller is proposed (Li et al., 2016), as shown in Fig. 7. The proposed control system is divided into the disturbance observer and controller. The disturbance observer aims to observe and measure the disturbance caused by the drag force of the blood flow. The controller calculates the optical trap position based on the cell position, destination, and estimated disturbance, with minimized steady-state error. Furthermore, the cell transportation trajectory can be updated online by adjusting the parameters of the observer. The change of trajectory caused by collisions with other obstacles can be corrected simultaneously.

According to Eqs. (21), (22), and (24), the following equation can be obtained:

$$\dot{p}_c = \frac{k_{OT}}{6\pi\mu r}d_o + v_l \tag{31}$$

Definition 10.1 The disturbance is defined as an equivalent offset of the optical trap caused by the blood flow, which can be expressed as $\varepsilon = \beta v_l$, where $\beta = 6\pi\mu r / k_{OT}$.

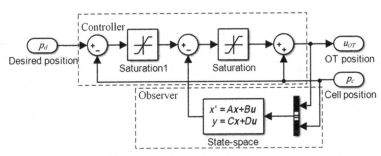

Fig. 7 Disturbance compensation control system.

With Definition 10.1, Eq. (31) can be rewritten as follows:

$$\dot{p}_c = \frac{1}{\beta}(d_o + \varepsilon) \tag{32}$$

Disturbance ε is determined by designing a disturbance observer, as follows:

$$\begin{cases} \dot{x} = Ax + Bu \\ \tilde{\varepsilon} = Cx + Du' \end{cases} \tag{33}$$

where

$$\begin{cases} A = -\alpha I \\ B = [(\alpha - \alpha^2\beta)I \quad -\alpha I] \\ C = I \\ D = [\alpha\beta I \quad 0] \end{cases} \tag{34}$$

$$u = \begin{bmatrix} \tilde{p}_c \\ u_{OT} \end{bmatrix} \tag{35}$$

$$u_{OT} = p_c + d_o \tag{36}$$

Note that x in Eq. (33) represents the observer status, and $\tilde{\varepsilon}$ is the observer output that represents the estimate of the disturbance ε. α is a positive real number.

The observer status can be initialized at the beginning of manipulation, as follows:

$$x = \tilde{\varepsilon}_0 - \alpha\beta\tilde{p}_c, \tag{37}$$

where $\tilde{\varepsilon}_0$ is the initial value of the estimated disturbance and is usually given as follows:

$$\tilde{\varepsilon}_0 = b\beta v_l, \tag{38}$$

where b is a positive real number. Parameter b affects the initial value of the observer, and further, the initial direction of the trajectory. In the current study, parameter b can be in the range of $[1, 5]$.

The disturbance compensation controller is then designed based on Eq. (17). The position of the optical trap can be given as follows:

$$u_{OT} = \tilde{p}_c + f_{r_0}\left(f_{r_0}(\tilde{e}) - \tilde{\varepsilon}\right) \tag{39}$$

3.3 The enhanced disturbance compensation controller

In order to improve the steady-state performance of the controller and the performance in high-speed blood flow, an enhanced disturbance compensation controller is designed (Li et al., 2017). Fig. 8 shows its structure.

To reduce the influence of the measurement error ω on the observer while ensuring the convergence speed of the observer, an observer with variable parameter is proposed. The structure of the observer remains unchanged, and the parameter α is changed to variable parameter, expressed as follows:

$$\alpha = \alpha_0 e^{-\gamma / \left\| \tilde{\boldsymbol{\varepsilon}} + d_o - \beta \ddot{\boldsymbol{p}}_c \right\|} \tag{40}$$

α is a variable parameter, which decreases as the disturbance estimate $\tilde{\boldsymbol{\varepsilon}}$ approaches the true value $\boldsymbol{\varepsilon}$. α_0 and γ are positive real numbers. In addition, the selection of parameters α_0 and γ affects the observer performance and the cell transportation trajectory. The larger the parameter α_0, the faster the convergence rate of the observer. A large parameter γ will help in reducing the steady-state error of the observer.

Based on Eq. (17), the enhanced disturbance compensation controller is designed as follows:

$$\boldsymbol{u}_{OT} = \tilde{\boldsymbol{p}}_c + f_{r_0}\left(f_{r_0}\left(\tilde{\boldsymbol{e}} \right) - f_{\varphi r_0}\left(\tilde{\boldsymbol{e}} \right) \right), \tag{41}$$

where $\varphi \geq 1$ is a positive real number.

3.4 Experiments

Simulations of moving single target cells in the blood vessel were performed first. The environmental parameters used in the simulations are as follows: trapping stiffness, $k = 1.5 \times 10^{-5}$ N/m; dynamic viscosity, $\mu = 4 \times 10^{-3}$ Pa s; cell radius,

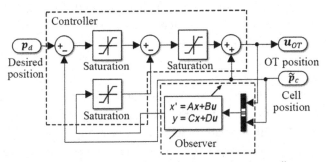

Fig. 8 Diagram of the enhanced disturbance compensation controller.

$r = 1 \times 10^{-5}$ m; and sampling time, $T = 0.05$ s. Based on Belharet et al. (2012), the speed of blood flow in the blood vessel is in parabolic distribution. In the simulation, the direction of blood flow is along the x-axis, and the velocity of blood flow has a parabolic distribution along the y-axis.

The countercurrent of the target cell with respect to the blood flow direction in in vivo environment makes the cell easily crash with other cells because of the high density of cells in the environment. The cell is not easy to move backward using optical tweezers if the maximum trapping force is less than $6\pi\mu r v_l$. Thus, the reverse movement of the cell should be avoided. Considering that blood flow runs faster in the middle than in the other parts of the vessel (Belharet et al., 2012), a stream of cells may be formed; and when a cell moves in the stream, it has a higher chance of colliding with other cells. Therefore, at the beginning of the transportation, the cell should be controlled to move perpendicular to the flow direction to escape from the cell stream, and then move along with the flow direction. In this way, the probability of collision can be significantly reduced.

The proposed controller allows the trajectory to be adjusted online with the disturbance observer. The disturbance may not be compensated perfectly when the disturbance observer estimates the disturbance, and this characteristic affects the trajectory generation. Three parameters α_0, b, and φ in the observer will play important roles in adjusting the trajectory. Parameter b is used to adjust the initial value of the disturbance observer, which affects the initial direction of the movement trajectory in getting away from the cell stream. Fig. 9A illustrates the cell transportation trajectories with different values of b when $\alpha_0 = 2$, $\varphi = 2$, and $\gamma = 10^{-6}$, where the velocity of blood flow is in parabolic distribution. An increase in b causes a large deviation of the cell movement trajectory from the blood flow. Parameter α is used to adjust the tracking speed of the observer and the noise rejection capability. This parameter is also used to adjust the curvature of the trajectory. Fig. 9B shows the transportation trajectories with different values of α_0 when $b = 2$, $\varphi = 2$, and $\gamma = 10^{-6}$. All the trajectories with different values of α have the same initial directions when b is fixed. As α increases, the curvature of the trajectory becomes smaller and the cell moves more directly toward the desired position.

Based on (Li et al., 2015a), the stability of the system cannot be definitely guaranteed unless the condition that $\|\boldsymbol{\varepsilon}\| < r_0$ always holds. Given that the drag force of blood flow is large and may not be compensated by the optical trapping force, the condition $\|\boldsymbol{\varepsilon}\| < r_0$ can only hold when the cell moves near the vessel wall. Fig. 9C illustrates the transportation trajectories with

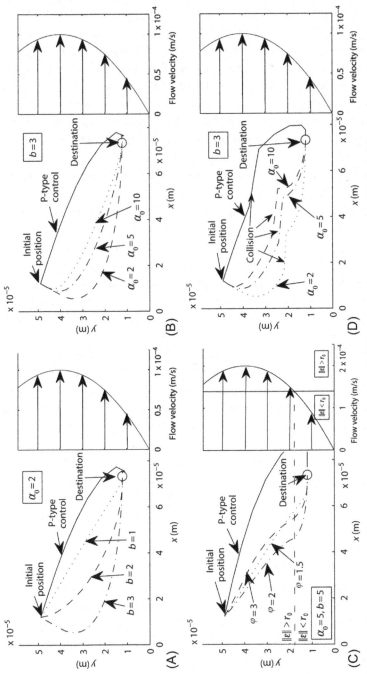

Fig. 9 The simulation result of cell transportation trajectories. (A) Cell transportation trajectories with different values of b. (B) Cell transportation trajectories with different values of α_0. (C) Cell transportation trajectories in rapid blood flow environment with different values of φ. (D) Cell transportation trajectories with different values of α in case of collision.

different φ, where $\|\boldsymbol{\varepsilon}\| < r_0$ holds around the destination but not at the initial position. The other parameters are $\alpha_0 = 5$, $b = 5$, and $\gamma = 10^{-6}$. Clearly, the proposed controller with all of the different parameters can drive the cell toward the destination, and when $\varphi = 2$, the cell can achieve $\|\boldsymbol{\varepsilon}\| < r_0$ more quickly. With the P-type control, however, the cell fails to reach the destination, because the P-type controller cannot help the cell to move away from the cell stream.

Collision with other cells and obstacles is inevitable during transport of the target cells in vivo. Furthermore, the controller should also be able to drive the cell back to the original trajectory after collisions. The influence of collision on cells is treated as disturbance and handled by the disturbance observer. The controller can compensate this disturbance to maintain the cell in its original trajectory. Fig. 9D shows the transportation trajectory when the collision exists with different values of α_0, where $b = 2$, $\varphi = 2$, and $\gamma = 10^{-6}$. The collision is simulated by adding an offset of 10^{-5} m in the x-axis direction on the target cell position. Evidently, the proposed controller can effectively compensate for the influence of the collision and drives the cell back to the original trajectory after collisions. However, the P-type controller (Li et al., 2015a) does not have such capability.

Steady-state error is an important criterion used to evaluate the performance of a position controller. The steady-state error of in vivo transportation is mainly due to the drag force caused by the blood flow (Li et al., 2015a). The proposed disturbance compensation controller can observe and compensate the drag force such that the steady-state error and the overshoot can be minimized. Fig. 10 illustrates the disturbance observer errors and the position errors, under the controller with the variable parameter observer, the controller with the fixed parameter observer, and the P-type controller. The measurement error is set as a white noise with a magnetite of 1 µm. The steady-state error under the proposed variable parameter observer is ~0.1 µm, which is less than that under the fixed parameter observer.

Experiments were then performed to track RBCs in 30-hpf zebrafish to demonstrate the effectiveness of the proposed controller. The RBCs of zebrafish before 48 hpf have a regular spherical shape, and can be easily identified and controlled. Fig. 11 shows the sinus venosus of zebrafish (with a depth ranging from 10 to 60 µm) where the cell movement experiment was performed. Given that zebrafish is transparent, the depth limit of the optical tweezers can reach ~120 µm, which is larger than that for mouse

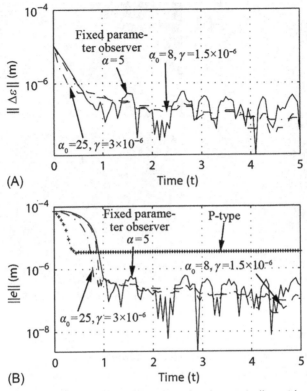

Fig. 10 Observer errors and position errors under the controller with the variable parameter observer, the controller with the fixed parameter observer, and the P-type controller. (A) Observer errors. (B) Position errors.

($\sim$60 μm) (Zhong et al., 2013b). Therefore, the cell transportation in zebrafish is immune to the depth limit of optical tweezers. The experimental process is described as follows: first, the program searches for the cell around the initial position; then, the identified cell is driven to the destination by the proposed controller and reaches a steady state; finally, the system goes back to the initial position for another trial. Each transportation trial was repeated for more than 30 times. The laser output power in all of the experiments was set to 3 W ($\sim$450 mW in the sample), which could provide an optical trapping force of 20 pN without causing significant heat damage to cells (Johansen et al., 2016; Gou et al., 2014). Two controllers, namely, the enhanced disturbance compensation controller and the P-type controller

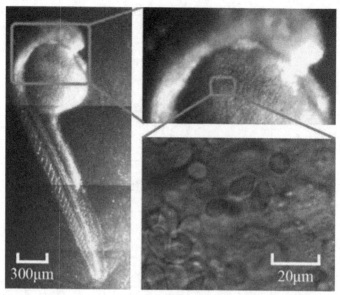

Fig. 11 Venue of the cell movement experiment in the yolk of zebrafish.

(Li et al., 2015a), were implemented in the experiments for comparison. Two groups of experimental results are reported.

The environmental parameters in the first group of experimental results were set as follows: flow velocity at the initial position, $\tilde{v}_l = (-0.9254, -0.9623) \times 10^{-5}$ m/s; estimate of the parameter, $\tilde{\beta} = 0.125$ s. The initial position of the trapped cell was $(0.7026, 0.4186) \times 10^{-4}$ m, the coordinate of destination was $(0.2730, 0.3256) \times 10^{-4}$ m, and the measure error was 1–2 μm. The parameters of the proposed controller were $b = 2$, $\alpha_0 = 0.5$, $\varphi = 2$, and $\gamma = 3 \times 10^{-7}$, and the parameter for the P-type controller was $k_p = 1$. The cell transportation was repeated for 10 times. Table 2 shows the average steady-state error and the travel time of each trial. The average steady-state error was calculated by $\int_\Sigma \|e\| dt / \int_\Sigma 1 dt$, where t represents the time and $\Sigma = \{t | \|e\| < 1 \times 10^{-5}\}$. With the proposed controller, nine cells successfully arrived in the destination, and one cell failed. With the P-type controller, only six cells arrived in the destination, and four cells failed. This result confirms that the proposed controller can achieve a high success rate in the transportation. Fig. 12 illustrates the position errors under the proposed controller and the P-type control of one trial, for comparison. The travel time under the proposed controller is 3.31 s, which is less than 4.78 s under the P-type controller. The average steady-state error under

Table 2 Average steady-state error and travel time.

| Experimental trial | Disturbance compensation controller | | P-type controller | |
	Average steady-state error (μm)	Travel time (s)	Average steady-state error (μm)	Travel time (s)
1	2.53	5.80	2.90	4.78
2	Fail	Fail	3.94	4.14
3	1.91	4.30	Fail	Fail
4	3.45	3.72	3.14	4.60
5	2.47	6.34	2.41	4.46
6	3.38	5.73	2.14	7.88
7	2.41	6.08	Fail	Fail
8	2.27	3.47	Fail	Fail
9	1.64	4.77	Fail	Fail
10	1.93	3.31	2.89	6.56
Mean	2.44	4.84	2.90	5.40

the proposed controller is 1.93 μm, which is less than 2.90 μm under the P-type controller. Fig. 13 shows the cell transportation trajectories with 10 experimental trials under the proposed disturbance compensation controller and the P-type controller (Li et al., 2015a), respectively. Evidently, the trajectory generated by the proposed controller was initially perpendicular to the flow direction and then turned toward to the destination. This phenomenon indicates that the proposed controller can make the cell move away from the cell stream at the beginning and then successfully drive the cell to the destination.

The environmental parameters in the second group of experiment are as follows: flow velocity at the initial position, $\tilde{v}_l = (-0.1389, -0.0851) \times 10^{-4}$ m/s; the estimate of the parameter, $\tilde{\beta} = 0.125$ s; initial position coordinate of the trapped cell, $(0.6450, 0.2105) \times 10^{-4}$ m; the position of destination, $(0.1334, 0.3905) \times 10^{-4}$ m. The proposed controller employed parameters of $\varphi = 2$ and $\gamma = 3 \times 10^{-7}$. Fig. 14 shows the experimental process of cell transportation with collision. The proposed controller, with parameters of $b = 2.5$ and $\alpha_0 = 0.2$, could manipulate the cell back to the original trajectory after colliding with other cells. Fig. 15 illustrates the cell transportation process under the proposed controller with three groups of parameters, namely, $b = 1.5$ and $\alpha_0 = 0.2$, $b = 1.5$ and $\alpha_0 = 1$, $b = 0.5$, and $\alpha_0 = 0.2$, and the P-type controller ($k_p = 1$). An increased value of b made the initial direction of the cell movement deviate more remarkably

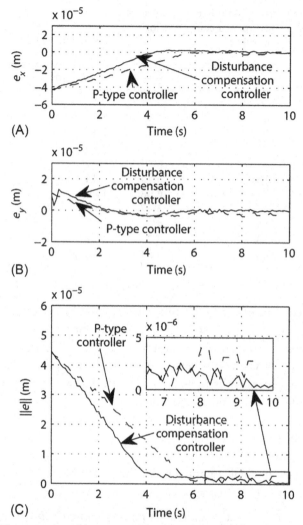

Fig. 12 Position errors in cell transportation experiments. (A) Position error in x-axis. (B) Position error in y-axis. (C) The norm of the position error.

from the flow direction, which can be seen through comparison of the initial trap force direction in Fig. 15C ($b = 0.5$) and that in Fig. 15A and B ($b = 1.5$). Furthermore, an increased value of α minimized the curvature of the trajectory. The steady-state errors of the proposed controller with the three groups of parameters are similar, which are all considerably lower than that under the P-type controller.

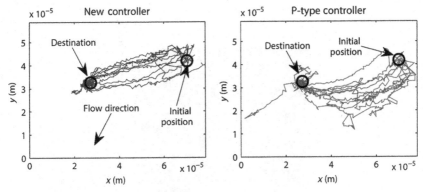

Fig. 13 Cell transportation trajectory.

4 Collision avoidance in vivo

To date, automatic in vivo cell transport has been achieved, and the effects of disturbances caused by blood flow have been minimized using disturbance compensation controller. However, a large number of RBCs exist in blood vessels. Collisions between the target cell and the other flowing cells are prone to occur during transport, resulting in transport failure. In this chapter, a collision–avoidance vector method is developed to solve the collision–avoidance problem in vivo (Li et al., 2018). Based on the method, the in vivo cell transport system exhibits collision–avoidance capability.

4.1 Collision-avoidance vector methods

The proposed collision-avoidance strategy aims to ensure that the trapped cell maintains a safe distance from other particles during its transportation to its destination. In previous studies, the position of the obstacle must be first identified using image processing. However, the identification of all the positions of obstacles is very time consuming and difficult, especially for chaotic in vivo images. A large amount of computational work will also affect the control performance of the system. Poor identification accuracy will cause the collision avoidance to fail, in turn causing the cell transportation task to fail.

To solve this problem, a novel concept, named the collision-avoidance vector, is proposed in this paper.

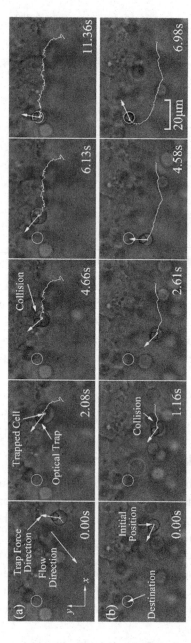

Fig. 14 Cell transportation with collision under (A) disturbance compensation controller with $b = 2.5$ and $\alpha_0 = 0.2$. (B) P-type controller.

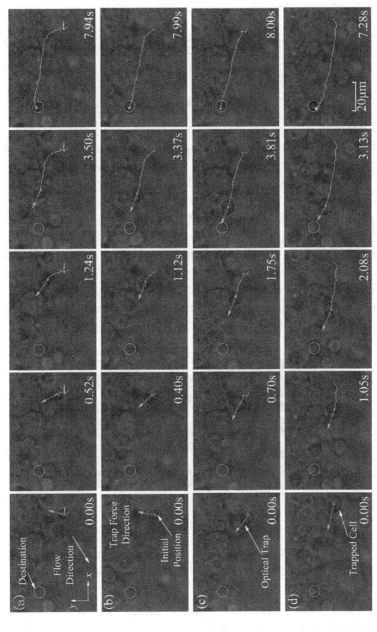

Fig. 15 Cell transportation with different parameters, under (A) disturbance compensation controller with $b = 1.5$ and $\alpha_0 = 0.2$; (B) disturbance compensation controller with $b = 1.5$ and $\alpha_0 = 1$; (C) disturbance compensation controller with $b = 0.5$ and $\alpha = 0.2$; and (D) P-type controller.

Definition 10.2 (Collision-avoidance vector)

The collision-avoidance vector is expressed as $\boldsymbol{G} = [G_x, G_y]^T$, with the components calculated as follows:

$$G_x = \sum_{\Delta x=-R_1}^{R_1} \sum_{\Delta y=-R_1}^{R_1} \delta_x(\Delta x, \Delta y) n_{p_{cx}+\Delta x, p_{cy}+\Delta y}(k), \qquad (42)$$

$$G_y = \sum_{\Delta x=-R_1}^{R_1} \sum_{\Delta y=-R_1}^{R_1} \delta_y(\Delta x, \Delta y) n_{p_{cx}+\Delta x, p_{cy}+\Delta y}(k), \qquad (43)$$

where R_1 denotes the maximum detection radius; $\Delta x = x - p_{cx}$ and $\Delta y = y - p_{cy}$ are the distances between the current position and the detected position $\boldsymbol{p}_c$ in the X and Y coordinate directions, respectively; and $\boldsymbol{\delta}(\Delta x, \Delta y) = [\delta_x(\Delta x, \Delta y), \delta_y(\Delta x, \Delta y)]^T$ denotes the collision-avoidance operator, which includes components in the X and Y directions.

Fig. 16 illustrates the process of calculating a collision-avoidance vector. The size of the collision-avoidance operator depends on image resolution and the radius of the obstacle detection region. Fig. 16 shows that the image of an obstacle is composed of a plurality of pixels with $n_{p_{cx}+\Delta x, p_{cy}+\Delta y}(k) \neq 0$. Therefore, avoiding collision from an obstacle is equivalent to avoiding

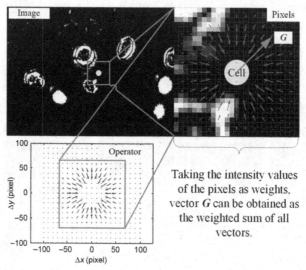

Fig. 16 Calculation of the collision-avoidance vector.

collision from those pixels in relation to the obstacle. The collision-avoidance operator $\delta(\Delta x, \Delta y)$ is used to guide the cell to avoid a single pixel at $(\Delta x, \Delta y)$. Taking the intensity value of the pixel as the weight, the collision vector can be calculated by summing all the vectors in the collision-avoidance operator by weight. With a different collision-avoidance operator $\delta(\Delta x, \Delta y)$, the proposed controller can employ different strategies without changing the control structure. Two operators for collision avoidance will be discussed here.

4.2 Collision-avoidance controller

Utilizing the collision-avoidance vector, a collision-avoidance controller is designed to transport the trapped cell to its destination while avoiding obstacles and compensating for the disturbance caused by blood flow. Fig. 17 shows the schematic of the proposed cell transportation control system with collision-avoidance capability.

The disturbance compensation controller was proposed to eliminate the effect of the drag force caused by blood flow. The disturbance observer can be used to adjust the moving direction of the target cell in the motion control. With this feature, a new disturbance compensation controller is designed in this section, aiming to avoid collision while eliminating the effect of drag force. Utilizing the collision-avoidance vector, the estimate

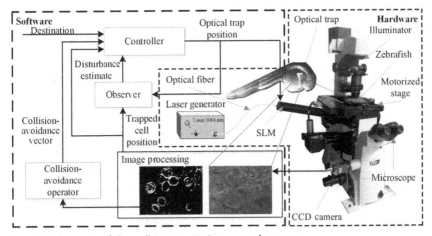

Fig. 17 Schematic of the cell transportation control system.

of disturbance, and the nonlinear saturation function, a controller that calculates the position of the generated optical trap is designed as follows:

$$\boldsymbol{u}_{OT} = \tilde{\boldsymbol{p}}_c + f_{r_0}\left(f_{r_0}\left(\tilde{\boldsymbol{e}}\right) - \tilde{\boldsymbol{\varepsilon}} + \gamma\boldsymbol{G}\right), \tag{44}$$

where γ denotes the weight of the collision-avoidance vector. Because the new controller is based on the disturbance compensation, the effect of the measure error $\boldsymbol{\omega}$ is also similar and was discussed in Section 3. In this section, the measure error can be ignored, and we have $\tilde{\boldsymbol{p}}_c = \boldsymbol{p}_c$ and $\tilde{\boldsymbol{e}} = \boldsymbol{e}$.

With the proposed controller (44), the positions of the other particles need not be identified, which significantly reduces the amount of online computational work.

4.3 Collision-avoidance operators

First, a so-called cotton operator is proposed by treating the trapped cell as if it is covered with a layer of invisible cotton to make the cell bounce off from obstacles.

Definition 10.3 (Cotton operator)
The cotton operator is defined as:

$$\delta_x(\Delta x, \Delta y) = \begin{cases} \dfrac{-\Delta x}{r} e^{-\frac{(r-R_2)^2}{2c^2}}, & R_2 < r < R_1, \\ 0, & \text{else} \end{cases} \tag{45}$$

$$\delta_y(\Delta x, \Delta y) = \begin{cases} \dfrac{-\Delta y}{r} e^{-\frac{(r-R_2)^2}{2c^2}}, & R_2 < r < R_1, \\ 0, & \text{else} \end{cases} \tag{46}$$

where R_2 denotes the minimum detection radius; $r = \sqrt{\Delta x^2 + \Delta y^2}$ is the distance between the current pixel (x, y) and the detected position $\boldsymbol{p}_c$ of the cell; and $c \in \mathbb{R}$ denotes the standard deviation.

Fig. 18A and B show the $\delta_x(\Delta x, \Delta y)$ and the $\delta_y(\Delta x, \Delta y)$ of the cotton operator, respectively. Fig. 18C illustrates the vector field of the cotton operator. Every pixel around the detected cell corresponds to a vector that is calculated by the collision-avoidance operator. The pixel with a smaller r corresponds to a longer vector length. However, when the pixel is extremely close to the trapped cell ($r < R_2$), it may be a part of the target cell image and should be ignored; in this regard, the corresponding vector length is zero. The pixel corresponds to a vector direction from the pixel position to the target cell position. Along this direction, the target cell will move away

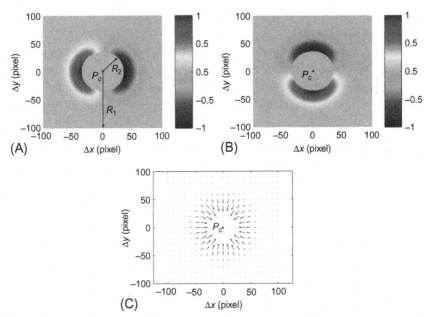

Fig. 18 Cotton operator. (A) The x-axis component of the cotton operator $\delta_x(\Delta x, \Delta y)$. (B) The y-axis component of the cotton operator $\delta_y(\Delta x, \Delta y)$. (C) The vector field distribution of the cotton operator.

from the corresponding pixel position. By considering all the pixels around the trapped cell, the collision-avoidance vector $\mathbf{G}$ can be obtained as the weighted sum of all the corresponding vectors with weights that are the intensity value $n_{p_{cx}+\Delta x, p_{cy}+\Delta y}(k)$ of pixels. Moving along the direction of the vector $\mathbf{G}$ will ensure that the cell avoids collision with other particles. Given that the cotton operator can be calculated offline, the online computational load for obtaining the vector is only $O(R_2^2)$, which is considerably lower than those of traditional obstacle detection algorithms.

The cotton operator utilizes a simple method to drive the trapped cell away from the other particles. With this strategy, the controller successfully avoids moving obstacles. However, when a stable obstacle exists in the environment and blocks the forward path of the controlled cells, the cotton operator has to drive the cells backward, which may not be the best method. Therefore, the cotton operator is more suitable for avoiding a moving obstacle but not for avoiding stable obstacles.

The effect of obstacle avoidance varies when different operators are used in the controller. Consider an obstacle with a neighborhood of pixels, with element number $m \in \mathbb{N}$ $(m \ll c^2)$ and its velocity relative to blood flow is expressed as $v_o \in \mathbb{R}^{2\times 1}$. The distance between the trapped cell and the obstacle is expressed as r_a.

Second, another collision-avoidance strategy that allows the trapped cell to bypass an obstacle in its forward path is proposed here. Given that this strategy makes the cell behave like a bar of soap that slips away from the obstacle, we designated this collision-avoidance operator as the soap operator:

Definition 10.4 (Soap operator)
The soap operator is defined as:

$$\delta(\Delta x, \Delta y) = \begin{bmatrix} j_x & -j_y \\ j_y & j_x \end{bmatrix} \delta_j(\Delta x, \Delta y), \tag{47}$$

where $j = [j_x, j_y]^T$ is a unit vector along the forward moving direction of the controlled cell and $\delta_j(\Delta x, \Delta y) = [\delta_{jx}(\Delta x, \Delta y), \delta_{jy}(\Delta x, \Delta y)]^T$ has the following expressions of the components:

$$\delta_{jx}(\Delta x, \Delta y) = \begin{cases} -(\cos\theta - 0.5)e^{-\frac{(r-R_2)^2}{2c^2}}, & R_2 < r < R_1, \\ 0, & \text{else} \end{cases} \tag{48}$$

$$\delta_{jy}(\Delta x, \Delta y) = \begin{cases} \dfrac{-\theta(\cos\theta + 1)}{2|\theta|} e^{-\frac{(r-R_2)^2}{2c^2}}, & R_2 < r < R_1, \\ 0, & \text{else} \end{cases} \tag{49}$$

where $\theta \in (-\pi, \pi]$ denotes the angle between vector j and vector $[\Delta x, \Delta y]^T$.

Unlike the cotton operator, the soap operator is not isotropic, and its utilization of cell motion in the forward direction makes the operator more purposeful. When an obstacle blocks the trapped cell from moving forward, the cell slightly changes its direction to continue its movement. Fig. 19A and B shows the components of the soap operator $\delta_j(\Delta x, \Delta y)$, which shares similar properties with the cotton operator. The pixel with a smaller r exerts a large influence when $r > R_2$, and has no influence when $r < R_2$. The direction of the soap operator, as shown in Fig. 19C, is different from that of the cotton operator. Therefore, the controller with the soap operator employs a collision-avoidance strategy that is different from that employed by the cotton operator. Although the computational load with the soap operator is more than that of the cotton operator because of the required extra coordinate transformation with Eq. (47), it is still within $O(R_2^2)$.

Note that the unit vector of the cell forward direction j, which is defined in Definition 10.4, determines the direction of the soap operator when the

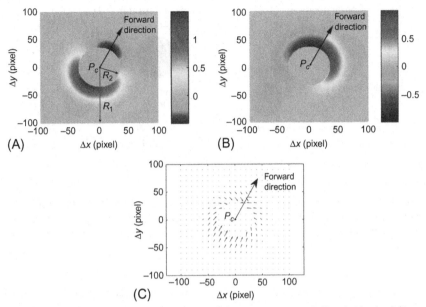

Fig. 19 Soap operator. (A) The component of the soap operator $\delta_{jx}(\Delta x, \Delta y)$ in the j direction. (B) The component of the soap operator $\delta_{jy}(\Delta x, \Delta y)$ in the $[-j_y, j_x]^T$ direction. (C) The vector field distribution of the soap operator.

collision-avoidance vector is calculated. This vector further affects the performance of controller in avoiding obstacles. In this study, two different forward directions are considered: the target direction $\left(f_{r_0}(e) - \tilde{\varepsilon}\right)/\left\|f_{r_0}(e) - \tilde{\varepsilon}\right\|$ and the instantaneous direction $\left(f_{r_0}(e) - \tilde{\varepsilon} + \gamma \mathbf{G}\right)/\left\|f_{r_0}(e) - \tilde{\varepsilon} + \gamma \mathbf{G}\right\|$. The target direction is fixed when the cell position error e and the estimate of the disturbance $\tilde{\varepsilon}$ are fixed. With the target direction, the moving direction of the trapped cell changes slightly when the cell encounters an obstacle, which helps the cell to keep moving forward to the destination in a highly cell-dense environment. With instantaneous direction, the collision-avoidance vector changes even when the position error e and the estimate $\tilde{\varepsilon}$ are fixed, which enables the trapped cell to continuously search the surrounding environment and escape from the local minimum.

Finally, the difference between the cotton operator and the soap operator is discussed. Although theoretically the controller with the cotton operator exhibits better obstacle avoidance performance than that with the soap operator, the controller with the soap operator behaves in a more intelligent manner in an obstacle-dense environment. Fig. 20 shows the difference between the cotton operator and the soap operator when the trapped cell

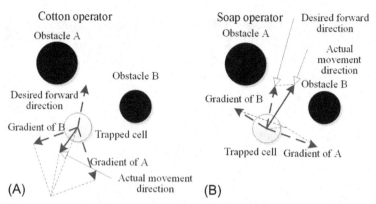

Fig. 20 Operators in handling two obstacles. (A) The collision-avoidance vector of the cotton operator pushes the cell to move away from the two obstacles to avoid a collision. (B) The collision-avoidance vector of the soap operator guides the cells to pass through between the two obstacles.

encounters two obstacles. Based on Eq. (44), the cell actually moves in the direction along the sum of vectors $f_{r_0}(e) - \tilde{\varepsilon}$ and γG. In Fig. 20A, the collision-avoidance vector of the cotton operator pushes the cell to move away from the two obstacles to avoid collision. In Fig. 20B, the collision-avoidance vector of the soap operator guides the cells to pass through between the two obstacles.

4.4 Experiments

For simulations, the target cells were transferred to a destination in a working environment with stable and moving obstacles. The simulation scenario was extracted from the actual in vivo environment of living zebrafish. Both the cotton operator and the soap operator with two different forward directions (namely, target direction and instantaneous direction) were employed in the simulations. The environmental parameters were selected as follows: trapping stiffness, $k = 1 \times 10^{-5}$ N/m; dynamic viscosity, $\mu = 6 \times 10^{-3}$ Pa s; cell radius, $r_c = 3 \times 10^{-6}$ m; maximum detection radius, $R_1 = 100$ pixel; minimum detection radius, $R_2 = 35$ pixel; image resolution, 1 pixel $= 1.224 \times 10^{-7}$ m; weight of the collision-avoidance vector, $\gamma = 0.15$; and sampling time, $T = 0.05$ s.

Figs. 21 and 22 show the simulation results for avoiding two different stable obstacles. In the simulations, the blood flow speed $\boldsymbol{v}_l$ was set to 0 to simulate the in vitro environment, and the standard deviation was $c = 15$. Fig. 21 shows that the trapped cell could avoid obstacles and finally arrive at the destination under the proposed controller with all the operators.

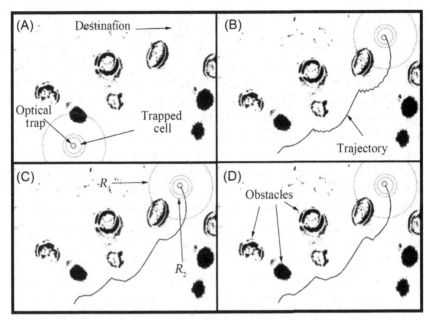

Fig. 21 Simulated trajectories with sparse stable obstacles. (A) Beginning scenario. (B) Cell movement trajectory under the cotton operator. (C) Cell movement trajectory under the target direction soap operator. (D) Cell movement trajectory under the instantaneous direction soap operator.

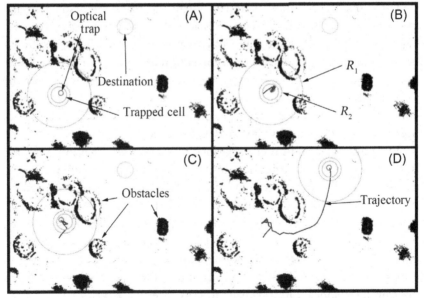

Fig. 22 Simulated trajectories with crowded stable obstacles. (A) Beginning scenario. (B) Cell movement trajectory under the cotton operator. (C) Cell movement trajectory under the target direction soap operator. (D) Cell movement trajectory under the instantaneous direction soap operator.

The cell movement trajectories with the target-direction soap operator (Fig. 21C) and the instantaneous-direction soap operator (Fig. 21D) were smoother, whereas the trajectory with the cotton operator (Fig. 21B) oscillated slightly when an obstacle was encountered. In Fig. 22A, the trapped cell is surrounded by obstacles at the beginning of the scenario. Only the use of the soap operator with instantaneous direction, as shown in Fig. 22D, could drive the trapped cell to the destination. Using the cotton operator (Fig. 22B) and the target direction soap operator (Fig. 22C) resulted in the trapped cell being blocked by the local minimum, which caused the cell transportation task to fail. Under the instantaneous-direction soap operator, as seen in Fig. 22D, the cell continued searching the surrounding environment and discarded the local minimum to move forward.

Fig. 23 shows the simulation results for the avoidance of moving obstacles. In this simulation, the blood flow speed was set as $v_l = (0.17, 0.1) \times 10^{-4}$ m/s, which simulated the in vivo environment, and the standard deviation was $c = 10$. At the times of 1.60–2.10 s, the trapped cell must move along the direction of blood flow to avoid collision with obstacle A. The use of both the cotton operator and the soap operator enabled the trapped cell to avoid the obstacles effectively. The controller behaved differently with different operators. Compared with the cotton operator (see Fig. 23A), the soap operator made the trapped cell move a longer distance to reach the destination and maintained a smaller distance to obstacle A at the time of 2.10 s (see Fig. 23B and C). When the cell density was high and avoiding all collisions became difficult, the controller with the cotton operator prioritized collision avoidance and tended to ignore cell transport, whereas the controller with the soap operator struck a balance between collision avoidance and cell transport. When the cell density was sparse, the cotton operator was more advantageous because of its stronger capability to avoid obstacles.

Experiments were further performed to transfer yeast cells in vitro and RBCs in 30-hour postfertilization (hpf) zebrafish in vivo. The environmental parameters used in experiments were set as follows: environmental parameter, $\beta = 0.125$ s; image resolution, 1 pixel $= 1.224 \times 10^{-7}$ m; maximum detection radius, $R_1 = 100$ pixel; minimum detection radius, $R_2 = 50$ pixel; weight of the collision-avoidance vector, $\gamma = 0.15$; and standard deviation, $c = 15$. The laser power in all experiments was set to 3 W ($\sim$450 mW in the sample), which provided an optical trapping force of 20 pN on cells without causing significant heat damage to cells (Clevers, 2011).

Fig. 24 shows the in vitro experimental results with the proposed controller. In this experiment, a yeast cell was controlled and the flow speed v_l

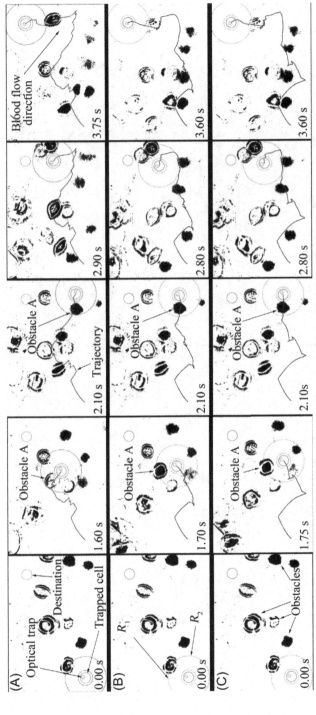

Fig. 23 Simulated trajectories with moving obstacles. (A) Cell movement trajectory under the cotton operator. (B) Cell movement trajectory under the target-direction soap operator. (C) Cell movement trajectory under the instantaneous-direction soap operator.

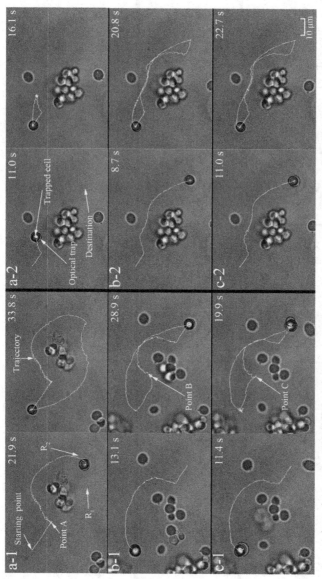

Fig. 24 Experimental results in vitro. (A) Cell movement trajectory with the cotton operator. (B) Cell movement trajectory with the target direction soap operator. (C) Cell movement trajectory with the instantaneous direction soap operator.

was zero. The yeast cell moved from the starting position to the destination, and then moved back to the starting position. Two different tracking tasks were implemented under the controllers with the cotton operator, the soap operator with target direction, and the soap operator with instantaneous direction. Fig. 24A-1, B-1, and C-1 shows the different performances of the operators in task 1, where the trapped cell was stuck at the local minimum (points A, B, and C). The trapped cell was stuck for 7 s at point A (Fig. 24A-1) and for 6 s at point B (Fig. 24B-1). This phenomenon occurred because the controllers with the cotton operator and the soap operator with the target direction could not escape from the local minimum. By contrast, the soap operator with the instantaneous direction could have the trapped cell search and escape from the local minimum; thus, the cell immediately left point C (Fig. 24C-1). Fig. 24A-2, B-2, and C-2 shows the different performance of the operators in task 2, where the trapped cell was required to pass between obstacles. As shown in Fig. 24B-2 and C-2, the soap operator guided the cell to pass between obstacles. By contrast, the cotton operator drove the trapped cell away as seen in Fig. 24A-2. Fig. 25 shows the position errors between the cell and the destination in these two tasks.

Figs. 26 and 27 present the results of an in vivo experiment performed on a living, 30-hpf zebrafish. The experiment was performed in the sinus venosus of the zebrafish, where the flow speed was $v_l = (-0.252, 0.09) \times 10^{-4}$ m/s. In this experiment, an RBC was controlled to move toward a destination. Fig. 26A–C shows the cell movement trajectories under the controller with the cotton operator, the soap operator with target direction, and the soap operator with instantaneous direction, respectively. Fig. 26 shows the corresponding position error $\|e\|$. All three operators helped the controller drive the trapped cell away from the other particles during transportation and arrive at its destination. As shown in Fig. 26A, the cotton operator maintained a larger distance between the target cell and the obstacles than the soap operator. This phenomenon indicated that the cotton operator emphasizes collision avoidance and optimizes the performance of the controller with the cotton operator when cell density is low. By contrast, the soap operator performs better when the cell density is high because it strikes a balance between collision avoidance and cell transport. As shown in Fig. 26B, the trajectory of the soap operator with target direction is the one with the smallest changes in its moving direction. This phenomenon makes the controller with the target direction soap operator more suitable for the rapid blood flow environment where the trap force is not very strong compared with the drag force. Under the soap operator with instantaneous

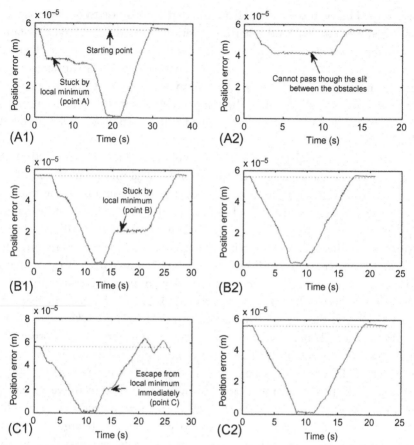

Fig. 25 Position errors of in vitro cell transportation experiment. (A) The cotton operation. (B) The target direction soap operator. (C) The instantaneous direction soap operator.

direction, the target cell had the most flexible movement, as seen in Fig. 26C. Therefore, the soap operator with instantaneous direction is more suitable for the slow blood flow environment where the trap force is sufficiently strong. Furthermore, the soap operator is able to handle the local minimum, which may appear in a slow (or static) blood flow environment.

5 Conclusion and future work

In vivo cell transport is a new research area vital for in vivo biological research and several potential applications, such as in determining the

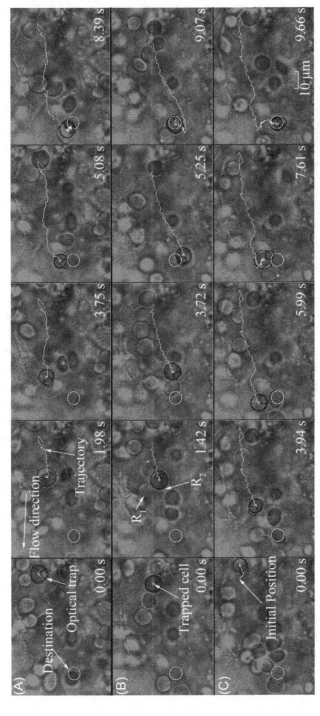

Fig. 26 Experimental results in vivo. (A) Cell movement trajectory with the cotton operator. (B) Cell movement trajectory with the target direction soap operator. (C) Cell movement trajectory with the instantaneous direction soap operator.

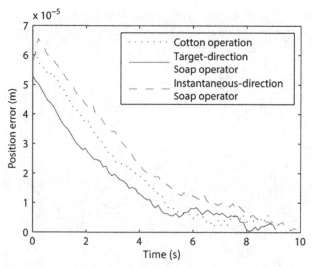

Fig. 27 Position errors in the in vivo cell transportation experiment.

mechanism of cancer metastasis, eliminating thrombus, targeted drug or cell delivery, screening nanoparticle-cell interactions for cancer therapy, and evaluation of tissue invasion in cancer and infection biology. This study initially proposes a close-loop control system for in vivo cell transport based on OTs technology. With this system, a disturbance compensation controller is developed to overcome the influence of blood flow on the cell. A collision-avoidance vector method is proposed to handle the collision problem during in vivo transport. The main contributions of this chapter can be summarized as follows:

(1) A close-loop control system is established to automatically transport cells in vivo by using a robot-aided OTs manipulation system. In this system, the cell position can be identified from the internal zebrafish images with image processing model and then tracked by position correlation method. A close-loop control system can be established based on the cell position by using a saturated P-type controller. Experiments are performed to demonstrate the effectiveness of the proposed system. The automatic in vivo cell transport was realized for the first time in this study, and the feasibility of the system was also demonstrated.

(2) A disturbance compensation controller is developed to overcome the influence of fluids (e.g., blood flow) on the cell. The proposed controller treats the influence of blood flow as a disturbance, which can be observed and compensated. The controller allows the trajectory to be adjusted online with the disturbance observer. When the trapped cell collides with other particles and deviates from its desired direction, the controller can enable the trapped cell to switch back to its original trajectory. Thus, the success rate of cell transportation is improved. The steady-state error of the cell motion can also be minimized by compensating for the disturbance caused by blood flow. Simulation and experimental results demonstrated that the proposed controller can efficiently and automatically transport a single trapped cell to the desired position.

(3) An automated cell transportation controller with collision-avoidance capability is developed to handle the collision problem during cell transport. A collision-avoidance vector method is proposed to avoid obstacles during transportation of the target cell. The proposed method integrates obstacle detection and collision avoidance into a single step, thereby reducing the duration of online processing while enhancing the accuracy of obstacle detection. With the proposed approach, different collision avoidance strategies are designed to suit for different transportation environments. The proposed approach exhibits the advantages of reduced online calculation, fast response, high accuracy, and disturbance compensation. Simulations and experiments were performed to demonstrate the effectiveness of the proposed controller.

This chapter may be used as basis for future studies.

(1) In this study, an automatic cell transport system is developed. However, the system only can handle the cell transportation in a fixed visual field, which is too small to the entire body. This kind of cell transport can be only named as a local cell transport. To realize a global cell transport, the visual field should follow the target cell and the optical tweezer needs to manipulate the cell uninterruptedly. Future work will involve the development of a global automatic cell transport system.

(2) With the proposed cell transport system, the target cell can be trapped and transported to the destination in zebrafish. This condition is sufficient for study of cancer metastasis. The cancer cell can be injected into zebrafish and transported to the target position with the transport system. In this technique, the location and initial cell number of the metastatic tumor can be precisely controlled. Therefore, future work will involve the study of cancer metastasis based on the proposed cell transport system.

(3) Magnetic manipulation can direct the magnetic microrobot in deeper tissues. Compared with the OTs, the structure of an in vivo magnetic manipulation system is similar and the dynamic equation of the control object is also the same. Therefore, several works in this chapter can be used in in vivo magnetic manipulation study. The proposed system can also be used as reference to build an in vivo magnetic manipulation system. Future work can focus on automatic in vivo microrobot manipulation with a magnetic manipulation system.

References

Arai, F., Yoshikawa, K., Sakami, T., Fukuda, T., 2004. Synchronized laser micromanipulation of multiple targets along each trajectory by single laser. Appl. Phys. Lett. 85 (19), 4301–4303.

Ashkin, A., 1970. Acceleration and trapping of particles by radiation pressure. Phys. Rev. Lett. 24 (4), 156.

Ashkin, A., Dziedzic, J., 1987. Optical trapping and manipulation of viruses and bacteria. Science 235, 1517–1521.

Ashkin, A., Dziedzic, J., Bjorkholm, J., Chu, S., 1986. Observation of a single-beam gradient force optical trap for dielectric particles. Opt. Lett. 11 (5), 288–290.

Ballar, D.H., 1981. Generalizing the Hough transform to detect arbitrary shapes. Pattern Recogn. 13 (2), 111–122.

Banerjee, A.G., Chowdhury, S., Losert, W., Gupta, S.K., 2012. Real-time path planning for coordinated transport of multiple particles using optical tweezers. IEEE Trans. Autom. Sci. Eng. 9 (4), 669–678.

Belharet, K., Folio, D., Ferreira, A., 2012. Simulation and planning of a magnetically actuated microrobot navigating in the arteries. IEEE Trans. Biomed. Eng. 60 (4), 994–1001.

Block, S.M., Blair, D.F., Berg, H.C., 1989. Compliance of bacterial flagella measured with optical tweezers. Nature 338, 514–518.

Chapin, S.C., Germain, V., Dufresne, E.R., 2006. Automated trapping, assembly, and sorting with holographic optical tweezers. Opt. Express 14 (26), 13095–13100.

Chowdhury, S., Thakur, A., Švec, P., Wang, C., Losert, W., Gupta, S.K., 2014. Automated manipulation of biological cells using gripper formations controlled by optical tweezers. IEEE Trans. Autom. Sci. Eng. 11 (2), 338–347.

Clevers, H., 2011. The cancer stem cell: premises, promises and challenges. Nat. Med. 17 (3), 313–319.

Cohen, M.A., Taylor, J.A., 2002. Short-term cardiovascular oscillations in man: measuring and modelling the physiologies. J. Physiol. (Lond.) 542 (3), 669–683.

Curtis, J.E., Koss, B.A., Grier, D.G., 2002. Dynamic holographic optical tweezers. Opt. Commun. 207 (1-6), 169–175.

Davies, P.F., Barbee, K.A., Volin, M.V., Robotewskyj, A., Chen, J., Joseph, L., Griem, M.L., Wernick, M.N., Jacobs, E., Polacek, D.C., Depaola, N., Barakat, A.I., 1997. Spatial relationships in early signaling events of flow-mediated endothelial mechanotransduction. Annu. Rev. Physiol. 59, 527–549.

Drummond, I.A., 2005. Kidney development and disease in the zebrafish. J. Am. Soc. Nephrol. 16 (2), 299–304.

Dufresne, E.R., Grier, D.G., 1998. Optical tweezer arrays and optical substrates created with diffractive optics. Rev. Sci. Instrum. 69 (5), 1974–1977.

Dufresne, E.R., Spalding, G.C., Dearing, M.T., Sheets, S.A., Grier, D.G., 2001. Computer-generated holographic optical tweezer arrays. Rev. Sci. Instrum. 72 (3), 1810–1816.

Dusenbery, D.B., 2009. Living at Micro Scale: The Unexpected Physics of Being Small. Harvard University Press.

Evensen, L., Johansen, P.L., Koster, G., Zhu, K., Herfindal, L., Speth, M., Fenaroli, F., Hildahl, J., Bagherifam, S., Tulotta, C., 2016. Zebrafish as a model system for characterization of nanoparticles against cancer. Nanoscale 8 (2), 862–877.

Finer, J.T., Simmons, R.M., Spudich, J.A., 1994. Single myosin molecule mechanics: piconewton forces and nanometre steps. Nature 368 (6467), 113–119.

Forbes, Z.G., Yellen, B.B., Barbee, K.A., Friedman, G., 2003. An approach to targeted drug delivery based on uniform magnetic fields. IEEE Trans. Magn. 39 (5), 3372–3377.

Fournier, J.M., Merenda, F., Rohner, J., Jacquot, P., Salathe, R.P., 2008. Comparison between various types of multiple optical tweezers. In: The International Society for Optical Engineering (SPIE), August 10, 2008–August 13, 2008, San Diego, CA.

Gou, X., Yang, H., Fahmy, T.M., Wang, Y., Sun, D., 2014. Direct measurement of cell protrusion force utilizing a robot-aided cell manipulation system with optical tweezers for cell migration control. Int. J. Robot. Res. 33 (14), 1782–1792.

Grier, D.G., 1998. Colloids: a surprisingly attractive couple. Nature 393 (6686), 621–623.

Gupta, G.P., Massagué, J., 2006. Cancer metastasis: building a framework. Cell 127 (4), 679–695.

Hu, S., Sun, D., 2011a. Automated transportation of single cells using robot-tweezer manipulation system. J. Lab. Autom. 16 (4), 263–270.

Hu, S., Sun, D., 2011b. Automatic transportation of biological cells with a robot-tweezer manipulation system. Int. J. Rob. Res. 30 (14), 1681–1694.

Johansen, P.L., Fenaroli, F., Evensen, L., Griffiths, G., Koster, G., 2016. Optical micromanipulation of nanoparticles and cells inside living zebrafish. Nat. Commun. 7, 10974.

Ju, T., Liu, S., Yang, J., Sun, D., 2014. Rapidly exploring random tree algorithm-based path planning for robot-aided optical manipulation of biological cells. IEEE Trans. Autom. Sci. Eng. 11 (3), 649–657.

Li, X., Liu, C., Chen, S., Wang, Y., Cheng, S., Sun, D., 2015a. Automated in-vivo transportation of biological cells with a robot-tweezers manipulation system. In: IEEE 15th International Conference on Nanotechnology (IEEE-NANO), pp. 73–76.

Li, X., Yang, H., Wang, J., Sun, D., 2015b. Design of a robust unified controller for cell manipulation with a robot-aided optical tweezers system. Automatica 55, 279–286.

Li, X., Liu, C., Chen, S., Wang, Y., Cheng, S.H., Sun, D., 2016. Automated in-vivo transportation of biological cells with a disturbance compensation controller. In: IEEE/RSJ International Conference on Intelligent Robots and Systems (IROS), pp. 2561–2566.

Li, X., Liu, C., Chen, S., Wang, Y., Cheng, S.H., Sun, D., 2017. In vivo manipulation of single biological cells with an optical tweezers-based manipulator and a disturbance compensation controller. IEEE Trans. Robot. 33 (5), 1200–1212.

Li, X., Chen, S., Liu, C., Cheng, S.H., Wang, Y., Sun, D., 2018. Development of a collision-avoidance vector based control algorithm for automated in-vivo transportation of biological cells. Automatica 90, 147–156.

Liesener, J., Reicherter, M., Haist, T., Tiziani, H., 2000. Multi-functional optical tweezers using computer-generated holograms. Opt. Commun. 185 (1), 77–82.

Molloy, J.E., Padgett, M.J., 2002. Lights, action: optical tweezers. Contemp. Phys. 43 (4), 241–258.

Nixon, M.S., Aguado, A.S., 2012. Feature Extraction & Image Processing for Computer Vision. Academic Press.

Oskarsson, T., Batlle, E., Massagué, J., 2014. Metastatic stem cells: sources, niches, and vital pathways. Cell Stem Cell 14 (3), 306–321.

Pries, A.R., Secomb, T.W., Gaehtgens, P., 2000. The endothelial surface layer. Pflugers Arch. – Eur. J. Physiol. 440 (5), 653–666.

Reicherter, M., Haist, T., Wagemann, E., Tiziani, H., 1999. Optical particle trapping with computer-generated holograms written on a liquid-crystal display. Opt. Lett. 24 (9), 608–610.

Reya, T., Morrison, S.J., Clarke, M.F., Weissman, I.L., 2001. Stem cells, cancer, and cancer stem cells. Nature 414 (6859), 105–111.

Stoletov, K., Klemke, R., 2008. Catch of the day: zebrafish as a human cancer model. Oncogene 27 (33), 4509–4520.

Tan, Y., Kong, C.-W., Chen, S., Cheng, S.H., Li, R.A., Sun, D., 2012. Probing the mechanobiological properties of human embryonic stem cells in cardiac differentiation by optical tweezers. J. Biomech. 45 (1), 123–128.

Tanaka, Y., Tsutsui, S., Kitajima, H., 2013. Design of hybrid optical tweezers system for controlled three-dimensional micromanipulation. Opt. Eng. 52(4).

Teng, Y., Xie, X., Walker, S., White, D.T., Mumm, J.S., Cowell, J.K., 2013. Evaluating human cancer cell metastasis in zebrafish. BMC Cancer 13 (1), 1.

Thakur, A., Chowdhury, S., Švec, P., Wang, C., Losert, W., Gupta, S.K., 2014. Indirect pushing based automated micromanipulation of biological cells using optical tweezers. Int. J. Robot. Res. 33 (8), 1098–1111.

Wang, M.D., Yin, H., Landick, R., Gelles, J., Block, S.M., 1997. Stretching DNA with optical tweezers. Biophys. J. 72 (3), 1335–1346.

Wang, G., Wen, C., Ye, A., 2006. Dynamic holographic optical tweezers using a twisted-nematic liquid crystal display. J. Opt. A: Pure Appl. Opt. 8 (8), 703–708.

Womersley, J.R., 1955. Method for the calculation of velocity, rate of flow and viscous drag in arteries when the pressure gradient is known. J. Physiol. 127 (3), 553.

Wu, Y., Sun, D., Huang, W., 2011. Mechanical force characterization in manipulating live cells with optical tweezers. J. Biomech. 44 (4), 741–746.

Wulff, K.D., Cole, D.G., Clark, R.L., Dileonardo, R., Leach, J., Cooper, J., Gibson, G., Padgett, M.J., 2006. Aberration correction in holographic optical tweezers. Opt. Express 14 (9), 4169–4174.

Xie, M., Wang, Y., Feng, G., Sun, D., 2015. Automated pairing manipulation of biological cells with a robot-tweezers manipulation system. IEEE/ASME Trans. Mechatron. 20 (5), 2242–2251.

Zhang, H., Liu, K.-K., 2008. Optical tweezers for single cells. J. R. Soc. Interface 5 (24), 671–690.

Zhong, M.C., Gong, L., Zhou, J.H., Wang, Z.Q., Li, Y.M., 2013a. Optical trapping of red blood cells in living animals with a water immersion objective. Opt. Lett. 38 (23), 5134–5137.

Zhong, M.C., Wei, X.B., Zhou, J.H., Wang, Z.Q., Li, Y.M., 2013b. Trapping red blood cells in living animals using optical tweezers. Nat. Commun. 4, 1768.

CHAPTER ELEVEN

Medical nanorobots: Design, applications and future challenges

Ahmad Taher Azar[a,b], Ahmed Madian[c], Habiba Ibrahim[c], Mazen Ahmed Taha[c], Nada Ali Mohamed[c], Zahra Fathy[c] and BahaaAlDeen M. AboAlNaga[c]

[a]Robotics and Internet-of-Things Lab (RIOTU), Prince Sultan University, Riyadh, Saudi Arabia
[b]Faculty of Computers and Artificial Intelligence, Benha University, Benha, Egypt
[c]School of Engineering and Applied Sciences, Nile University, Giza, Egypt

1 Introduction

Nanoscience is the field of science concerned with the study of material with sizes ranging between 1 and 100 nm. The term nano comes from the Greek word for dwarf; so a nanometer is the length scale of 10^{-9} m. Nanotechnology can be defined as the study or the development of products done on the nanoscale using nanomaterials. Furthermore, describing any minor field of technology with nano refers to applying or developing this field at the nanoscale (Ramsden, 2018). The major scientific revolutions in the twentieth century made it possible to merge information technology and biotechnology with nanoscience and nanomaterials which made it possible to other fields to emerge (Kumar and Kumar, 2014).

Robots—or robota, or mechanical laborers—is an idea introduced by Karel Čapek and Isaak Asimov in the theatrical play "Rossum's Universal Robots" in 1920; this play paved the way for robots as an idea to emerge. Tracking the inspiration for robots would lead to the seventeenth century in Japan, during the Edo period, where the karakuri ningyo toys were considered as robots for their mechanical properties and for their usage for tasks such as tea-serving (Hornyak, 2006). The characteristics needed nowadays to describe a device as a robot are the actuator, sensing element, controller, and frame. The actuator provides the robot with its degrees of freedom (DOF) while the sensing elements measure the properties of the surroundings. Additionally, the controller makes an input-based decision to modify the actuator mode relying on the controlling operation. Finally, the frame houses the components for safety purposes (Tsitkov and Hess, 2017).

Control Systems Design of Bio-Robotics and Bio-mechatronics
with advanced applications
https://doi.org/10.1016/B978-0-12-817463-0.00011-3

Assimilation of previously mentioned fields, robotics and nanotechnology, led to the synthesis of what is known as nanorobotics. Nanorobotics is an interdisciplinary compartment that supports applications in different fields such as biomedical engineering (Azar, 2011, 2012).

The rise of nanorobots was on December 29, 1959, when Richard Feynman introduced the idea of single atom/molecule manipulation using nanotechnology in his talk "There's Plenty of Room at the Bottom" (Kumar et al., 2014). Moving to 1974, Professor Norio Taniguchi illustrated nanotechnology and concepts of material deformation to output a separate cell (Kumar et al., 2014). Starting in the 1980s, Dr. Eric Drexler advertised nanoscale equipment in his published articles (Drexler, 1989, 1992). In 1987, Dr. Eric Drexler published the first book on nanotechnology that deals with the self-replication property of nanorobots (Drexler, 1987).

Through the rise of nanorobots, it was necessary to note the comparison between nanorobots and micro-robots as it shapes a main concerning conflict. Although both micro-robots and nanorobots are included in the class of small-scale devices, they have major differences that include scale, physics, and fabrication (Sitti, 2007). The primary difference is that components scale ranges from 1 to 100 µm (0.1 mm) for micro-robots while nanorobots components range from 1 to 100 nm (0.1 µm) (Ramsden, 2018). Regarding current achievements, micro-robots are taking advanced steps extended towards their manufacturing, but nanorobots are in the early stage of research and design. Besides, technologists rely on optical microscopes during working on micro-robots, which is not equipment compatible with nanorobots as they are too small to be seen. Therefore, electron microscopes have been used while developing nanorobots as they are dealing with an atom or a molecule at once. Additionally, different strategies are applied during fabrication of Nano and micro-robots; Bottom-up strategy is applied on the scale of nanorobots while micro-robots rely on top-down techniques of manufacturing (Ramsden, 2018). The collective field for micro-robots and nanorobots is the key applications in which both of them are anticipated to be utilized in healthcare, micro-, and nano-assembly (Bogue, 2010).

There is no doubt that the complexity of system design is directly proportional to the requirements that the system must fulfil. From this concept, the nanorobotics field emerges to cross new boundaries in robotics and broaden the domain of its applications. The smaller the size of the robot, the higher its capability to be used in critical size-limiting environments, and the more crucial the functionality of the robotic system (Tsitkov and Hess, 2017). In addition, here lies the distinction and importance of nanorobots from other known forms of industrial robots. The uniqueness of nanorobotics makes them the only solution for a set of challenging applications, especially in medicine. According to Kumar et al. (2014), research is conducted to fabricate nanorobots that help performing surgeries in vivo using molecular nanotechnology (MNT), doing diagnosis, delivering treatment to patients of serious diseases such as AIDs and cancer, and manipulating genes by chromosomal genetic replacement therapy. Nanorobotics not only has contributions in medicine, but also has great applications in other fields such as in mechanical structure of small elements to be used in future nanomachines, such as Drexler's bearing design (Drexler, 1992). Despite thorough interest being given to the research about nanorobotics, certain gaps have not been addressed well over the years. Obviously, previous work in the field provides a solid ground for the evolution of nanorobotics in different fields, but it lacks providing a complete conceptual design including all robot specifications and manufacturability. This insufficiency of design components is the reason why manufacturing nanorobots still a challenge under the given technological capabilities. Accordingly, the main contribution of this chapter is to provide a sufficient review on components of nanorobots and how they can be integrated. In addition, applicable drivers are discussed, with the aim of clarifying several working principles covering their navigation and control, and applications of nanorobotics in medicine is shown in brief in order to form a solid view about nanorobotics impact in such fields.

The remainder of this chapter is organized as follows. Section 2 is the literature review presenting all previous work with their achievements, and their limitations. Sections 3 and 4 are proposing the framework of nanorobots. The third section includes full information about the possible elements such as sensors, actuators, and body parts followed by the fourth one that covers the current applicable designs. Then, Section 5 is dedicated to succeeding nanorobots with recent applications in the medical field. Section 6 is an exhaustive discussion that represents current limitations and future opportunities for research in this field. Then, conclusion about all main lines is conducted in Section 7.

2 Related work

Believing in the vision that one day nanorobots would be able to repair human tissues, Drexler (1992) proposed to freeze his body, cryogenically, upon death so that one-day human technology would be advanced enough in order to bring life back to his tissues using nanorobots

(Drexler, 1992). With advances in science and technology, the process of fabricating nanorobots follows one of two main strategies: bottom-up and top-down strategies. The former is based on construction of nanorobots from individual molecules and atoms while the latter is based on reducing the sizes of microscale objects by chemical or mechanical methods into nanoscale structures (Petrina, 2012).

As far back as 2008, research was conducted to suggest and formulate a general algorithm or plan of how nanorobots can be manipulated within the human body, even in narrow blood vessels. A proposal for using flagellated magnetotactic bacteria in directing the nanorobot in vivo was presented by Martel et al. (2008). It uses computers to prove the idea that hybrid nanorobots—which are composed of both synthetic and biological components—can be controlled and tracked inside the body to deliver agents to tumor cells. In addition, another group of researchers applied an optimization method based on particle swarm optimization (PSO) algorithms in a simulation of multiple nanorobots investigating throughout the human body. The study aims to let the nanorobots organize and control motion on their own in order to have full monitoring over a coverage area (Hla et al., 2008). These robots can be used to discover the existence of a certain disease, rather than attacking a tumor cell or delivering a drug. Concluded from both researches, their approaches focus on propulsion, and algorithms to control movement and coordinate the robots rather than manufacturing them.

In 2009, research focused again on the propulsion of nanorobots, and more specifically, on using flagella as a method for movement. A proposal for the use of helical flagella as a propulsion method for nanorobots is shown in Subramanian et al. (2009). In this contribution, mathematical modeling and MATLAB simulation was performed to verify the efficiency of this propulsion method. Changes in the geometry of the helical pattern of flagella have shown that when the amplitude of the helical increases linearly, the model experiences faster propulsion and reduced internal strain at the flagella end than shown in a constant amplitude model. Additionally, a contribution by Wautelet (2009) described the construction problems of nanorobots starting from required precision to assemble such devices reaching the risk taken in approaching a full avoidance of surrounding disturbances that may cause malfunction.

Proceeding to 2010, a study by Merina (2010) was done to discuss the potential of using nanorobots in heart transplant surgeries. The study covers medical knowledge about the surgery as well as suggested approaches to fabricate nanorobots. This contribution could be considered a key for opening the door for further exploration in using nanorobots in medical applications. At this point, it is obvious that researchers are going deep into exploration of what nanorobots can do, how to benefit from the emerging field, as well as how to use current technologies to apply nanorobots in reality.

In 2011, a proposal by Martel (2011) was presented to discuss the use of flagellated bacteria as a propulsion for hybrid nanorobots until reaching the target area in the human body. This time, experimental results were provided for using magnetotactic bacteria in blood and in water to control its directional motion and steering. Fig. 1 shows the microscopic observation of the movement of hybrid microrobot using flagella. Fortunately, the advances of research that is concerned with nanorobotics in 2011 expanded to include another important aspect. A contribution was made to suggest an algorithm by which nanorobots can reach their target using the nerve signals in vivo. Block diagrams for directional control (Fig. 2A) and speed control (Fig. 2B) are proposed to describe the process (Quader et al., 2011).

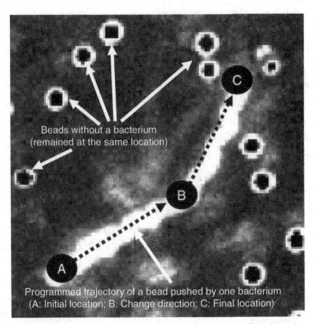

Fig. 1 Experimental results with optical microscope for a single flagellated magnetotactic bacterium propulsion for hybrid microrobot. *(From Martel, S., 2011. Flagellated bacterial nanorobots for medical interventions in the human body. In: Surgical Robotics. Springer, Boston, MA, pp. 397–416, with permission.)*

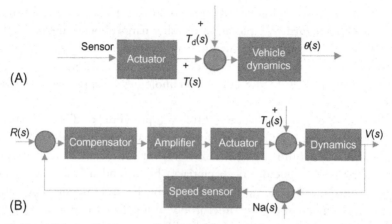

Fig. 2 Block diagrams for proposed directional control (A) and velocity control (B) for a nanorobot guided by nerve signals (Quader et al., 2011).

In 2012, the goal of using nanomedicine as treatment for diabetes encouraged scientists to find an expected path that the nanoparticles (NPs) would take to deliver the insulin in vivo. However, they faced some obstacles due to the metabolic changes, intestinal barrier functions, and differences in blood flow that prevented them from proposing any initial design (Krol et al., 2012). Moreover, in the same year, a study proposed to show the potential of using nanorobots propelled by external electromagnetic field to detect tumors in the human breast. The nanorobot is tracked and monitored using differential microwave imaging (DMI), and single walled carbon nanotubes (SWCNTs) are used in detection (Kosmas and Chen, 2012). It is exciting and interesting how research conducted in 2012 had encountered new areas for application of nanorobots in medicine such as diabetes and nervous system. There is no doubt that in 2013, researchers tackled the topic of nanorobots from different points of view. The revolution and development of nanomaterials had paved the way for Aguilar, in 2013, to propose theoretical considerations for using nanoparticles as targeted drug carriers and the factors affecting the process such as surface charge, particle size, and distribution (Aguilar, 2013). Additionally, the suggestion of nanorobots use in a new area of application was proposed. A simulation on nanorobots use for repairing injured blood vessels walls was done based on PSO algorithm and Herschel-Bulkley fluid model to represent the non-Newtonian blood (Trihirun et al., 2013). In this study, a rigid tube model for the blood vessel is used with the aim to use an elastic tube model in further research.

Furthermore, propulsion of nanorobots using flagella was, for the third time, a topic for study. In 2013, a proposal was made for the use of artificial flagellum in nanorobots to provide a propulsion mechanism for the robot inside the blood vessels. In other words, propulsion driven by external rotational magnetic field. Then, a motion MATLAB simulation is done on the proposed model where the flagellum is the tail of the robot (Xu, 2013). Also, a new approach of controlling the movement of nanorobots within the body is proposed to assist in performing surgery. This approach uses optical tweezers and light pulses emission to guide the nanorobot towards its destination (Thammawongsa et al., 2013). In the same marvelous year of 2013, fabrication of nanorobots came to reality using lasers. A fabrication process along with manipulation technique for a robot, whose components are within the nanoscale but the overall body size is in the microscale, is proposed; fabrication uses a laser beam from optical tweezers. Also, by using holographic optical tweezers, translation speed of 100 μm/s and rotation speed of 1140 deg/s is achieved (Fukada et al., 2013). The setup of this device by which the nanorobot was manufactured is shown in Fig. 3.

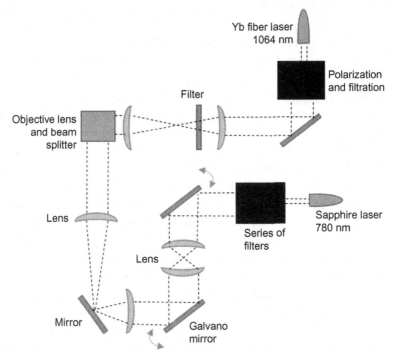

Fig. 3 Setup of holographic optical tweezers (Fukada et al., 2013).

As expected after the great achievements of 2013, 2014 came with greater enthusiasm about nanorobots and wider coverage of related topics. First, for nanorobotics manipulation in single cell structure, analysis done by Shen and Fukuda (2014) discussed two ways of cell injection: by hand (an inefficient way) and automatically (dependent on robotics technology). For cell positioning, different techniques are represented using micro tools and magnetic nanorobots. The manipulation system will also be responsible for giving info about cell mechanical and electrical properties and behavior with the surrounding environment. Another research direction was taken in 2014, which was drug delivery to cells using nanotechnology. Kumar and Kumar (2014) proposed a variety of emerging nanomaterials along with the methods to use them as drug delivery systems (DDS). The book discussed the ability to use NPs for neurological disorder, cancer treatment, cardiovascular diseases, and stroke treatment. The book faced some challenges that were presented as the toxicity of materials and introduction of different particles when implemented in vivo (Kumar and Kumar, 2014). The use of flagella was revisited in 2014 with some advances. This study is unique in simulating and comparing efficiency of nanorobots with different body to flagellum ratios (BFR). Results showed that for single actuation, BFR greater than 0.2 is recommended, while for multipoint actuation, BFR less that 1 is recommended (Jia et al., 2014).

Also in 2014, two different researches were conducted to provide communication techniques between a group of nanorobots to reach their target cells. As shown in the work done by Elsayed et al. (2014), the communication technique between a swarm of nanorobots is approached based on a modified PSO algorithm as well as blood cells detection and avoidance algorithm. In other words, the proposal intends to aid a group of nanorobots in reaching their destination cells faster (Elsayed et al., 2014). Using different strategies, another communication technique among multiple nanorobots is studied based on evolution strategy (ES). Simulations among three strategies—straight strategy, swap strategy, and high strategy—verifies that high strategy showed higher effectiveness in arriving at the target area (Ahmed et al., 2014). The background provided from both contributions can help in finding the best strategy among all the methods mentioned.

The development regarding fabrication of nanorobots never reaches an end. In 2014, two fabrication techniques were also presented in two researches. Focused ion beam (FIB) technique was used by researchers in Bao et al. (2014) to fabricate a nanorobot from platinum. Experiments of the movement of nanorobots in hydrogen peroxide solution were carried out to test the effect of geometry on robot motion. As a result, the robot shows bidirectional movement behavior, on the contrary with previous studies showing unidirectional movement. Propulsion of this robot is done chemically, as shown in Fig. 4. In addition to FIB method, researches had proposed a process of fabrication and manipulation of nanorobots using laser exposure. Nanorobots made from carbon nanotubes (CNTs) were fabricated, and a cell puncture using infrared laser to generate heat and make a puncture point is experimented with (Hayakawa et al., 2014). This work showed the new use of laser in cell puncture, rather than limiting its use to fabrication and manipulation.

Heading to 2015, a communication algorithm based on using acoustic signals was proposed to aid a swarm of nanorobots to coordinate themselves in non-Newtonian fluid with respect to target cancer cells. Based on the simulation results, the ratio of missed target decreased from 0.456 to 0.019 after implementing the algorithm (Zhao et al., 2015). Additionally, Loscri and Vegni (2015) proposed the use of acoustic signals for communication between nanorobots to reach the cell, destroy the tumor, and then inform the rest of the nanorobots, and simulate the behavior. Then, for advances in fabrication, nanorobots fabrication method using FIB and plasma sputtering technique (PST) was implemented in a new manner, but this time using platinum and gold elements. Robot movement was studied in water with different hydrogen peroxide concentrations while chemical reactions were the means of self-driving of the nanorobot; the higher the hydrogen peroxide concentrations, the higher the robot speed (Chen et al., 2015). The design of the nanorobot is shown in Fig. 5.

Additionally, manipulation of NPs using a nanorobot probe was discussed in 2015. The paper published by Korayem et al. (2015) mainly

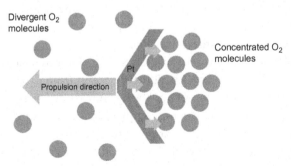

Fig. 4 The chemical propulsion mechanism of a catalytic platinum nanorobot in hydrogen peroxide solution (Bao et al., 2014).

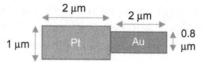

Fig. 5 The design of the nanorobot fabricated from Au and Pt using FIB (Chen et al., 2015).

discussed manipulation of NPs moving them by an atomic force microscope (AFM) nanorobot probe. They made their experiments on particles of different materials assuming desired position. The results aimed to limit the force that should be applied to reach a specific location in a typical time. In detail, the paper discusses the simulating and testing process supported by graphs.

For construction of nanosized objects and components of nanorobots, a study developed by Daulbaev et al. (2017) described the process of obtaining polymer films by the electrospinning method. The idea to use a 3D printer in combination with an electrospinning machine was newly proposed, which made it possible to print the simplest nano size objects using polymeric fibers.

Keeping up to date, in 2018, Farazkish (2018) discussed different reliability models of nanorobots, checking if their applicability is worthy or not. It also discusses the advantages of having trusted nanorobots and the barriers that should be overcome in order to get benefit from them. Moreover, a new aspect of research has come into play, which is the software for small-scale robots. In Tiemerding and Fatikow (2018), some software packages were evaluated in a survey with regard to multiple parameters and queried for advantages and disadvantages. Accordingly, any of them can be chosen based on the presented issue.

To sum up all the valuable contributions to the field of nanorobotics manufacturing, manipulation, and control, Table 1 shows the highlighting researches in a brief, organized timeline.

Although efforts made in nanorobotics-related research have been significant, there is still a gap between all the theoretical and experimental work presented and the possibility of applying nanorobotics in reality. This gap rises from the fact that there is no complete framework of the fabrication, manipulation, control, and simulation of nanorobots in vivo. Accordingly, contributions made that addresses the topics from different aspects are not homogenous with each other in a manner that allows their merging in one complete nanorobot. This gap is the motive for this chapter with the

Table 1 Summary of nanorobotics-related work.

Year	Authors	Contribution
2008	Martel et al.	Proposal to use flagellated magnetotactic bacteria in controlling and tracking hybrid nanorobots until reaching tumors
	Hla et al.	Proposal to detect diseases by a swarm of nanorobots guided by PSO
2009	Subramanian et al.	Modeling MATLAB simulation to verify the effectiveness of flagella propulsion in hybrid nanorobots
2010	Merina	Proposal to use nanorobots in heart transplant supported by suggestions for nanorobot fabrication and control
2011	Martel	Experimental results for using magnetotactic bacteria as a propulsion method to control movement of nanorobots
	Quader et al.	Proposal of algorithm for enabling control and manipulation of nanorobots depending on nerve signals
2012	Krol et al.	Path planning of nanoparticles to deliver insulin in order to help in treating diabetes
	Kosmas and Chen	Proposal of using external electromagnetic field to manipulate a robot to detect human breast cancer
2013	Aguilar	Proposal of theoretical considerations for using nanoparticles as targeted drug carriers analyzed for many factors
	Trihirun et al.	Simulation on using nanorobots for repairing injured blood vessels walls using PSO
	Xu	Motion MATLAB simulation for artificial flagellum as a means of propulsion for nanorobots
	Thammawongsa et al.	Proposal for using nanorobots in surgery and controlling their movement within the body by using optical tweezers and light beams
	Fukada et al.	Fabrication and manipulation of nanorobots using laser beam from optical tweezers with experimental results
2014	Shen and Fukuda	Proposal of methods for cell injection, cell positioning, and cell manipulation based on nanorobots and nanomaterials science
	Kumar and Kumar	Proposal of variety of emerging nanomaterials along with the methods to use them as DDS
	Jia et al.	Simulation and analysis of using flagella in propulsion of nanorobots for varying body to flagella ratios to test the effect

Continued

Table 1 Summary of nanorobotics-related work—cont'd

Year	Authors	Contribution
	Elsayed et al.	Proposal for communication technique between nanorobots using modified PSO algorithm as well as blood cells detection and avoidance algorithm
	Ahmed et al.	Proposal for comparing three evolutionary strategies—straight strategy, swap strategy, and high strategy—for communication between nanorobots
	Bao et al.	Fabrication of nanorobots composed of platinum using FIB technology
	Hayakawa et al.	Fabrication, manipulation, and cell puncture of nanorobots made from carbon nanotubes using laser exposure
2015	Zhao et al.	Simulation of communication algorithm using acoustic signals to aid a swarm of nanorobots to coordinate themselves in non-Newtonian fluid with respect to the cancer cells target
	Loscri and Vegni	Simulation of using acoustic signals in communication between robots that destroy the tumor cells
	Chen et al.	Fabrication of nanorobots from gold and platinum using FIB and PS technique
	Korayem et al.	Experimental results for manipulation of nanoparticles moving by AFM nanorobot probe
2017	Daulbaev et al.	Fabrication of nanorobot components by 3D printing using polymer films by the electrospinning method
2018	Farazkish	Proposal of different reliability models of nanorobots checking if their applicability is worthy or not
	Tiemerding and Fatikow	Evaluation and comparison for software packages for small scale robots regarding multiple parameters

determination to provide a complete review about manufacturing nanocomponents, applicable designs of drivers, and previously applied missions. This work will be the bridge taking nanorobots from research labs to reality.

3 Medical nanorobotic components design and selection

Since the rise of the primary models of robot manipulators (Stewart, 1965), the control issue of such frameworks has drawn the consideration of various scientists (Ammar and Azar, 2020; Azar et al., 2017, 2020a, b, 2018a, b, 2019a, b; Barakat et al., 2020; Azar and Serrano, 2019). The fabrication of nanorobots is, in fact, an assembly practice for different nanorobots subsystems; for example, electronic chips, sensors, actuators, and drivers. These subsystems are manipulated by control systems to ensure that all subsystems work in harmony so that the objective behind the fabrication of nanorobots is achieved. Obviously, different types of those subsystems should be chosen according to size, environment, and the task to be done. Surprisingly, within the medical nanorobots category itself, different subsystems are used based on the task (i.e., disease detection or treatment), the targeted area within the body (i.e., limbs, heart, etc.), the mechanism of diagnosis (i.e., a swarm of nanorobots or an individual one), the duration that the nanorobots will spend inside the body, and finally the disease itself. All these factors combined make it necessary to analyze thoroughly all the different methods that can be used to manipulate nanorobots and then design a specific-purpose nanorobot that can be assembled by integrating all these subsystems in a single assembled package.

With the increasing interest in nanorobots among different fields, especially medical applications, many methods have been suggested to fabricate nanoelectronics systems, nanosensors, nanoactuators, and nanodrivers. However, these methods have two concerns. The first concern is that not all of the materials used in the literature are bio-oriented to be used in vivo. The second concern is that, with the high specialty of research works, individual development of each method is proposed without extensive focused framework to aid nanorobot manufacturing.

3.1 Nanoelectronic chips in nanorobots

In the last century, advances in electronic chips were enormous and spanned multiple areas of development, such as new functionalities and smaller sized components. The microelectronic chip was first produced in 2000, which opens the door for researchers to think about nanoelectronic chips and making efforts in order to minimize the chip size. The transition between microelectronics and nanoelectronics happens by 2010, when researches made a nanoelectronic chip with length 32 nm while also maintaining most of the functionality they planned to have (Hoefflinger, 2012).

3.1.1 Nanomaterials-based nanoelectronics

Nanomaterials is a broad term that include many materials with a wide range of applications. While designing an electronic system to function in a nanorobot, it is wise to check the opportunities provided by nanomaterial for having electronic components with such small size. For example, nanoparticles from alloying tin and silver are proposed to be used in lead-free nano interconnects in electronic chips. Another example is using polymer nanocomposites for making embedded capacitors in circuits. Last but not least, nanomaterials have been proposed frequently to be used in nano interconnects and circuit adhesion and connections with the substrate (Wong et al., 2010), as shown in the process in Fig. 6 (Li et al., 2010).

Nanopacking is the process of placing and integrating an electronic component on a nanochip. In nanopacking technology, some materials face limitation in being applied in nanosized chips. For example, copper in nano size below 22 nm is subjected to electro migration, so it fails on such a level. Accordingly, interests had been given to nonconventional materials to see their possibility for implementation in nanosized chips. One of these unconventional methods is using graphene nanoribbons (GNRs) and CNTs in fabricating nano interconnects for electronic chips. Additionally, integration between these nano interconnects and complementary metal-oxide semiconductor (CMOS) systems are possible, where the carbon-based interconnects open new opportunities for CMOS technology to minimize the size of chips and implement high frequency CMOS oscillators with reduced component heating (Chiariello et al., 2013). Vertical interconnects design is shown in Fig. 7 while horizontal interconnects design is shown in Fig. 8.

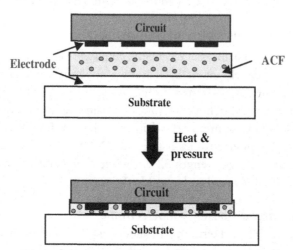

Fig. 6 Bonding of nanocircuits using nanomaterials based anisotropic conductive films. *(From Li, Y., Moon, K.., Wong, C., 2010. Nano-conductive adhesives for nano-electronics interconnection. In: Wong, C., Moon, K.S., Li, Y. (Eds.), Nano-Bio-Electronic, Photonic and MEMS Packaging. Springer, Boston, MA, with permission.)*

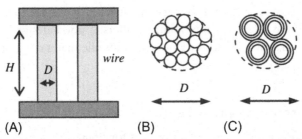

Fig. 7 Vertical interconnects in (A) the structure of interconnects, (B) the cross section of an interconnect with single walled CNTs, and (C) the cross section of an interconnect with multi-walled CNTs. *(From Chiariello, A.G., Maffucci, A., Miano, G., 2013. Temperature effects on electrical performance of carbon-based nano-interconnects at chip and package level. Int. J. Numer. Model.: Electron. Netw. Devices Fields 26 (6), 560–572, with permission.)*

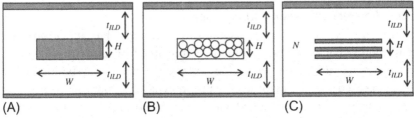

Fig. 8 Horizontal interconnects in (A) the structure of interconnects, (B) the interconnects made by CNTs, and (C) the interconnects made by GNRs. *(From Chiariello, A.G., Maffucci, A., Miano, G., 2013. Temperature effects on electrical performance of carbon-based nano-interconnects at chip and package level. Int. J. Numer. Model.: Electron. Netw. Devices Fields 26 (6), 560–572, with permission.)*

3.1.2 Nano optomechanical systems for nanoelectronic chips

Briefly, optomechanics is a field of science concerned with the effect and interaction between optical radiation and mechanical systems through radiation pressure. Particularly, the relation between a laser beam and mechanical vibrations is an example of optomechanics applications that is applied at micro and nano scale resonators. As medical nanorobots experience the need to measure mechanical movement of other particles in the swarm or particles inside the body, an optomechanical system integrated on the chip for sensing that mechanical displacement provides a good mechanism to implement robot functions and tasks. Without optomechanical systems, electrical methods of measuring the displacement would have been the available solution even with its limitation for nanorobot size. Nevertheless, the external optomechanical systems—when they are not integrated on chip—suffer from diffraction of laser while being in use in nanoelectromechanical systems (NEMS). Thus, the on–chip solution offers a good design for medical nanorobots that aims to measure movement while maintaining its small size (Diao et al., 2017).

3.2 Nanosensors in nanorobots

For fabricating a bio-nanomedical robots for operating in vivo, high accuracy in functionality, operation, and communication between the robot and the driving mechanism is needed. Indeed, all these targets are to be achieved using sensors. Sensors are essential to locate the robot with respect to other particles or other robots in a swarm of nanorobots.

3.2.1 Polymer clusters as nanosensors

Nano switches can be a great option for nanosensors in bio-applications and nanorobots. They have been proposed in the literature for usage in ion concentration measurements in cells. Particularly, nanosized polystyrene clusters are being used as switches for sensing purposes based on optical manner. This optical detection is accompanied with the use of florescent dyes to establish a detection and recognizing mechanism as shown in Fig. 9 (Méallet-Renault et al., 1999).

Although being a promising and good sensing methodology that is special for cells, the size of polymer nanoparticles used for making the nanocluster in literature is 100 nm. Further research and trials to make the size smaller is recommended before implementation in the detection and diagnosis nanorobot.

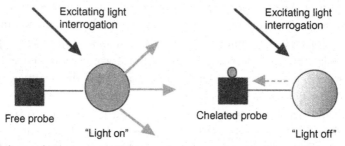

Fig. 9 Polymer clusters nano switches used as nano sensors in living cells for ion detection. *(From Méallet-Renault, R., Denjean, P., Pansu, R.B., 1999. Polymer beads as nanosensors. Sens. Actuators B Chem. 59 (2–3), 108–112, with permission.)*

3.2.2 Silver-based nanocluster nanosensors

Nanoclusters (NCs) are the combination of several NPs and atoms of one or more elements in one cluster or package. Throughout the years, interest had been given to fabricating silver nanoclusters for their high sensitivity in detecting lead ions in solutions and living cells, which gives them great potential for usage as nanosensors in medical nanorobots. However, the rapid development has faced a challenge in the instability and long preparation time for fabricating silver-based NCs. Fortunately, a proposal made by Wang et al. (2017) illustrated a novel method for fabrication with shorter time and high stability of clusters in terms of time, metal and pH stability. That method depends on using chemical agents in order to speed the chemical process of nanocluster aggregation.

3.3 Nanoactuators in nanorobots

Nanoactuators are mechanisms for imparting movement and actuation in nanostructures. Based on this concept, they are crucial parts in nanorobots used in medical applications in order to allow the robot to move and interact with living cells and particles. As wide as the range of applications for nanoactuators in the medical field, various research areas and methods are discussed in literature for fabricating nanoactuators. Actuators, in general, are classified according type of motion as rotary or translational actuators. On the nano scale, actuators are mainly dependent on a reaction or factor level like moisture, temperature, or PH level.

3.3.1 Electrostatic force-based nanoactuators

Gripper is a kind of end effector that can be used and implemented in nanorobots. In the work done by Fujiwara et al. (2016), layers of graphene are folded using a nanomanipulator for being used as a nano gripper driven via electrostatic force. This process is in fact divided into two sub processes: gripper fabrication and actuation implementation. The gripper is made by forming a pattern at first then making cuts in the pattern and folding them. Afterwards, energizing an electrode probe and the graphene gripper by different polarities results in the movement and manipulation for the gripper. The two-step process is shown in Fig. 10, where major stages are highlighted. This complete design can be an add-on to the structure of any nano medical robot for permitting the robot to interact and grab cells and particles in vivo.

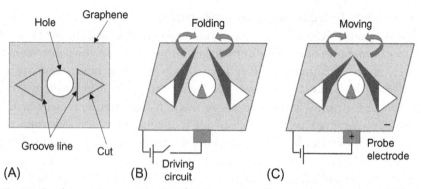

Fig. 10 Graphene actuator with electrostatic actuation illustrated as in (A) the unfolded graphene pattern, (B) the folded graphene pattern, and (C) the actuation method (Fujiwara et al., 2016).

3.3.2 CNTs-based nanoactuators

Surprisingly, in addition to being used in fabricating nanoelectronic chips and nanosensors, CNTs play a role in fabricating nanoactuators. The design of this nanoactuator is based on a cantilever beam design for a manipulator reinforced by CNTs and actuation using electromechanical coupling, as shown in Fig. 11 (Yang et al., 2014).

3.3.3 Viral protein-based nanoactuators

Focusing on getting use of biomaterials and biocomponents in the fabrication of nanoactuators, a proposal for a medical nanorobot for viral detection was made done by Dubey et al. (2003). The effective and significant method has the same actuation concept as the envelope proteins change of a virus while it tries to penetrate a cell. The proposed actuation method depends

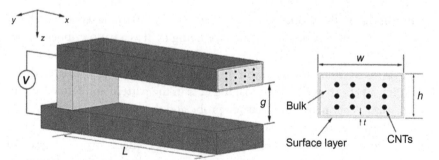

Fig. 11 Structure of CNTs-reinforced nanomanipulator actuated by electromechanical coupling. *(From Yang, W.D., Wang, X., Fang, C.Q., Lu, G., 2014. Electromechanical coupling characteristics of carbon nanotube reinforced cantilever nano-actuator. Sens. Actuators A Phys. 220 (2014), 178–187, with permission.)*

on using similar proteins and controlling the conformational changes in the actuator movement using environmental and experimental conditions, which can cause a linear movement of 10 nm (Dubey et al., 2003).

3.3.4 Prefoldin-based nanoactuators

From a biological point of view, prefoldin is a type of protein used as a molecular chaperone (proteins aiding in processing molecular structures) in single-celled microorganisms as well as multicelled microorganisms. Structure of prefoldin consists of multiple coils of proteins with a cavity at the middle of the structure. In a proposal made by Ghaffari et al. (2010), a design for a bio-nanoactuator used to hold bodies in the nano size intended to use the protein itself as an actuator, which presented a great solution for bio-safety and healthiness when used in vivo. Using simulation, the research established that some segments in the prefoldin are more flexible than others, a result that can further assist in controlling the actuator movement (Ghaffari et al., 2010).

3.3.5 Focused ion beam manufactured, thermally driven nanoactuators

This specific type of actuator is used to drive a bimorph. In fact, bimorph is a cantilever shaped structure consisting of several layers and, by design, it is allowed to move at one end. This bimorph is made using a focused ion beam-chemical vapor deposition (FIB-CVD) process, in which fabricating nano structure is done using ion energy for localizing and adding structure. The actuation method of the bimorph is done using thermal energy where the actuation process itself is a result of thermal strain. The actuator is connected to a plate that is heated electrically, and transfers heat energy to the actuator. Controlling the current in the circuit gives a direct control over the plate heat and thus the manipulator strain and movement (Chang et al., 2007). Fig. 12 demonstrates the setup of the manipulator for thermal actuation, while Fig. 13 illustrates the structure of a FIB manipulator with its materials.

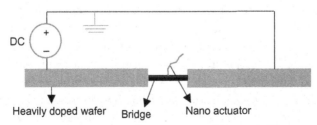

Fig. 12 Setup for thermal actuation of nanostructured bimorph (Chang et al., 2007).

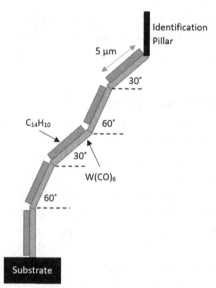

Fig. 13 Structure of nanosized bimorph actuator (Chang et al., 2007).

4 Applicable designs

As a proof of nanorobots' achievements, several designs have been approached and experimentally tested for its validation. Indeed, many of those designs have been conducted for nanorobots to serve medical field requirements and desires. The following are the designs that are most common in the recent progress.

4.1 Drug delivery system using hydrogel bilayer

One of these designs is self-folding hydrogel bilayer, which is especially suited to drug loading, encapsulation, and transport, as it is able to absorb several times more drug solution than its dry weight (Huang et al., 2016). The research done by Huang, Petruska, and Nelson handles the problem that hydrogels perform drug prerelease before reaching the targeted area of cells, due to high water content; it then introduces a mechanism preventing this phenomenon without affecting drug loading efficiency (Huang et al., 2016). Hydrogels have many advanced properties such as self-folding capability through stress-induced bending, exceptional drug loading ability, forming 3D structures from 2D patterns, and ability to control drug release rate (Fusco et al., 2015). This process is only compatible with drugs that are insensitive to UV-light.

The proposed design is based on a hydrogel bilayer that is composed of a drug-loaded layer and supporting layer. The first is a thermally responsive hydrogel with a relatively high-swelling property and is responsible for drug carry and temperature dependent folding control. The second degradable magnetic nanocomposite representing the supporting layer and concerns about folded shape and enabling the tube to be magnetically manipulated. Encapsulation occurs by raising the temperature to body temperature of 37°C, causing the drug-loaded layer to be isolated from the external surrounding environment. Then, the tube is magnetically guided to the target area without leaking drug on the way. Drug release is introduced once the supporting layer is degraded causing the tube to unfold (Huang et al., 2016).

4.1.1 Hydrogel bilayer fabrication

The bilayer is fabricated through several steps using photolithography. First, a magnetic field of 10 mT is applied in the planar direction for 1 min to align the magnetic NPs in the nonresponsive pregel solution. The direction of the applied magnetic field determines the structure folding direction. A non-thermally responsive pregel solution is polymerized by UV light (365 nm, 3 mW/cm^2) between a glass mask and a 10-μm thick spacer for 1.5 min to form a supporting layer. The spacer is then removed, and the thermally responsive pregel solution is introduced into the space between the photomask and the new 30-μm thick spacer substrate. The thermally responsive layer below the supporting one is polymerized by UV-exposure for 2 min. After UV curing, the cell is opened, and the bilayers attached to the photomask are released. Note that they are folded into tubes once immersed in water.

Regarding the analysis of hydrogel nanocomposites, characterization is achieved by weighing them in different temperatures starting from 22°C, increasing to 40°C. Each layer goes through the cycle of polymerization, drying, and then hydration using deionized (DI) water at room temperature, then it is kept in a water bath with a controllable temperature. This process provides enough information about dried and swollen weights to determine the weight swelling ratio (WSR) of hydrogel nanocomposite using Eq. (1):

$$\text{WSR} = \frac{M_s - M_d}{M_d} \qquad (1)$$

where M_d is dried weight, M_s is swollen weight, and WSR is weight swelling ratio.

Another analysis for drug release at room temperature and body temperature is performed as a part of testing criteria. Thermally responsive hydrogel disks, with a diameter of 7 mm and thickness of 400 µm, are, accordingly, prepared to analyze the drug encapsulation and leakage. Between two glass slides that are separated by a stacked cover glass, hydrogel polymerization is sustained for 2 min. Afterwards, it is immersed in DI water for two days to allow hydrogel to detach from the glass slides. After cutting it in defined diameters, gel disks are exposed to air for a day and again immersed in a model drug solution (Brilliant Green, BG in PBS solution, 1 mM) for a day at room temperature. Disks are removed from the solution and again immersed in DI water for 10 s to clean up any residue and immersed in 300 of PBS solution within an Eppendorf; release is monitored at room temperature for a day and then all samples are left in monitoring at room temperature for 10 days—the assumed period for a sample to release all drug quantity contained. The PBS solution and any release of BG is collected at defined periods and measured by UV-VI spectroscopy at the 622 nm wavelength and new solution is added to achieve best sink conditions.

Doing the drug release analysis at room temperature requires following steps of immersing bilayers in water bath at a temperature of 45°C for 1 min. As a result of placement in a temperature that is higher than the lowest-critical-solution temperature (LCST) of NIPAAm, the internal water is pressed out and removed with tissues. Bilayers are then placed in a 1 mM BG solution and left to absorb the drug for a day, after which, they are immersed in DI water for 10 s, rinsing off any residuals. Finally, they are introduced to 300 µL PBS solution within an Eppendorf and release progress is monitored for a week considering the collection, measuring, and replacement of the solution each 5 min.

Regarding analysis at body temperature, the same process applies, with an additional step of heating bilayers to body temperature of 37°C for an hour and then removal from the BG solution. Degradation characterization is performed for individual bilayers and supporting layers. They are immersed in 1 mM concentrated sodium hydroxide (NaOH) solution to boost progress of the process and kept at body temperature for four days, mimicking the in vivo environment. According to monitoring and observation of previously conducted analysis, the behavior of the hydrogel mechanism is provided. It applies the property of self-folding due to coupling two material layers—drug-loaded layer and supporting layer—that have different WSR. In this context, the supporting layer has the low WSR while the drug-loading layer has a relatively high ratio. Changing temperature affects the behavior of the mechanism in different cases. Generally, WSR of the drug-loaded layer decreases as temperature increases, causing an unfolding process to occur. When temperature is set to a value higher than the LCST, the hydrogel bilayer folds in an opposite sense, giving constant WSR for the two layers. Moreover, having ambient temperature that is much lower than the LCST causes the bilayer to fold in a cylindrical manner, forming a tube with the drug-loaded layer exposed to the surrounding environment; this scenario is advantageous in the drug release process. In contrast, when temperature is higher than LCST, the bilayer folds in the opposite way, hiding the drug-loaded layer inside and preventing high rate leakage of the drug; this scenario is beneficial for motion until the device reaches the targeted area. Fig. 14 shows a hydrogel bilayer capsule at different temperatures, giving three different behaviors. Self-folding occurs at room temperature while refolding happens at body temperature.

Thermally responsive release of hydrogel drug shows a step of success as it decreases the leakage to approximately 20% at high temperature, although it approaches 50% at room temperature. In other words, leakage and amount of loaded drug depends on pores size, which slightly changes with temperature. Accordingly, hydrogels initially release an amount of drug in high temperatures that is higher than the amount released at room temperature due to pores squeezing. Hydrogels release 70% of loaded drug in two days when left at room temperature; however, they releases only about 30% of it when left at body temperature. Actually, encapsulation of drugs can be

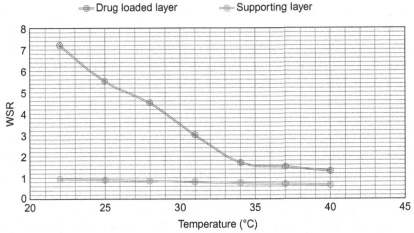

Fig. 14 Hydrogel capsule at different temperatures.

enhanced by isolating the drug-loaded layer such as implemented in the behavior of refolding or by adding an extra supporting layer to limit diffusion rate of drug. Degradation of layers differs according to which layer is meant to degrade. For supporting layers, it takes only 1 h to completely degrade, while drug-loaded layer takes about four complete days to completely degrade, which is about 72 times the duration required for the supporting layer.

4.2 Artificial bacterial flagella

The studies conducted on the micro/nano organisms that live in the low Reynolds number fluids proved that those organisms have a wide range of swimming techniques. The two most reliable techniques for micro/nano organisms' nonreciprocal motion are the flexible oar motion and the corkscrew motion. The flexible oar motion is mainly found in the motion of eukaryotes as they propel themselves; once activated, a propagated wave is generated by an active flagellum. The corkscrew motion is mainly presented in some prokaryote bacteria. This motion contains a group of flagella driven by helical shaped motors. Given the recent advancement in nanotechnology, scientists are working to fabricate devices that are able to propel in the micro and nano scales. The main obstacles that face the scientists during working at these scales are the power sources and actuations methods. Operating on nano scale made it impossible for the traditional power supplies to be compatible with the size of the robot that needed to be directed in biomedical applications. Seeking an acceptable solution to overcome this obstacle, two main approaches were improvised: depending on external power supply or harvesting energy from the surrounding environment. Both approaches proved to be efficient on the micro and nano devices that operates in low Reynolds number liquids. Those swimmers are using chemical locomotion to apply the approach of harvesting the energy from chemical components in the environment. Another type of swimmer benefits from being driven by a nanomotor; these are also known as artificial swimmers, as they increase their swimming velocity to reach 100 body lengths per second, compared to the fastest animal on earth, the cheetah, whose speed reaches approximately 25 body lengths per second. Using external power supply introduces the artificial swimmers with the advantage of being able to function with the least energized fluids (Zhang et al., 2009).

The first introduced remotely controlled helical swimmer was in 1996 and it was 0.15 mm with a permanent magnet fixed on one end, then it was developed based on the motion of spermatozoa to be a micro swimmer with a paramagnet that enables its attraction to a blood cell. After being attached to the blood cell, oscillating magnetic waves are applied to perform a propelling motion around the cell similar to the motion of eukaryotic flagellum. The modern advancement in helical swimmers is based on the same concept of motion, yet with the diameter of 200 nm and length of 2 μm, which enables the swimmers to operate on the cellular level with accurate position control and high velocity. Another actuation method for micro/nano swimmers can be harnessing the natural bacteria and using it as a power source. The required force to move the swimmer can be generated by molecular motors attached to these bacteria, this force can drive the swimmer randomly until it is controlled by external excitation methods such as light and magnetic fields. Focusing on the artificial bacterial flagella (ABF) as nanorobots, there are several concerns about employing them in sensitive applications such as surgeries and different biomedical fields. Those concerns can be defined as the precise motion control, separation of the individual swimmer from the associated group, accurate force and torque control of ABF, and the optimization of the swimmer's shape to fit the desired application. Due to the sensitivity of the application that uses nanorobots, missing or losing control of one of those parameters can result in serious consequences. Fig. 15 illustrates the major steps to use the ABFs in different applications.

4.2.1 Fabrication and magnetic actuation
Controlled fabrication
The recent ABFs have a nanoscale helical propeller that is directly actuated by an external magnetic field and are able to rotate around their helical axis. Using a magnetic field as an actuation technique makes the design simpler, as it eliminates the need to create a unique relative motion between the tip and the tail of the ABF. In spite of the difficulties of fabrication in the micro and nanoscales, due to the 2-D patterning of the lithography technique, a variety of nanohelix fabrications have been achieved based on the bottom–up fabrication approach such as the coiled CNTs, ZnO semi conductive nanohelices, and helical shape polymer fibers. However, although those fabrications are widely used, their geometry cannot be controlled accurately. New fabrication techniques were adopted to generate complicated 3-D nanostructures using direct laser writing, yet it was found not to be suitable with batch processing of 3-D structures. Another ABF fabrication technology has proved to be beneficial in generating 3-D nanostructures from

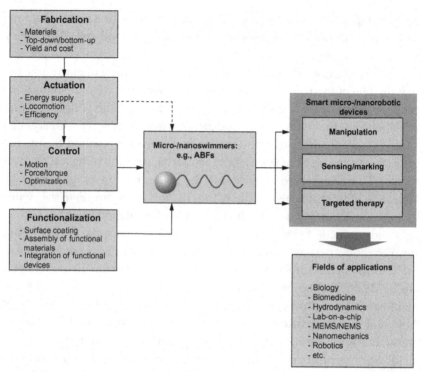

Fig. 15 The roadmap for ABF. *(From Zhang, L., Peyer, K.E., Nelson, B.J., 2010. Artificial bacterial flagella for micromanipulation. Lab Chip 10 (17), 2203–2215, with permission.)*

planar thin films. This technology is based on the top–down fabrication technique combined with a self-organizing step to introduce what is called the self-scrolling technique. The self-scrolling technique is known for its flexibility as the 3-D dimensions of the generated structures can be controlled and scaled up or down. Self-scrolling technique also enables the changes in the geometrical parameters of the helix to be controlled precisely. Moreover, a wide range of materials can be implemented in the structures produced by this technique, so it is considered as an adaptive fabrication technique (Zhang et al., 2009; Peyer et al., 2014).

Self-scrolling technique is used to fabricate two different types of ABFs: An InGaAs/GaAs semiconductor bilayer tail and InGaAs/GaAs/Cr semiconductor metal tri-layer tail. The studies revealed that InGaAs/GaAs/Cr tail design is preferable for its mechanical properties and flexible manipulation and motion. The design consists of a magnetic head that uses Cr/Ni/Au as a composite of the thin metal layer forming the head. The thickness of the three-metal composite Cr/Ni/Au is 10/180/10 nm, respectively. Another fabrication technique used to introduce another type of artificial helical nano propeller is the glancing angle deposition (GLAD). The technique is used to fabricate ABF helical tail by evaporating permanent magnet film on one side of the swimmer. The GLAD technique requires specific materials to be deposited by the vapor deposition, and it is mandatory that the helical tail of the nano propeller is coated with the magnetic film, which results in minimizing the area of the active surface. Despite the high yield of GLAD technique, the self-scrolling technique guarantees more flexibility in material selection (Peyer et al., 2014).

Magnetic actuation

Due to the self-propelling motion of the ABF, the soft magnetic head of the ABF experiences a continuous torque given by Eq. (2):

$$\tau_m = \mu_0 V \vec{M} \times \vec{H} \qquad (2)$$

where V is the volume, M is the magnetization, H is the applied field, and μ_0 is the permeability of the free space.

In order to actuate the ABF, the magnetic field needs to be rotated perpendicularly on the helical axis. As the head of the ABF is made of nickel, it is magnetized along the diagonal axis, which generates an increasing torque related to the angle between the plate and the applied field. The maximum magnetic field is calculated based on Eq. (3):

$$\left|\vec{B}\right| = \mu_0 \left|\vec{H}\right| \qquad (3)$$

where $\vec{B}$ is the field vector.

The amount of field strength generated based on Eq. (3) will be approximately 2 mT. In order to apply higher torque on the helical axis, stronger magnetic material is needed which will be beneficial in controlling and steering the swimmer.

Motion control

Steering precision Due to the motion behavior of ABF and the corkscrew-like movement, moving backward and forward is performed by the rotary motion of the spiral tail and can be switched by reversing the direction of rotation or by changing the direction of the magnetic field. Changing the angle of direction of the effecting rotating field B will steer the propeller ABF in the desired direction. The steering technique is based on changing the perpendicularity of the magnetic field, which will, in turn, reposition the ABF to be perpendicular to the rotating magnetic field (Zhang et al., 2009).

Shape optimization Seeking the optimum swimming performance, many experiments were deducted to optimize the shape of the helical artificial swimmer. Recent studies have proved that even with the same input frequency and by reducing the diameter of the ABFs from 2.8 to 0.8 μm and the length of the propeller from 30 to 8 μm, the swimming relative velocity of the ABFs will increase. Table 2 shows the dependency of the swimming velocity on the diameter and the length of the ABFs. The three ABFs were given the same input frequency, equal to 10 Hz (Zhang et al., 2009).

As shown in Table 2, both the relative velocity and the ratio of the transitional displacement increase with the smaller dimensions of the helical swimmer, which results in more efficient propelling ABF. More experiments were conducted, and it was proved that an ABF with a larger head will swim more slowly because of the higher fluidic drag; in contrast, it has larger maximum swimming velocity due to the magnetic torque produced by the input frequency. With specific tail length, the relation between increasing the magnetic torque and decreasing the fluidic drag can be adjusted to gain the maximum velocity (Zhang et al., 2009).

Another model for the ABF can have a spherical head and helical tail with diameters of d and 2σ, respectively. In order to study the effective parameters on the motion of the ABF, the helical tail is considered to be the reason behind the generated forward propulsion of the swimmer. As the tail consists of incredibly small cylindrical elements, the equilibrium force lays between the drag force and the driving force, which affects the swimmer's tail. The second effective parameter is the spherical head of the swimmer, which creates a torque and a resistive drag force that varies linearly with the rotational speed and the forward velocity of the ABF, respectively. Eq. (4) relates the maximum velocity with the maximum torque induced by the ABF.

Table 2 Comparison between velocities of different ABFs.

	Number of turns (n)	Length (L/μm)	Diameter (d/μm)	Velocity (v/μm s^{-1})	Relative velocity (v/L)	Ratio (v/Lf)
ABF 1	3.5	30	2.8	6	0.2	0.02
ABF 2	3.5	8	0.8	4	0.5	0.05
ABF 3	4	2	0.2	2	1	0.1

$$v_{\max} = \frac{b}{b^2 - ac} \tau_{\max} \qquad (4)$$

where $a = 1.5 \times 10^{-7}$, $b = -1.6 \times 10^{-14}$, and $c = 2.3 \times 10^{-19}$ are the parameters of propulsion.

4.3 Rotating nickel nanowire

Another approach for manipulating micro objects in the biological field is to use rotating nickel nanowires actuated by a rotating magnetic field. This approach enables the nickel nanowires to propel near solid surfaces using tumbling motion, as shown in Fig. 16. Manipulation on the cellular level consists of two main strategies: noncontact manipulation and contact manipulation. The first strategy is based on remote manipulation and steering using wireless techniques such electric field effect, directed laser beam, or acoustic force implementation, while the second contact manipulation strategy is based on direct contact between the end effector and the micro objects. Applying those strategies on the cellular level ensures precise object control and manipulation; however, they require complex calculations and accurate experimental setups. The complexity of applying the contact approach on the nano or even the micro scale relies on the effect of gravitational forces on that scale and the Van Der Waals force; it is a distance-dependent force generated by the interactions between the molecules. On the other hand, applying noncontact manipulation on that scale is simpler due to the absence of the mechanical contact between the end effector and the targeted cell. Additionally, the noncontact approach is more applicable on the cellular level as it provides smooth and gentle handling for the targeted cells using fluid flow control techniques. Controlling the flow is mainly done by inducing magnetic field with low strength using magnetic nanowires (NWs), specifically nickel nanowires (Ni NWs). It was also proved that Ni NWs can be used in producing hypothermia in living cells as they can be heated by radio frequencies that reach 810 MHz (Zhang et al., 2012).

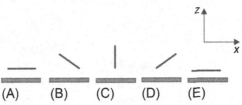

Fig. 16 Sequence of tumbling motion (Zhang et al., 2012).

4.3.1 Fabrication and characterizations

The idea of Ni NWs is developed by using a template-assisted electrochemical deposition technique. The template is fabricated from anodic aluminum oxide (AAO) filament that contains channels with pore sizes of 100 and 200 nm. The field emission scanning electron microscopy (FESEM) of a Ni NW array implanted in AAO template is shown in Fig. 17. The characterization of those implanted arrays can be simply performed using a vibrating sample magnetometer.

Magnetic actuation

Seeking accurate control and steering for the NWs, three orthogonal Helmholtz pairs of coils are used in order to obtain a uniform magnetic field. The induced magnetic field strength reaches approximately 15 mT, which is more than enough to perform manipulation processes on the cellular level. The actuation process starts with turning on the rotating magnetic field, which in turn rotates the Ni NWs within the same plane of the generated magnetic field. The rotation of the NWs occurs due to the driven torque (τ_m) that tends to align the axis of the NWs with the long axis of the induced field. The relation between the uniform field and torque was illustrated in Eq. (1). As for the directing the NWs, adjusting the rotation plane of the magnetic field is sufficient to obtain precisely steered NW (Zhang et al., 2012).

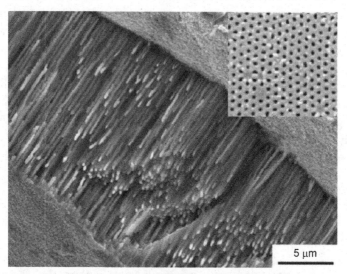

Fig. 17 Micrograph of Ni NW arrays implanted in an AAO template. *(From Zhang, L., Petit, T., Peyer, K.E., Nelson, B.J., 2012. Targeted cargo delivery using a rotating nickel nanowire. Nanomed.: Nanotechnol. Biol. Med. 8 (7), 1074–1080, with permission.)*

Unlike the contact manipulation technique, applying the noncontact manipulation on the cellular level is simple, as it enables pushing and delivery processes to be performed based on less complicated calculations. In order to push and manipulate a micro object without direct contact between it and the end effector, equal velocities of both the end effector and the micro object are essential. Reducing the interaction forces and friction forces between the end effector and the object results in a smooth pushing process with smaller forces applied on the micro object. Performing noncontact manipulation is based on two main parameters: the rotational speed of the NW and the interaction force applied on the manipulated object. Controlling those parameters can be done by adjusting the rotation velocity of the NW to generate the desired force on the micro object through generating proportional fluid flow field. Given this relation, adjusting the rotational velocity of NW is crucial to determine the preferable force to be applied on the micro object in order to move it with the same velocity as the NW. Applying this principle requires the axis of the NW to be aligned with the axis of the targeted object which introduces some challenges due to the surface roughness of the targeted micro object and other conditions that make it difficult to avoid drifting. On the other hand, performing noncontact pulling manipulation results directly in nearly equal velocities of the NW and the micro object, with fewer imperfect boundaries (Zhang et al., 2012).

4.4 Positioning and control

4.4.1 Control by gradient field

The magnetic torque and force are dependent on the magnetization and volume of the object being analyzed (Schuerle et al., 2013). It is possible to remove the dependence on volume and material properties to yield normalized magnetic force positioning of the nanorobot, as in Fig. 18A, described by:

$$\mathbf{F_m} = (\mathbf{m} \cdot \nabla)\mathbf{B} \qquad (5)$$

$$\mathbf{F_m} = \begin{bmatrix} \dfrac{\partial \mathbf{B}_x}{\partial x} & \dfrac{\partial \mathbf{B}_y}{\partial x} & \dfrac{\partial \mathbf{B}_z}{\partial x} \\ \dfrac{\partial \mathbf{B}_x}{\partial y} & \dfrac{\partial \mathbf{B}_y}{\partial y} & \dfrac{\partial \mathbf{B}_z}{\partial y} \\ \dfrac{\partial \mathbf{B}_x}{\partial z} & \dfrac{\partial \mathbf{B}_y}{\partial z} & \dfrac{\partial \mathbf{B}_z}{\partial z} \end{bmatrix} \begin{bmatrix} \mathbf{m}_x \\ \mathbf{m}_y \\ \mathbf{m}_z \end{bmatrix} = \begin{bmatrix} \mathbf{m}^T\mathbf{B}_x \\ \mathbf{m}^T\mathbf{B}_y \\ \mathbf{m}^T\mathbf{B}_z \end{bmatrix} \mathbf{I} \qquad (6)$$

This aligns the nanorobot magnetic moment vector $\mathbf{m}$ with the magnetic flux density vector $\mathbf{B}$ at current vector $\mathbf{I}$ in the coils. In Eq. (5), all spatial

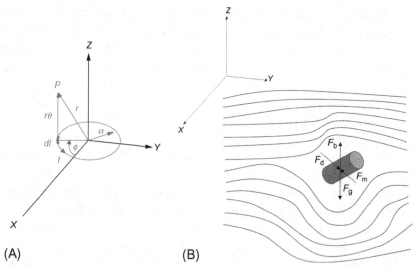

Fig. 18 (A) A circular loop carrying a current **I** of a magnetic field evaluated at point $P(x, y, z)$. (B) Illustration of forces acting on the nanorobot flow in a non-Newtonian fluid (Ghanbari et al., 2014).

derivatives of the magnetic flux density are evaluated at the location of the nanorobot. The components of the magnetic field can be transformed to the global coordinate system where the magnetic field of n coils is superimposed according to:

$$\mathbf{B}(\mathbf{P}) = \sum_{i=1}^{n} \mathbf{B}_i(\mathbf{P}) \mathbf{I}_i \tag{7}$$

where P is the position vector.

Combining Eqs. (6), (7) gives:

$$\begin{bmatrix} \mathbf{B} \\ \mathbf{F_m} \end{bmatrix} = \begin{bmatrix} \mathbf{B}(\mathbf{P}) \\ \mathbf{m}^\mathsf{T}\mathbf{B}_x(\mathbf{P}) \\ \mathbf{m}^\mathsf{T}\mathbf{B}_y(\mathbf{P}) \\ \mathbf{m}^\mathsf{T}\mathbf{B}_z(\mathbf{P}) \end{bmatrix} \mathbf{I} = \mathbf{C}(\mathbf{m}, \mathbf{P}) \mathbf{I} \tag{8}$$

For a desired magnetic field and magnetic force, one can solve Eq. (8) using the inverse of matrix **C** to obtain currents through coils for exact positioning as:

$$\mathbf{I} = \mathbf{C}(\mathbf{m}, \mathbf{P}) \begin{bmatrix} \mathbf{B} \\ \mathbf{F_m} \end{bmatrix} \tag{9}$$

If the desired force, magnetic moment, and position of nanorobot are determined, the current in the coils can be calculated by Eq. (9). This equation generally describes the positioning and the manipulation of the nanorobot by any coil configuration exerting magnetic force.

A nanorobot can be positioned by a magnetic force in the three translational DOF in a non-Newtonian fluid, where the equation of linear motion for motion is obtained by using conservation of linear momentum as:

$$F_m + F_g + F_{drag} + F_b = ma \tag{10}$$

where the applied magnetics force F_m, the fluid hydrodynamic drag F_{drag}, the gravitational force F_g, and the buoyancy force F_b on the nanorobot shown in Fig. 18B are equivalent to its mass m times its acceleration a.

The nanorobot's drag force varies linearly with its velocity and acts in the opposite direction of the velocity. Accordingly, the magnetic force positioning the nanorobot using the position vector $\mathbf{P}$ will be:

$$F_m = m\ddot{P} - F_g - F_{drag} - F_b \tag{11}$$

where the second time derivative of the nanorobot position $\ddot{P}$ represents its acceleration.

All terms that contain uncertainty, such as the contact forces, van der Waals forces, and electrostatic forces, can be represented by the parameter H as:

$$H = (m - m_0)\ddot{P} - \left(F_g + F_{drag} + F_b\right) \tag{12}$$

By substituting with Eq. (11) into Eq. (12):

$$H = F_m - m_0\ddot{P} \tag{13}$$

Considering the time delay estimation methodology (TDE), which can control systems with nonlinearities and uncertainties by a straightforward gain design (Jin and Chang, 2009), the parameter H can be estimated based on the previous time-delayed value of the system variables, as the time delay δ tends to zero, which is given as follows:

$$\hat{H}(t) = H(t - \delta) = F_m(t - \delta) - m_0\ddot{P}(t - \delta) \tag{14}$$

Based on Eq. (13), the input control force F_c is given as:

$$F_c(t) = m_0 V + \hat{H} \tag{15}$$

where V is a function of the desired position vector $\mathbf{P}_d$, the proportional gain K_p, and the derivative gain K_v, given as:

$$V = \ddot{p}_d + K_v\left(\dot{P}_d - \dot{P}\right) - K_P(\mathbf{P}_d - \mathbf{P}) \tag{16}$$

where the position error can be defined as $e = \mathbf{P}_d - \mathbf{P}$.

Substituting by Eqs. (15), (14) in Eq. (16) will yield:

$$F_c(t) = m_0\left[\ddot{p}_d + K_v\dot{e} - K_Pe\right] + F_m(t - \delta) - m_0\ddot{P}(t - \delta) \tag{17}$$

The gains, K_p and K_v, are diagonal matrices designed individually in three directions to cope with the dynamics of the nanorobot in the three directions, giving an error of a desired response. As shown in Fig. 19, the control system requires continuous position information $P(t)$, and its time derivatives until acceleration $\ddot{P}(t)$.

OctoMag

This device, shown in Fig. 20, can navigate micro and nanorobots in bodily fluids to enable a number of minimally invasive diagnostic medical procedures. The OctoMag was designed to optimize the manipulability of the magnetic field, as it has five degrees of freedom (5-DOF) wireless magnetic control, which decomposes to 3-DOF position and 2-DOF pointing orientation. Accurate force control will be able to levitate the robot against its own weight or to push with a specific force. On the other hand, torque is needed to rotate the robot, and a low magnitude of torque applied in the correct direction will give the correct control effort.

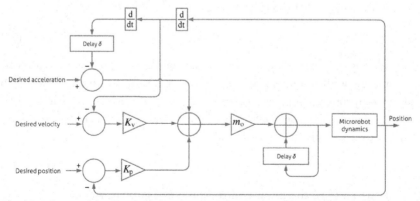

Fig. 19 Scheme of the TDE control system used in the simulation, where m_0 was tuned for optimum performance of the TDE controller (Ghanbari et al., 2014).

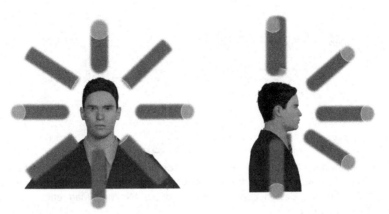

Fig. 20 OctoMag system for intraocular microrobots control (Kummer et al., 2010).

The OctoMag electromagnetic system was proposed for the control of intraocular microrobots (Fig. 20) with a camera fitted down the central axis to image the robot in the eye, where an eyeball is at the center of the system's workspace (Kummer et al., 2010). Using an OctoMag prototype, a robot has been injected into the lapine vitreous humor, where a disposable planoconcave vitrectomy lens is placed on the eye to increase visibility (Fusco et al., 2014). Increasing the size of the prototype system to accommodate a human head requires more powerful current amplifiers to generate the same magnetic field strength, because the strength of the field gradients attenuates with a factor by which the system is scaled (Kummer et al., 2010).

4.4.1.1 MiniMag

The MiniMag is a magnetic actuation system for manipulating micro and nano swimmers with translational motion along their orientation axis. It consists of eight electromagnetic coils focused on the center, and arranged in the form shown in Fig. 21A and B. By passing a current through the coils, a magnetic field is given within the set workspace, and assuming ideal coils, the individual magnetic fields giving the resultant magnetic flux density could be superimposed as in Eq. (7).

The MiniMag is designed to restrict the locations of the electromagnetic coils to a single hemisphere (Schuerle et al., 2013), which is accomplished by moving the intersection point of the coils orientation axis. Less physical restriction is achieved in the workspace in addition to compatibility with an inverted microscope. The MiniMag's reachable workspace is larger

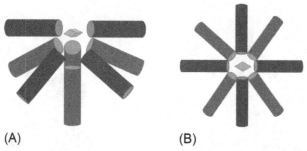

                    (A)   (B)

Fig. 21 Schematic diagram showing the electromagnetic coils of the magnetic actuation system surrounding the workspace. (A) Isometric view, (B) top view (Ghanbari et al., 2014), (C) MiniMag design configuration, and (D) MiniMag Magnetic workspaces (Schuerle et al., 2013).

due to the smaller working distance, where the difference in workspaces configuration is due to the tilted geometry.

Using the MiniMag TDE control configuration requires determining the desired error dynamics by selecting a natural frequency ωn, and a damping ratio ζ, and according to Ghanbari et al. (2014), the gains were selected to $K_p = \omega_n^2$ and $K_v = 2\zeta\ \omega$. The value of m_0 should be tuned according to the given implementation as it depends on the sampling frequency. As for m_0, one can start with a value close to or less than the nanorobot mass and vary the value based on the performance (Jin et al., 2008). The non–Newtonian fluid containing the robot is placed on the magnetic positioning system, where the position of the nanorobot was determined using the two cameras, one giving a top view and the other giving a side view.

The TDE Minimag requires a high damping ratio to overcome the overshoot of the step response (Ghanbari et al., 2014), which occurred for the dependence of the controller performance on the time delay. Increasing the sampling rate (by increasing the camera speed) will reduce the time delay, which results in less overshoot and faster controller response. In addition, the inherent nonlinearities of the magnetic actuation were not included in the simulation. These nonlinearities associated with the unmodeled forces, including stiction at the initial position when the robot is in contact with the container surface, are sources of the overshoots observed in the experiment, which do not exist in the simulations (Ghanbari et al., 2014).

Control by rotating field
This method targets mainly swimming magnetic helical micro and nanorobots in non–Newtonian fluids. The swimmer dynamics view

applying an external force to the robot, as it reaches a steady-state movement instantaneously, where the drag forces F_d equal the external forces F_{ext} (Peyer et al., 2014), as:

$$F_{ext} + F_d = 0 \qquad (18)$$

In addition, drag torques T_d equal to external torques T_{ext}, as:

$$T_{ext} + T_d = 0 \qquad (19)$$

The drag forces and torques relates to the robot's linear and rotational velocity by the resistance matrix $\mathbf{D}$, given by:

$$\begin{bmatrix} F_d \\ T_d \end{bmatrix} = -\mathbf{D} \begin{bmatrix} U \\ \Omega \end{bmatrix} \qquad (20)$$

The translational velocity U and rotational velocity Ω of the robot are given in Eq. (20), in relation to the magnetic torque that is applied externally. Magnetic forces are neglected, as the external magnetic field existing is uniform. The gravitational forces are also neglected, as the movement due to the helical propulsion is the main physical interest (Mahoney et al., 2011). The resistance matrix D can be calculated by using the resistive force theory (RFT), which is a method to capture the swimming behavior of micro and nano scaled subjects (Lauga et al., 2006). Consequently, the magnetic torque T_m applied to the swimmer is given by:

$$T_m = m \times B \qquad (21)$$

The magnetic property m is expressed as the magnetic dipole $m = VM$. Given that B is a static field, the robot would align for M and B to be collinear and remain stationary. To achieve a rotation of the robot, B is rotated at a constant speed:

$$B(\omega t) = \begin{bmatrix} B\cos(\omega t) \\ B\sin(\omega t) \\ 0 \end{bmatrix} \qquad (22)$$

The magnetization M is constant for hard magnetic material M_h, which describes the direction and magnitude of the magnetization. Soft-magnetic material magnetization M_s can be expressed by a relation between the apparent susceptibility tensor χ_a and the external magnetic field $H = B/\mu_0$, where μ_0 is the permeability of free space (Abbott et al., 2007):

$$M_s = \chi_a \cdot R \cdot H \qquad (23)$$

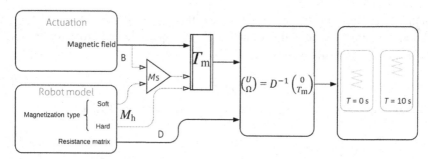

Fig. 22 Micro and nanorobots locomotion model using rotating field (Peyer et al., 2014).

where R represents a rotational matrix transforming the external magnetic field into the local coordinate system in which M and χ_a are expressed.

The robot locomotion model is shown in Fig. 22, where the robot's motion is a function of the resistance matrix and the applied torque. The torque applied in the system depends on the magnetization of the robot, which is constant for hard-magnetic material or a function of the applied magnetics field in the case of a soft-magnetic material. In experimentations by Peyer et al. (2014), the continuous rotation of the helix around its axis results in a constant advancement forward along the helical axis. Depending on the angle between M and B, either a positive or a negative torque is exerted onto the robot, which causes the robot to switch between clockwise and counterclockwise rotation.

Helmholtz One of the main devices for micro and nano swimmers' positioning using a rotating field is the Helmholtz coil, shown in Fig. 23, whose

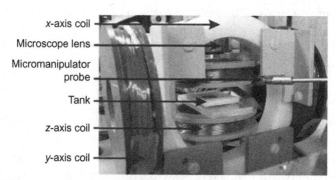

Fig. 23 Helmholtz coils experimental setting. *(From Zhang, L., Peyer, K.E., Nelson, B.J., 2010. Artificial bacterial flagella for micromanipulation. Lab Chip 10 (17), 2203–2215, with permission.)*

magnetic workspace is a union of the predicted magnetic workspace for all combinations of field and gradient orientations. It uses three pairs of orthogonal electromagnetic coils to generate a uniform rotating magnetic field to actuate and control mostly ABFs (Zhang et al., 2010). The Helmholtz coils witness singularities when the field and force are not aligned, and thus it does not show the ability to exert force harmonically. A Helmholtz system can generate gradients using an unequal current input through the coils, if oriented along the same axis as the field. A singularity in the system is caused by using an orthogonal system to orient the device along the x-axis and move it along the y-axis. An ABF is rotated along its axis and self-propelled by applying a torque emerging from a uniform rotating magnetic field acting on the ABF soft-magnetic metal head, which is given by Eq. (5). Motion direction can be switched by reversing the rotating direction of the magnetic field, which is much more straightforward and time-efficient than reversing motion by turning the swimmer 180°. The velocity depends on the rotation frequency, as shown in Fig. 24A, where the curves show that velocity

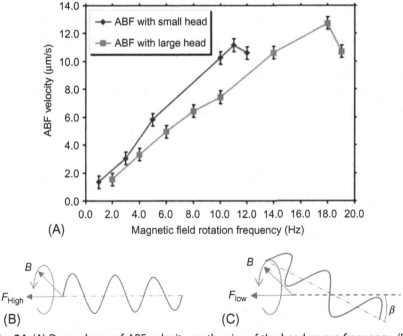

Fig. 24 (A) Dependence of ABF velocity on the size of the head versus frequency. (B) High actuation frequencies the wobbling decreases. (C) At low frequency, the precession angle β is large (Peyer et al., 2014). ((A) From Zhang, L., Peyer, K.E., Nelson, B.J., 2010. Artificial bacterial flagella for micromanipulation. Lab Chip 10 (17), 2203–2215, with permission.)

increase if the frequency increases, and decrease as the available magnetic torque is no longer sufficient to keep the ABF motion (Zhang et al., 2010). At low frequencies, the experimental results show that, although the ABF rotated in sync with the input field, it wobbled about its helical axis with a constant precession angle. The wobbling angle increases significantly at lower frequencies, as in Fig. 24C, so for precise steering, these effects must be taken into account, as they influence the propulsion direction of the swimmer (Peyer et al., 2014).

5 Biomedical applications

Nanorobotics shows a promising future in medicine and its applications as its applications offer new types of tools in treatment and improvising human biological systems (Manjunath and Kishore, 2014; Azar and Eljamel, 2012). The potential starts from drug delivery to several targets inside human body by loading the drug dose to the drug delivery system (pharmacyte). Then, it can transport safely to the target and release the dose by nanoinjection or progressive cytopenetration. A second major use of nanorobots is body surveillance or continuous monitoring of vitals, this provides a frequent response for a detected abnormal behavior. In addition, they are beneficial in dentistry as they can induce oral analgesia or manipulate tissues by aligning them in a desired form. Moreover, applications of nanorobots raise the success rate of delicate and difficult surgeries such as surgeries of the retina and its surroundings. For instance, fetal surgery is considered as one of the most risky surgeries due to high mortality rate of either the mother or the baby, but it will become much easier using nanorobots, as they will target the area desired with high precision and minimal trauma (Bhat, 2014). Nanorobots provide gene therapy and artificial substitutes of human blood. In gene therapy, it gets allowable to compare the molecular structure of DNA and proteins found in cell and do a replacement for the found defective. On the other side of providing artificial blood substitutes, three different nanorobots are introduced.

5.1 Surgical nanorobots

5.1.1 Nanotechnology in surgical tool

The recent advancements in nanotechnology have made it possible for it to be implemented in the most vital applications like surgery. Implementing nanotechnology in surgery introduces the concept of nanobiotechnology, which deals with injuries, curing diseases, and using the required tools to perform surgeries on the cellular level. The demand for less invasive surgeries with less damage to healthy tissues has made it necessary for scientists to develop nanobiotechnology in an attempt to fulfill this need. Due to the fact that even the smallest scalpels are thousands of times the size of a cell, performing an accurate operation within a single cell is impossible for surgeons without the existence of nanoscaled tools. Furthermore, scientists are considering the ability of using this technology to enhance the abilities of human beings by applying DNA modifications, and new coronary arteries delivery. Applying this technology requires advanced and reliable diagnostic abilities in order to perform precise surgery and to provide effective treatment. This section will present different applications of nanorobots in surgery, treatment, and even diagnosis (Mali, 2013).

5.1.2 Nanocoated blades

As cutting blades are the main instrument in performing any surgery, introducing new structures at the nanoscale for surgical blades contributes to developing more precise surgeries with less traumatic effects and less invasive wounds (Roszek et al., 2005). Coating the micro-structured metal with nanolayers of diamond is beneficial in the field of ophthalmic surgery and neurosurgery. The recent manufacturing technologies have enabled the production of cutting blades with diameters of 5 nm to 1 μm (Mali, 2013).

5.1.3 Suture nanoneedles

Forms of surgery currently being enhanced include plastic and ophthalmic surgeries, as they improve the appearance and the ability for better vision; improving those surgeries requires precise tools and minimal invasive operations, which can be provided by using nanoneedles. These needles are made of stainless steel with a particle size of 1–10nm quasicrystals by applying the techniques of thermal ageing. Using silicon to prepare these needles improves their ductility, corrosion resistance, and strength. Taking into consideration that operating on the cellular level requires the least deformation of cells, as the minimal undesired mechanical reaction will interfere with the result of manipulation, the needles are manufactured at a size of 200–300 nm in diameter and 6–8 μm in length. The continuous improvement of nanoneedles has made it possible to use them in the immobilization techniques of DNA, and chemicals or proteins loading. Nanoneedles also support other surgery techniques on the cellular level that use stem cells to make differentiations that help to produce healthy and functional cells for donations (Roszek et al., 2005).

5.2 Optical nanosurgery

5.2.1 Optical tweezers

Optical tweezers are also known as the optical trap, which is a powerful technology that is widely used in noninvasive manipulation for objects in the micro-scale. The technique is based on using light beams to control the motion and the formulation of particles. Once a continuous laser beam is directed on an object, repositioning of the object occurs due to steering and positioning of the light beam. The tweezers are made of CNTs attached to the electrode, which can bend, under specific force, to grab the desired molecule. This technology is precise and noninvasive as it enables the surgeons to perform delicate operations on a single cell; the diameter of a nanotweezer is 50 nm and can be excited at 8.5 V (Mali, 2013; Roszek et al., 2005).

5.2.2 Femtosecond laser neurosurgery

A femtosecond laser can emit ultrashort optical pulses with a duration measured in femtoseconds (1 femtosecond equals 10^{-15} seconds). Because of their ultrashort and ultralong pulses, ease in usage, and precision, femtosecond lasers are widely used in molecular structure manipulation and localization. They are also used in ophthalmic surgery to correct human vision. Using femtosecond near-infrared laser pulses with low energy helped to cut cell organelles without harming any surrounding tissues; this application is called "nanoscissors." Based on the properties of the ultrashort pulses, they are capable of performing accurate operations on nerve cells and cut through nano-cell structures. Those operations are performed by heating the targeted area then cutting thorough it, but this can form a threat to the surrounding tissues, so nanoscissors provide the optimum solution for this problem.

5.3 Nanocoated implant surfaces

An implant is an artificial replacement for damaged body functional parts such as hip and knee joints, bone fracture treatment, and temporomandibular joints. An implant is usually made of titanium, polymers, or cement, which are biocompatible with the human body. The life of implants is affected by various factors such as poor growth on the surface of the implant. However, the life time for an implant can be enhanced by using materials that stimulate the formation of the fractured bones and by fixing the implant with an adjacent bone. The most preferable implants for joints are metallic, like stainless steel and cobalt chrome, due to their mechanical properties. On the other hand, the high Young's modulus for those metals may cause stress shielding, as they will not be as elastic as the surrounding living bones and joints. Seeking solutions for this challenge, scientists proved that the adsorption of proteins that mediate specified osteoblast adhesion are improved on materials' nanophase. The bioactivity of proteins is reformed on the nanophase materials based on their surface energetics, and the surface features that are close to the size of proteins (Roszek et al., 2005). Given the fact that the hydrophobicity or hydrophilicity of a surface can directly affect the behavior of cells and that the wettability of the material can contribute in categorizing this material with regards to hydrophobicity/hydrophilicity categories, implants or biomaterial composition with the presence of the required nanofeatures and immobilization of chemical agents can alter the wettability and enhance the behavior of cells on that surface.

5.4 NPs for wound dressing

Based on the antiinfective properties of silver, it has been beneficial for wound dressing by formulating bilayers of silver-coated polyethylene and an adsorptive core. It was demonstrated that using a coat of nanocrystalline silver on nonhealing wounds like burns and ulcers has sustained release that help keep antibacterial and fungicidal activity on the surface of those wounds. The silver nanoparticles start functioning by entering the wound through the body fluids, and then start to kill bacteria within 30 min. The wound surface can withstand the coating layer for several days, taking into consideration the amount of silver in the composite to avoid any toxic side effects. Wound care applications also benefit from the electrospinning technique that enables the production of polymer nanofibers. First-aid wound care products are developed based on creating a new delivery platform, which can be used as a powerful and efficient hemostatic agent in wound care. The spinning technique was developed by nanofiber technology companies to produce nonwoven materials for a wide range of biomedical, chemical, microelectronic, and industrial applications (Roszek et al., 2005).

5.5 Tissue engineering

Tissue engineering is the technology concerned with the replacement of the injured, damaged, or even missing tissues in the human body. The core three components to provide an alternative tissue are isolated cells, signal molecules, and extracellular matrix. The contribution of nanotechnology in tissue engineering presents in providing wide range of options for the extra-cellular matrix, which is also known as the scaffold. This scaffold is primarily contributing in three roles, which are, promoting the localization process of cells in the human body, supporting the regeneration process of new well-structured tissues, and provide the required guidance for this regeneration. The interaction between the cells and the extracellular matrix has a direct effect on the function of the new generated tissues, so the process of con-trolling the porosity of the suggested nanomaterials is crucial to obtain the required properties. Additionally, controlling the mechanical character-istics can help generate well-functioning tissue-engineered products (TEPs). The role of nanotechnology presents itself by providing the most successful candidates for TEPs application. The first candidate is nanofiber-structured tissue, which is prepared using the electrospinning technique. Applying this technique, knowing the physical properties for nanofiber matrices such as the interconnective pores, high porosity, and high surface area make this candidate perfect for developing the scaffold of TEPs. The second option is CNT, with its ability to guide the live biological cells and the ability of forming neural networks in vitro that play a great role in the repair and regeneration processes of the central nervous system. Those processes are essential for brain or spinal cord injuries. (Mali, 2013, Roszek et al., 2005)

5.6 Nanorobots for cellular-level surgery

Digging through the field of nanotechnology, the AFM has proven to be the best instrument to study the nanoworld. AFM is a tool that enables imaging of living samples and measuring a variety of mechanical properties. An AFM-based nanorobot is a powerful candidate for the critical need of oper-ating on living cells. As the AFM has a tip that is approximately 20 nm or less, which can operate on living-cells samples, an AFM-based nanorobot can use this tip as an end effector that will conduct precise operations. Given the several advantages of AFM technology, such as high-resolution imaging and vacuum-free working environment, developing AFM-based nanorobots will be capable of generating images on nanoscale, providing accurate measures of mechanical properties, manipulating samples on nano-scale, and conducting accurate operations on living cells, due to the precise motion control of the nanorobot. Due to the complexity of cellular-level surgeries, and the need for real-time monitoring to guarantee precise oper-ation, three techniques were developed to overcome these challenges. (Song et al., 2012)

The three developed techniques present the operating principles of AFM-based manipulators. First, the interface developed by augmented reality systems enables the operator to control the position of the nanorobot using a joystick. Second, the end effector of the nanorobot can be equipped with specialized tips to perform various types of surgeries and drug delivery operations. Finally, the whole operation should be monitored using real-time system as a feedback to insure the accuracy of the operation. This online monitoring is critical to update the operator with the most recent images of the currently running operation by taking discrete images of the current position of the nanorobot or by producing continuous imaging for the full work surface. Increasing the rate of real-time monitoring is beneficial for the accuracy of the operations, so the scientists developed two methodologies to achieve this goal. The first is online sensing, which enables generation of accurate calculations of the force exerted on the tip of AFM due to the cutting pressure and depth. The other methodology is developing an adaptive scan for the area under operation that will generate more reliable real-time topography. The following sections will discuss the three main techniques of the working principle of AFM-based manipulators (Song et al., 2012).

5.6.1 Augmented reality system

The augmented reality system is designed to provide an active interface that shows the user or the operator the actual position and motion of the AMF tip. The joystick is used to control the directions of the nanorobot along with providing it with three-dimensional (3-D) forces. Along with controlling the motion, an augmented reality system in AMF provides updated images with the high quality and high frame rate (video rate), which keep the user updated with the real-time position of the nanorobot. In spite of the continuous development in position control of the nanorobots, two factors cause critical errors regarding position control. The first factor is the thermal drift, which is the drift in actuator outputs caused by temperature change. Controlling the environment surrounding the manipulator will result in stable thermal drift, which will indicate the accurate position of the tip. The thermal drift can be controlled by using local scan mechanisms that detect the initial position of the tip before starting the operation and then calculate the effect of thermal drift, which will make it easy to predict the error caused by this factor. The second factor is the cantilever bending, which can cause a sideways displacement along the cantilever length. The effect of this factor can be avoided by controlling the stiffness of the probe, and the position algorithm (Song et al., 2012).

5.6.2 Local drug delivery

The AFM tip can be used in local drug delivery by attaching antibodies or ligands to it. This application can enhance the researches concerned with the effects of those chemicals in vitro local drug delivery. There are two main methods for loading a functionalize AMF tip with specific chemicals. The directly used method is dipping the cantilever or the tip in the specified chemical solution. The widely known method is to use a linker molecule between the AMF tip and the targeted molecule, which will increase accurate targeting in 3-D. The principle of this method is based on creating a covalent bond between the chemical and the linker molecule (Song et al., 2012).

5.6.3 Online monitoring for nanosurgery

As discussed before, the monitoring system is responsible for providing real-time video of the position of the operating nanorobot. The online monitoring process has several steps that presents the functioning technique. The first step is to calculate the required force on the AMF tip to cut through the targeted surface. Determining if the tip has cut through the sample surface or not is considered as the critical step to start the tip motion planner. Once the tip goes through the surface and the motion planner starts working, the visual feedback is generated and sent to the operator to detect the current position of the nanorobot. The motion planner also performs accurate scans that generate precise directions and ranges of the cutting tip. The initial scan range is quarter of the original real image size, and it can be customized to fit several nano surgeries.

Another important term in online monitoring is compressive scan technology, which is a fast imaging strategy that samples fewer data and then generates a new reconstructed image. This technique enhances the capturing rate of the changes occurring during the nanosurgery, which results in more precise and accurate operation. The compressed sensing can be controlled through the motion planner, which controls the exact location and time of the new compressive scan. After capturing and updating the required images, the video is then displayed on the online monitoring system to guide the operator or the user (Song et al., 2012).

5.7 Cancerous tumor killing using nanorobots

The first type of nonrobots that simulate red blood cells in the functionality of oxygen and carbon dioxide exchange are respirocytes. They are typical spheres of 1 µm diameter and constructed of 18 billion atoms arranged in diamondoid pressure tanks. These cells can store about 3 billion oxygen molecules and deliver at a rate 236 times higher than natural red blood cells. They include three rotor types for storing oxygen while travelling, capturing carbon dioxide from the blood stream and releasing at the lungs, and taking in glucose from blood as a fuel source (Frcitas, 1998; Arpita et al., 2013).

Second type of nanorobots simulating blood are microbivores, also known as nanorobotic phagocytes. They are made up of diamond and sapphire with a measurement of 3.4 μm in diameter along the major axis and 2.0 μm diameter along the minor axis. It mainly contains 610 billion structural atoms that are arranged in a perfect form. Microbivore aims mainly to perform phagocytosis process. They consist of four fundamental components: an array of reversible binding sites, an array of telescoping grapples, a morcellation chamber, and a digestion chamber to do the whole digestion cycle. These robots are completely safe in removal as they exit the body through the kidneys and are then excreted in urine (Bhat, 2014).

The last type of nanorobots that simulates a part of human blood are clottocytes. The main function of clottocytes is doing the hemostasis process, the process of blood clotting by platelets once damage to the endothelium cells of blood vessels is detected. Nanotechnology shows an impressing success in this field, as it reduces clotting time to be 100,000 times faster than the natural hemostatic system. Furthermore, it avoids the use of drugs with side effects trying to treat the irregularity in the hemostasis process in some bodies. These drugs were creating bad side effects such as hormonal secretion that could damage lungs and allergic reactions (Boonrong and Kaewkamnerdpong, 2011). Clottocytes are 2 μm in diameter, powered by serum-oxyglucose, and contain fiber mesh folded onboard; this mesh can be biodegradable. Although clottocytes show a huge benefit, there is a major risk of the mechanical platelets triggering the disseminated intravascular coagulation due to their extra-activity. This may result in multiple micro thrombi (Eshaghian-Wilner, 2009).

In fact, the motivation behind the great development of nanorobots and related technologies was the desire of finding the optimal solution for cancer treatment. Cancer can be presented as a group of diseases that grow rapidly in an uncontrollable way, causing tumors. Now, it can be treated by combining the current knowledge and medication with the technology of nanorobots and its advanced DDS. These systems are beneficial as they can perform diagnosis and treatment of cancer by carrying large amounts of a drug to tumors while decreasing side effects of chemotherapy (da Silva Luz et al., 2016). Nanorobots as drug carriers for dosages allow maintaining the chemical compounds for a longer time, as necessary, into the bloodstream circulation. Fuzzy control is applied to identify and treat cancerous tumors in patients' bodies. The process starts by detecting the exact position of tumors or defective cells getting benefit of its magnetic property. Defective cells identification is performed based on a given data set to the controlled and built-in membership functions; it depends mainly on the shape and temperature of the cell in comparison to normal conditions. Then, the stage of cancer cell is determined according to gathered information about size and shape of nucleus and cytoplasm. Lastly, the drug carried by the nanorobot is delivered exactly to the diseased cells, which is beneficial in two ways: it avoids drug wastage by delivering the entire amount to the tumor, and prohibits the drug from interacting with surrounding healthy cells, avoiding side effects of chemotherapy (da Silva Luz et al., 2016; Karan and Majumder, 2011).

Another way is to use magnetic nanocatalysts in cancer therapy. A research proposed by scientists discusses how it is efficient to use magnetic nanorobots to activate chemotherapeutic prodrugs in the treatment process. In fact, it has been shown that the only approach to use anticancer drugs is transforming chemotherapeutic agents into latent prodrugs; this is only effective upon catalytic activation. Prodrugs are divided into two sections. The first group relies mainly on the metabolic activation via enzymes, which requires insertion of foreign enzymes to the human body. Consequently, it develops an error of hitting the targeted tumor or diseasing surrounding healthy cells. The second class are called biorthogonal prodrugs; these are more physiologically stable precures as they can get activated through biocompatible heterogeneous catalysts by means of biorthogonal organometallic reactions. Despite focusing on transiting metals to catalysts, part of these transitions causes toxic side effects. Palladium was the best choice as it exhibits strong catalytic activity.

A nanowire was previously proposed as a composition of magnetic iron (Fe) and palladium (Pd), which is a very well-known biorthogonal catalyst. This composition, added with Pro-5-FU, led to reduced viability of cells. The magnetic field property of cells attracts the nanorobots to the area of disease, which demonstrates a high ability of hitting the target. This research was proved by injecting the proposed nanowire into cancer tumor xenografts. This resulted in retarding the tumor without causing any significant side effects (Hoop et al., 2018).

5.8 Laparoscopic cancer surgery using nanorobots

Nanorobots not only have a promising future in medicine, they are also beneficial for the field of surgery. In specific, nanorobots have an important role in laparoscopic cancer surgery. This implies the concept of integration between surgical teleoperations and nanorobotics contribution in the medical field. Although cancer can be treated using the current medicines and technologies of tumor elimination, it has crucial factors that determine the survival period of cancer patients. For example, both the precision of the surgeon in eliminating diseased cells and the growth behavior of the cancer give an indication about treatment success. Avoiding human error in such surgeries, different methods are being used, although their results are disappointing. Although clearer, magnetic resonance imaging (MRI) was limited in specificity and sensitivity. Retroperitoneal lymph node dissection (RPLND) showed an acceptable result as it is directly related with determining areas associated with tumor cell invasion. Accordingly, nanorobots are inserted to gather reliable information about targeted area, forming a map to deal with the medical procedure in a precise and professional way. In general, laparoscopic systems can depend on different tools, such as sound, to control steps of robotic arm motion with an aim of improvising smooth behavior.

The DaVinci is the most advanced master-slave system developed so far, and is responsible for a huge number of delicate surgeries (Guillou et al., 2005). This system has seven DOF, while normal laparoscopic instruments allow only four DOF. The DaVinci consists of three main components:

(a) A surgeon console, which is responsible for controlling the movement of the arm although being placed away. It considers motion scaling to eliminate tremor and get very precise motion steps.

(b) Patient-side cart has the arm mounted on it to provide the three-dimensional view desired in the progress.

(c) Image insufflation stack, which has the camera-control units allowing vision for surgery assistants and carrying the rest of the units such as image-recording devices, laparoscopic insufflator, and monitor.

(d) Nanorobots, which is the new part added to the DaVinci system to provide it with an accurate mapping of the area required to be treated or eliminated without harmful side effects; it uses high precision transducers in allocating these areas in the real time domain.

Nanorobots' use in medical treatment and surgery may be enhanced in a way that it demarks areas that are more likely to be infected or get a new growing tumor (Guillou et al., 2005).

Nanorobots also push the boundaries of treating oral cancer. It requires a prestep using saliva diagnostic medium and different highlighters to define the infected cells, then nanoshells, which are miniscule beads that are specific tools in cancer therapeutics, take the responsibility of destroying infected cells. Another method, which is under experimental testing, is nanoparticle-coated. This method uses radioactive resources to destroy the tumor by being placed near to or even inside it. Although it seemed like an impossible mission in the past, it has reached the phase of experiment testing relying on the conceptual idea of nanorobots that swim in vivo.

5.9 Cell cutting using nanorobots

One medical application of nanorobots that concerns many researchers is cell manipulation in vivo. This manipulation can be represented by many operations such as cell cutting, cell treatment, cell DNA accessing and manipulation, and drug delivery. All of these areas of research have been given great interest over years. Being one of the promising fields with a lot of usage in surgeries and diseases treatment, cell cutting using nanorobots has been analyzed multiple times by researchers over the years.

In a work done by Shen et al. (2010), the nanorobotics system was represented by a nanoknife that was designed and manufactured to perform cutting of a single cell. Making advancement to the technology used at that time, the researchers were capable of fabricating a nanoknife with an edge angle of 5°. Not only advances to the design were proposed, but changes in the materials were also introduced in that research. The use of CNTs in the operation of cell cutting came as a replacement to the conventional technique of using glass and diamond. Fabrication was done by welding using FIB etching technique and AFM.

Moving a step forward to the control of the designed cutting nanoknife, environmental scanning electron microscopy was used as the equipment for manipulation, taking advantage of its capability to allow live tracking of biological systems in high humidity. This manipulation, along with theoretical and numerical analysis, was used to calibrate the knife for doing its cutting task. The cutting force used in performing the cell cutting by the Nano knife was calculated based on the deformation in the shape of the nanoknife.

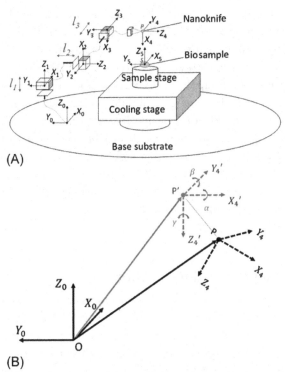

Fig. 25 The model of the nanorobot system for single cell cutting using conventional robotics kinematics. *(From Shang, W., Lu, H., Wan, W., Fukuda, T., Shen, Y., 2016. Vision-based nano robotic system for high-throughput non-embedded cell cutting. Sci. Rep. 6, 22534, under the Creative Commons license.)*

In research done by Shang et al. (2016), another successful proposal for the use of nanorobots in cell cutting was made. What distinguishes this proposal is the development of an automatic system that works exactly like other familiar robots in measuring the distance between the knife and the target cell, using homogenous transformations as well as calibration for ensuring the results to guide the automatic manipulator to perform its cutting task, as shown in Fig. 25. Again, environmental scanning electron microscopy was used due to its features in observing the operation and controlling of the nanorobot in different stages. Success of this fabrication and manipulation proposal was shown by the ability of the nanorobot to cut within 1–2 min. It becomes superior to the use of AFM, which is able to perform cutting in the normal conditions of the cell, in the capability to cut a single cell, which is not possible with AFM. Additionally, the shape edge and accuracy of cutting has a great effect in reducing cell compression during the cutting operation and preventing cell damage after being cut. The analysis of this proposal was done using yeast cells, which are very similar to human cells and are already used by many other researchers to study genetics. The cell was cut into two pieces with cutting force of 30.3 µN.

These advances in the applications of nanorobots open the door for more interest to be given to that emerging and promising field to be more capable for application in real life operations. In fact, there is an obvious defect between the successful results of research and the implementation in real life. This was obvious by recognizing that although a proposal for a nanoknife was made in 2010 by Shen et al. (2010), the same conventional methods of cell cutting—which all have limitations in single cell cutting in normal cell conditions—were still used in 2016 at the time of the new nanonknife proposal by Shang et al. (2016).

5.10 Bacteria propelled nanorobots

When nanorobots' usage in medical applications was suggested, many proposals for the method of propulsion inside the body were made. One of the most suitable, widely researched, and promising methods of propulsion is the propulsion by bacteria or flagella. This method represents a good integration between the fabricated artificial parts of the nanorobot along with biological systems in order to perform a certain task. Advances in research about propulsion is of great importance, because it is essential for all nanorobots in medical application, no matter what task the nanorobot will perform or what kind of disease the nanorobot contributes in treating. This generality has caused this research area to attract the attention of researchers over the years. Additionally, the interest in bacteria propulsion stems and rises from the scientists' trials to reduce the power sources for on-board nanorobotic bodies for minimization of robot size to fit inside small blood vessels. The smaller the body of the robot, the more flexibility in doing medical operations, and the more potential to use a swarm of many nanorobots communicating together to achieve a task.

In a research done by Martel (2008), a nanorobot propelled by the two-diameter MC-1 magnetotactic bacterium was proposed. In order to control these nanorobots, magnetosomes embedded in the propelling bacteria were the means of propulsion. Magnetosomes are structures present in the membrane of the bacteria, and they contain magnetic particles rich in iron. By knowing their structure, control and manipulation of them is used to affect the movement of bacteria. In other words, torque induction was applied over the magnetosomes to control the direction of bacterial propulsion, and thus of the driven nanorobot. Scientifically speaking, microelectronic circuits were used to deliver current to generate the necessary torque on the magnetic particles. The achievements of this driving mechanism using magnetotactic bacterium in this proposal were a speed of 300 μm/s of the motion for unloaded bacteria and a thrust force of 4 pN per bacteria. These experimental results are promising, since they can be of great suitability to propel a medical nanorobot inside the human body to either detect or investigate cellular functions or treat a disease. This proposed technique can be used in other applications, such as nanoactuators or nanosensors. The dependence on bacteria provides a good platform for many applications regarding medical fields, and it makes propulsion easier than designing some artificial bio-friendly devices to perform this driving task.

In a work done by Martel et al. (2009), the use of magnetotactic bacterium in nanorobotics propulsion is proposed again, with advances in the research. The core of this research was the desire to enable the nanorobot to go through the smallest capillaries. An external—not on-board—controller for the motion of the nanorobot was used to control the swarm of bacteria for propulsion through controlled magnetic field intensity. The gap which this research points out was the fact that previous literature controlled the bacteria motion precisely in terms of controlling when the motion will start and when it will end, with low control over the directional movement of the bacteria-propelled nanorobot body. In this research, a mathematical equation for finding the terminal velocity of the bacteria was proposed. Analysis for the relation between the terminal velocity and the body temperature was carried out to prove that higher body temperature affects the velocity negatively by decreasing it as a result of changes in the viscosity of the blood. As a part of propulsion system designing, the loading method for the bacteria was studied in this research. Four loading methods were discussed, as shown in Fig. 26. The loading method done by attaching nanoparticles to the cell is one of the manners with the high promising potential to be used in many medical applications.

Not only the body temperature affects the propulsion speed of the Nano robot, but also the drag force induced by the small-sized diameter of the capillary in which the motion will take place. More importantly, the loading method of the nanorobot or NPs will further affect the drag force and the speed. Here is the area where all design parameters interconnect and affect each other. DC magnetic field was used to control the motion of the bacteria, while a proposal for the use of chemical agents was also

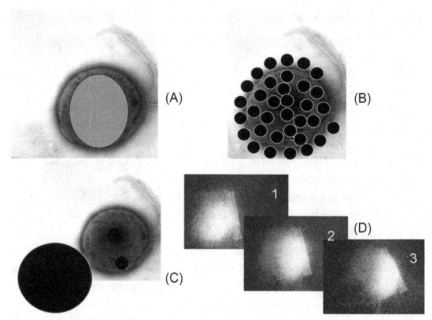

Fig. 26 Loading methods for bacteria propelling nanorobots: (A) The deliverables are loaded inside the cell of bacteria. (B) The deliverable nanoparticles are loaded by attaching them to the cell. (C) The nano or micro body is pushed by the means of bacteria. (D) A large body was propelled by a swarm of bacteria. *(From Martel, S., Mohammadi, M., Felfoul, O., Lu, Z., Pouponneau, P., 2009. Flagellated magnetotactic bacteria as controlled MRI-trackable propulsion and steering systems for medical nanorobots operating in the human microvasculature. Int. J. Robot. Res. 28 (4), 571–582, with permission.)*

proposed within the research. This proposal was accompanied by two limitations. The first one is the potential effect of this agent on the place where it is used—which is in vivo. The second limitation is the fact that chemical agents cannot be controlled through the electronic circuits. In a nutshell, manipulating the nanorobot movement through the bacteria by the use of chemical agents will not give us full control over the movement. Furthermore, researchers have investigated the possibility of nanorobot propulsion using artificial flagella. This technique of bio-mimicking stems from the idea of using bacteria as the propulsion driver while using artificial parts in order to have a higher control, and to allow usage in a wider field of applications.

5.11 Heart surgery using nano robots

Heart disease, heart attacks, and heart transplant are all important areas that concern many people because of the criticality of having a serious problem with the heart. The application of nanorobots in solving heart problems is one of the promising fields that carry hope to many people, and thus it is tackled by many researches from different points of view. These points of view range from using nanorobots in regular health checks to detect problems at an early stage, to the use of nanorobots in heart surgery, to even the use of them for after surgery care while they are maintained in vivo.

In a work done by Cavalcanti et al. (2006), nanorobots were proposed to be used in stenosed coronary occlusion (SCO) treatment. SCO is the medical case of obstruction of blood flow to the coronary artery, which results in heart attack. Chemical and thermal treatment will be provided by the means of nanorobots. Accordingly, the components of the Nano robot are sensing elements—temperature and chemical sensors, energy supply and actuator. Although real fabrication is not done in this research, simulation is done using the clinical data and biological information to ensure the possibility of fabricating this nanorobot and manipulating it. In addition, valuable data about blood velocity, blood streamlines, and blood temperature are discussed, as they are the key design inputs in the process of manufacturing a nanorobot that will go through the blood stream—especially, if it is planned to manipulate heart issues.

In another work, carried out by Merina (2010), a novel proposal was made for using nanorobots for after surgery care in heart transplant cases. From a medical point of view, although heart transplant surgery itself is useful in helping patients with heart failure to survive, a major problem of cell rejection to the new heart is common among patients; with at least 40% of heart transplant survivors having at least one rejection case during the first year after the transplant surgery. In many cases, symptoms of this rejection do not appear in the patient for a long time. In this research, the design of a nanorobot to help in detecting and controlling the presence of infection or heart rejection is developed. Although no trial for fabricating the nanorobot was done in this work, framework and design elements ranging from main components to materials are discussed. Additionally, for fabricating the Nano robots, self-assembly and external-assembly were proposed. The former is done by letting the robot assemble itself from cells and components, while the latter is a human-made assembling operation.

In research conducted by Wamala et al. (2017), nano soft deformable robots have been proposed for usage in surgery as automated surgical tools. The main focus of this work was the novel use of soft robots, which are necessary and have preference over other rigid robots in surgeries related to the heart. The reason behind this is the fact that these robots will not damage tissues while doing their task or while moving inside the blood vessels. Their ease of deformation and their fabrication from bio-friendly materials makes their motion more flexible with fewer obstacles. Furthermore, this proposed soft robot will depend on its control over biological systems and physical properties, instead of many external circuits and motors. This adjustment to the design from other conventional designs of nanorobots is perfect to suit the application in which the nanorobot will be used.

The challenge that still impedes the implementation of such nanorobot designs in real life and in performing real surgeries is the precise propulsion. A combination between this design for soft robots with the bacteria propulsion system proposed by Martel et al. (2009) will probably build the link between the proposals and the implementations. Experiments on models as well as simulations should be conducted to ensure that the interference of the nanorobot with the heart is well and healthy for the functionality of the heart.

Technological advances are what distinguish earlier efforts on the topic of using nanorobots as a treatment for cardiac issues from efforts made by researchers nowadays. Access to laboratories with higher technologies and equipment to help in fabricating and testing the proposed nanorobot is shaping the future of research and implementation in this area. Precision and accuracy are of great importance in this specific application due to the criticality associated with operating or performing surgery related to the heart.

6 Discussion

Regarding the medical field, nanorobots present impressive solutions and enhancements for problems, diseases, and even systems. However, the review conducted about nanorobots components and their existing designs, fabrication methods, and motion scientific bases was limited by the currently available technology. In other words, most of the discussed designs and their fabrication depend mainly on the magnetic field in their motion, whether it is gradual, rotary, or even reciprocal. One of the limitations that faces the upgrade done by nanorobots is providing a suitable powering source for the designed system. Actually, researches have tried, in a theoretical way, to provide a solution by presenting an artificial molecular pump (Cheng et al., 2015). This pump can perform its function by transportation of small positively charged rings from a bulk solution, considered as an initial environment, to high-energy multicomponent assembly representing the new environment. The pump functions against local concentration gradient so as to form potential energy. Additionally, the design underneath represented a mechanism that choreographs relative molecular forces, controlling the noncovalent forces in a very precise way (Cheng et al., 2015). If this mechanism goes the way of proof of concept and applicability testing, it can add a new solid layer of applications to be added to benefits of nanorobots in the medical field.

One of the promising actuation methods for nanorobots is piezoelectricity, which is a physical property that is widely known; it is the polarization experienced by a material when being subjected to mechanical stress. It has been encountered in research many times for being used in electric power generation. In the framework of medical nanorobots, piezoelectricity forms a method of actuation for nanomanipulators. Researches done by Sinha et al. (2009) and Li et al. (2016) discussed the use of piezoelectric properties of materials for two-dimensional drive for nanoactuators. The results of both researches have shown that the former had reached the speed of 0.3 mm/s, while the latter had reached vertical deflection of 40 nm under actuation by 2 V. the drawback in the latter research is that the layer thickness, which was fabricated from aluminum nitride, is 100 nm; this thickness represents a challenge when trying to make a complete nanorobot within the nanoscale (Li et al., 2016). Additionally, considering power for a nanorobot that targets in vivo missions is still a challenge.

Research and advances in electronic engineering have been directed towards minimizing the size of passive components, such as capacitors and inductors, in order to fit into nanoelectronic chips. Apart from fabricating small components and mounting them on the surfaces on electronic chips, including these components in substrate of chip in embedded form improves the size target as well as the functionality of components (Khan et al., 2018).

Continuing with the challenges, the conventional classical laws of circuit theory cannot be applied anymore due to tunneling phenomenon that occurs in such a scale, and their behaviors are better described by quantum theory. With shifting the concentration from classical theory to the quantum one, designs for single electron transistors (SETs) as well as resonant tunneling diodes (RTDs) are proposed, which show how electric components that are used in everyday life, such as transistors and diodes, can work under new concepts and governing rules (Forshaw et al., 2004).

The CMOS process is considered an ideal technology to create pH-sensitive arrays due to their significant scalability and low-power operation. This unmodified ISFET type is created using an extended metal gate that comes with a sensing membrane above; it allows the CMOS to use this layer as a pH sensor. The detection of this type is based on using the sensing membrane as a double layer of SiO_2-Si_3N_4 (Rasooly and Herold, 2009). The structure of the floating-gate FET (FGFET) consists of two main gates: sensing and control. Those two gates are joined in a capacitive manner to a common floating gate; therefore, any change in the potential in either of them adjusts the potential of the floating gate. In fact, this type is widely used in DNA charge molecule detection as the principle based on the assumed charge generated by DNA molecules (Rasooly and Herold, 2009).

The extended-gate FET (EGFET) is a simple combination of the merits of both ISFETs and the coated-wire technology. This type is distinguished by the possibility to separate between the dry and wet environments. The working principle of the EGFET is that the sensor generates the signal physically from high to low impedance environments. The EGFET simple structure is used in specific sensing cases such as a pH sensing, and BHV-1 antibodies detection based on sensing the electrical enzyme-linked immunosorbent assay. The specific structure for such type is also useful in sensing extracellular $K+$ concentration with pad fabricated on the micro-scale (Rasooly and Herold, 2009). The dual-gate sensor (DGFET) shares a similar structure to the FGFET sensor. The dual-gate type is widely used in pH sensing and it can be used with minor scale of DNA detection. The structure of this type implements additional gates that affect the sensor response positively (Rasooly and Herold, 2009).

Regarding future technology for nanorobots, a general optimism should be spread as there is a good solid ground to build on. Despite the huge risk taken in each step, introducing the internet of nano things (IoNT) to the nanorobots world will cover a good portion of tedious tasks and applications. Nevertheless, additional researches conduction and development is a must. IoNT helps in having a complete view of the internet of everything (IoE) when it provides the incorporation of nanosensors in diverse object by the help of nanonetworks (Miraz et al., 2018). It is concerned with the connection between the four pillars of people, process, data, and things, instead of giving all concern only to "things." Actually, the ability of practical implementation of IoNT has been proved theoretically when transceivers succeeded in being implemented using Tetrahertz frequencies. Accordingly, the future work should focus on developing the interface between IoNT and actually existing micro and nano devices. It can be even taken from different perspectives of industry and the medical arena. Moreover, standardized software and protocols of work can be achieved by the agreement of institutions developing such researches to implement the concepts of IoNT. The software to be used should be designed on optimal architecture criteria. Main requirements of implementation of IoT, such as power requirements and energy-aware routing, should be considered.

Regardless the focus on control and electronic solution growth, it is a wonderful idea to consider material engineering and science in nanorobot material selection procedures. This field can get benefit from the rise of material biodegradability concepts to avoid taking additional steps. For instance, using biodegradable material in manufacturing nanorobots may help save drugs inside the robot for a specific time period until the material is completely dissolved, or it may help in getting rid of the nanorobot after completing its mission without the need of surgery to eliminate it from the human body. Therefore, material scientists should give some concern to trying to help in this particular area, helping in the great impact of nanorobots.

7 Conclusion

The scientific revolution allows the merging of different fields of information technology, biotechnology, medicine, and robotics. This integration helps greatly in different fields, i.e., the medical field has reached an impressive phase whereby manufacturing of robots in the nanoscale starts to provide significant support in disease identification, drug-delivery, and even treatment. Previously conducted researches, from 2008 to 2018, reached solid steps in the field with highlighted discoveries. Field previous achievements are summarized in the proposal done to use flagellated magnetotactic bacteria in the control and tracking subfield of nanorobots, another to detect diseases relying on the scientific base of PSO algorithm, and a proposal to use nanorobots in several applications such as heart transplant or DDS. Modeling and simulations have been done to validate the efficiency of using flagella propulsion in hybrid nanorobots. Another is done for nanorobots that repair damaged blood cells. These achievements were continued by researches discussing the fabrication of nanorobots based on several perspectives in 2014 and 2015. Researches in 2018 tried to validate if manufacturing nanorobots for discussed benefits is worthy or not.

Robot main components like actuators, sensors, electronic ships, and controllers are presented as discussed in different previous researches to suit several aims of manufacturing a nanorobot. Moreover, discussed in brief are the designs that have already been applied in the real world and shows a success in motion in vivo such as flagellum bacteria, delivering drug to cancerous tumor, or even taking a sample from an infected cell to be tested and analyzed outside the body. These collections of designs open the door for imagination of specific integration between designs and components producing a nanorobot that is responsible for a specific target.

Several applications are discussed as a proof of the assistance of nanorobots in the medical field. These applications cover different areas starting from detecting diseases by noticing the generation of specific enzymes around the infected area or in blood, passing by drug delivery, including different diseases such as diabetes or even cancer, and reaching surgical targets of nanorobots in tumor elimination without harming the surrounding area. Nanorobots also help traditional ways of tumor detection, treatment, or elimination by giving full data/map to the area to be handled and give help in monitoring human body behavior in normal conditions as a check-up for being healthy. In a nonstoppable way, nanorobots show an impressing functionality of simulating blood components and providing artificial blood that helps in treatment of several diseases resulting from shortage of one of the blood components.

Looking forward to new discoveries and enhancements regarding nanorobots, the integration with the internet of nano things to control the nanorobot from outside the body would be a great addition to the continuous success of nanorobots and a new level of integration between technology, intelligence, and control. Furthermore, use of biodegradable materials in manufacturing of nanorobots gives an advantage of forming easily destroyed components that cause no harm or side effects in the human body. In addition, it saves cost and eliminates the risk of components' elimination by surgery or any other way.

References

Abbott, J.J., Ergeneman, O., Kummer, M.P., Hirt, A.M., Nelson, B.J., 2007. Modeling magnetic torque and force for controlled manipulation of soft-magnetic bodies. IEEE Trans. Robot. 23 (6), 1247–1252.

Aguilar, Z., 2013. Nanomaterials for Medical Applications. Elsevier, Amsterdam.

Ahmed, S., ElAraif, T., Amin, S.E., 2014. A novel communication technique for nanorobots swarms based on evolutionary strategies. In: 2014 UKSim-AMSS 16th International Conference on Computer Modelling and Simulation, 26-28 March 2014, Cambridge, UK, IEEE, pp. 51–56. https://doi.org/10.1109/UKSim.2014.72.

Ammar, H.H., Azar, A.T., 2020. Robust path tracking of mobile robot using fractional order PID controller. In: The International Conference on Advanced Machine Learning Technologies and Applications (AMLTA2019). AMLTA 2019. Advances in Intelligent Systems and Computing, vol. 921. Springer, Cham, pp. 370–381.

Arpita, J., Hinali, T., Atanukumar, B., Krunali, T., 2013. Nanotechnology revolution: respirocytes and its application in life sciences. Innovare J. Life Sci. 1 (1), 8–13.

Azar, A.T., Zhu, Q., Khamis, A., Zhao, D., 2017. Control design approaches for parallel robot manipulators: a review. Int. J. Model. Ident. Control (IJMIC) 28 (3), 199–211.

Azar, A.T., 2011. Biomedical engineering specialization and future challenges. Int. J. Biomed. Eng. Technol. (IJBET) 6 (2), 163–177.

Azar, A.T., 2012. Overview of biomedical engineering. In: Joel, J.P.C., de la Torre Diez, I., de Abajo, B.S. (Eds.), Telemedicine and E-Health Services, Policies and Applications: Advancements and Developments. IGI Global, USA, ISBN: 978-1466608887, pp. 112–139.

Azar, A.T., Eljamel, M.S., 2012. Medical robotics. In: Sobh, T., Xiong, X. (Eds.), Prototyping of Robotic Systems: Applications of Design and Implementation. IGI Global, USA, ISBN: 978-1466601765.

Azar, A.T., Ammar, H.H., de Brito Silva, G., Razali, M.S.A.B., 2020a. Optimal proportional integral derivative (PID) controller design for smart irrigation mobile robot with soil moisture sensor. In: The International Conference on Advanced Machine Learning Technologies and Applications (AMLTA2019). AMLTA 2019. Advances in Intelligent Systems and Computing, vol. 921. Springer, Cham, pp. 349–359.

Azar, A.T., Serrano, F.E., Vaidyanathan, S., Albalawi, H., 2020b. Adaptive higher order sliding mode control for robotic manipulators with matched and mismatched uncertainties. In: The International Conference on Advanced Machine Learning Technologies and Applications (AMLTA2019). AMLTA 2019. Advances in Intelligent Systems and Computing, vol. 921. Springer, Cham, pp. 360–369.

Azar, A.T., Hassan, H., Razali, M.S.A.B., de Brito Silva, G., Ali, H.R., 2019a. Two-degree of freedom proportional integral derivative (2-DOF PID) controller for robotic infusion stand. In: Proceedings of the International Conference on Advanced Intelligent Systems and Informatics 2018. AISI 2018. Advances in Intelligent Systems and Computing, vol. 845. Springer, Cham, pp. 13–25.

Azar, A.T., Ammar, H.H., Barakat, M.H., Saleh, M.A., Abdelwahed, M.A., 2019b. Self-balancing robot modeling and control using two degree of freedom PID controller. In: Proceedings of the International Conference on Advanced Intelligent Systems and Informatics 2018. AISI 2018. Advances in Intelligent Systems and Computing, vol. 845. Springer, Cham, pp. 64–76.

Azar, A.T., Serrano, F.E., 2019. Fractional order two degree of freedom PID controller for a robotic manipulator with a fuzzy type-2 compensator. In: Proceedings of the International Conference on Advanced Intelligent Systems and Informatics 2018. AISI 2018. Advances in Intelligent Systems and Computing, vol. 845. Springer, Cham, pp. 77–88.

Azar, A.T., Ammar, H.H., Mliki, H., 2018a. Fuzzy logic controller with color vision system tracking for mobile manipulator robot. In: The International Conference on Advanced Machine Learning Technologies and Applications (AMLTA2018). AMLTA 2018. Advances in Intelligent Systems and Computing, vol. 723. Springer, Cham, pp. 138–146.

Azar, A.T., Kumar, J., Kumar, V., Rana, K.P.S., 2018b. Control of a two link planar electrically-driven rigid robotic manipulator using fractional order SOFC. In: Proceedings of the International Conference on Advanced Intelligent Systems and Informatics 2017. AISI 2017. Advances in Intelligent Systems and Computing, vol. 639. Springer, Cham, pp. 57–68.

Bao, J., Yang, Z., Nakajima, M., Shen, Y., Takeuchi, M., Huang, Q., Fukuda, T., 2014. Self-actuating asymmetric platinum catalytic mobile nanorobot. IEEE Trans. Robot. 30 (1), 33–39.

Barakat, M.H., Azar, A.T., Ammar, H.H., 2020. Agricultural service mobile robot modeling and control using artificial fuzzy logic and machine vision. In: The International Conference on Advanced Machine Learning Technologies and Applications (AMLTA2019).

AMLTA 2019. Advances in Intelligent Systems and Computing, vol. 921. Springer, Cham, pp. 453–465.

Bhat, A.S., 2014. Nanobots: the future of medicine. Int. J. Manage. Eng. Sci. 5 (1), 44–49.

Bogue, M., 2010. Microrobots and nanorobots: a review of recent developments. Ind. Robot: Int. J. 42 (2), 98–102.

Boonrong, P., Kaewkamnerdpong, B., 2011. Canonical PSO based nanorobot control for blood vessel repair. Int. J. Biomed. Biol. Eng. 5 (10), 428–478.

Cavalcanti, A., Rosen, L., Shirinzadeh, B., Rosenfeld, M., 2006. Nanorobot for treatment of patients with artery occlusion. In: Proceedings of Virtual Concept, November 26th–December 1st, 2006. Mexico, Cancun.

Chang, J., Kim, J., Min, B.K., Lin, L., 2007. Thermally driven bimorph nano actuators fabricated using focused ion beam chemical vapor deposition. In: TRANSDUCERS 2007-2007 International Solid-State Sensors, Actuators and Microsystems Conference, 10–14 June 2007, Lyon, France, IEEE, pp. 541–544. https://doi.org/10.1109/SENSOR.2007.4300187.

Chen, K., Chen, T., Liu, H., Yang, Z., 2015. A Pt/Au hybrid self-actuating nanorobot towards to drug delivery system. In: 10th IEEE International Conference on Nano/Micro Engineered and Molecular Systems, 7–11 April 2015, Xi'an, China, IEEE, pp. 286–289. https://doi.org/10.1109/NEMS.2015.7147428.

Cheng, C., McGonigal, P.R., Schneebeli, S.T., Li, H., Vermeulen, N.A., Ke, C., Stoddart, J.F., 2015. An artificial molecular pump. Nat. Nanotechnol. 10 (6), 547–553.

Chiariello, A.G., Maffucci, A., Miano, G., 2013. Temperature effects on electrical performance of carbon-based nano-interconnects at chip and package level. Int. J. Numer. Model.: Electron. Netw. Devices Fields 26 (6), 560–572.

da Silva Luz, G.V., Barros, K.V.G., de Araújo, F.V.C., da Silva, G.B., da Silva, P.A.F., Condori, R.C.I., Mattos, L., 2016. Nanorobotics in drug delivery systems for treatment of cancer: a review. J. Mater. Sci. Eng. A 6 (5-6), 167–180.

Daulbaev, C.B., Dmitriev, T.P., Sultanov, F.R., Mansurov, Z.A., Aliev, E.T., 2017. Obtaining three-dimensional nanosize objects on a "3D Printer + Electrospinning" machine. J. Eng. Phys. Thermophys. 90 (5), 1115–1118.

Diao, Z., Sauer, V.T., Hiebert, W.K., 2017. Integrated on-chip nano-optomechanical systems. Int. J. High Speed Electron. Syst. 26 (01n02), 1740005.

Drexler, K.E., 1987. Engines of Creation: The Coming Era of Nanotechnology. Anchor Library of Science, New York, ISBN: 978-0385199735.

Drexler, K.E., 1989. Biological and nanomechanical systems: contrasts in evolutionary capacity. In: Langton, C.G. (Ed.), Artificial Life. Addison-Wesley, Redwood City: CA.

Drexler, K.E., 1992. Nanosystems: Molecular Machinery, Manufacturing, and Computation. Wiley & Sons, Chichester, UK.

Dubey, A., Mavroidis, C., Thornton, A., Nikitczuk, K., Yarmush, M.L., 2003. Viral protein linear (VPL) nano-actuators. In: 2003 Third IEEE Conference on Nanotechnology, 2003. IEEE-NANO 2003, 12–14 Aug. 2003, San Francisco, CA, USA, USA, IEEE, vol. 1, pp. 140–143. https://doi.org/10.1109/NANO.2003.1231735.

Elsayed, S., Amin, S., Alarif, T., 2014. Assessment of applying path planning technique to nanorobots in a human blood environment. In: 2014 European Modelling Symposium, 21–23 Oct. 2014, Pisa, Italy, IEEE, pp. 45–51. https://doi.org/10.1109/EMS.2014.37.

Eshaghian-Wilner, M., 2009. Nano-Scale and Bio-inspired Integrated Computing. Nature-Inspired Computing Series. Wiley-Blackwell, ISBN: 978-0470116593.

Farazkish, R., 2018. Robust and reliable design of bio-nanorobotic systems. Microsyst. Technol. 25 (4), 1519–1524.

Frcitas, R.A., 1998. Exploratory design in medical nanotechnology: a mechanical artificial red cell. Artif. Cells Blood Subst. Biotechnol. 26 (4), 411–430.

Forshaw, M., Stadler, R., Crawley, D., Nikoli, K., 2004. A short review of nanoelectronic architectures. Nanotechnology 15 (4), S220–S223.

Fujiwara, T., Nakajima, M., Ichikawa, A., Ohara, K., Hasegawa, Y., Fukuda, T., 2016. Electrostatic actuation of folded multi-graphene structure for nano-gripper. In: 2016 IEEE 16th International Conference on Nanotechnology (IEEE-NANO), 22–25 Aug. 2016, Sendai, Japan, IEEE, pp. 34–35. https://doi.org/10.1109/NANO.2016.7751477.

Fukada, S., Onda, K., Maruyama, H., Masuda, T., Arai, F., 2013. 3D fabrication and manipulation of hybrid nanorobots by laser. In: 2013 IEEE International Conference on Robotics and Automation, 6–10 May 2013, Karlsruhe, Germany, IEEE, pp. 2594–2599. https://doi.org/10.1109/ICRA.2013.6630932.

Fusco, S., Ullrich, F., Pokki, J., Chatzipirpiridis, G., Özkale, B., Sivaraman, K.M., Ergeneman, O., Pane, S., Nelson, B.J., 2014. Microrobots: a new era in ocular drug delivery. Expert Opin. Drug Deliv. 11 (11), 1815–1826.

Fusco, S., Huang, H.W., Peyer, K.E., Peters, C., Häberli, M., Ulbers, A., Spyrogianni, A., Pellicer, E., Sort, J., Pratsinis, S.E., Nelson, B.J., Sakar, M.S., Pané, S., 2015. Shape-switching microrobots for medical applications: the influence of shape in drug delivery and locomotion. ACS Appl. Mater. Interfaces 7 (2), 6803–6811.

Ghaffari, A., Shokuhfar, A., Ghasemi, R.H., 2010. Design and simulation of a novel bio nano actuator by prefoldin. In: 10th IEEE International Conference on Nanotechnology, 17–20 Aug. 2010, Seoul, South Korea, IEEE, pp. 885–888. https://doi.org/10.1109/NANO.2010.5697837.

Ghanbari, A., Chang, P.H., Nelson, B.J., Choi, H., 2014. Magnetic actuation of a cylindrical microrobot using time-delay-estimation closed-loop control: modeling and experiments. Smart Mater. Struct. 23 (3), 35013.

Guillou, P.J., Quirke, P., Thorpe, H., Walker, J., Jayne, D.G., Smith, A.M., Heath, R.M., Brown, J.M., MRC CLASICC Trial Group, 2005. Short-term endpoints of conventional versus laparoscopic-assisted surgery in patients with colorectal cancer (MRC CLASICC trial): multicentre, randomised controlled trial. Lancet 365 (9472), 1718–1726.

Hayakawa, T., Fukada, S., Arai, F., 2014. Fabrication of an on-chip nanorobot integrating functional nanomaterials for single-cell punctures. IEEE Trans. Robot. 30 (1), 59–67.

Hla, K.H.S., Choi, Y., Park, J.S., 2008. Mobility enhancement in nanorobots by using particle swarm optimization algorithm. In: 2008 International Conference on Computational Intelligence and Security, 13-17 Dec. 2008, Suzhou, China, IEEE, vol. 1, pp. 35–40. https://doi.org/10.1109/CIS.2008.108.

Hoefflinger, B., 2012. Chips 2020. Springer, Berlin.

Hoop, M., Ribeiro, A.S., Rösch, D., Weinand, P., Mendes, N., Mushtaq, F., Chen, X.Z., Shen, Y., Pujante, C.F., Puigmartí-Luis, J., Paredes, J., 2018. Mobile magnetic nanocatalysts for bioorthogonal targeted cancer therapy. Adv. Funct. Mater. 28 (25), 1705920.

Hornyak, T., 2006. Loving the Machine the Art and Science of Japanese Robots. Kodansha International, Tokyo.

Huang, H.W., Petruska, A.J., Sakar, M.S., Skoura, M., Ullrich, F., Zhang, Q., Pané, S., Nelson, B.J., 2016. Self-folding hydrogel bilayer for enhanced drug loading, encapsulation, and transport. In: 2016 38th Annual International Conference of the IEEE Engineering in Medicine and Biology Society (EMBC), 16–20 Aug. 2016, Orlando, FL, USA, IEEE, pp. 2103–2106. https://doi.org/10.1109/EMBC.2016.7591143.

Jia, X., Li, X., Lenaghan, S.C., Zhang, M., 2014. Design of efficient propulsion for nanorobots. IEEE Trans. Robot. 30 (4), 792–801.

Jin, M., Chang, P.H., 2009. Simple robust technique using time delay estimation for the control and synchronization of Lorenz systems. Chaos Solitons Fractals 41 (5), 2672–2680.

Jin, M., Kang, S.H., Chang, P.H., 2008. Robust compliant motion control of robot with nonlinear friction using time-delay estimation. IEEE Trans. Ind. Electron. 55 (1), 258–269.

Karan, S., Majumder, D.D., 2011. Molecular machinery—a nanorobotics control system design for cancer drug delivery. In: 2011 International Conference on Recent Trends in Information Systems, 21–23 Dec. 2011, Kolkata, India, pp. 197–202. https://doi.org/10.1109/ReTIS.2011.6146867.

Khan, M.I., Dong, H., Shabbir, F., Shoukat, R., 2018. Embedded passive components in advanced 3D chips and micro/nano electronic systems. Microsyst. Technol. 24 (2), 869–877.

Korayem, A.H., Korayem, M.H., Taheri, M., 2015. Robust controlled manipulation of nanoparticles using the AFM nanorobot probe. Arab. J. Sci. Eng. 40 (9), 2685–2699.

Kosmas, P., Chen, Y., 2012. Possibilities for microwave breast tumor sensing via contrast-agent-loaded nanorobots. In: 2012 6th European Conference on Antennas and Propagation (EUCAP), 26–30 March 2012, Prague, Czech Republic, IEEE, pp. 185–189. https://doi.org/10.1109/EuCAP.2012.6206533.

Krol, S., Ellis-Behnke, R., Marchetti, P., 2012. Nanomedicine for treatment of diabetes in an aging population: state-of-the-art and future developments. Maturitas 73 (1), 61–67.

Kumar, N., Kumar, R., 2014. Nanotechnology and Nanomaterials in the Treatment of Life-threatening Diseases. William Andrew, Oxford. ISBN: 9780323264334.

Kumar, R., Baghel, O., Sidar, S., 2014. Applications of nanorobotics. Int. J. Sci. Res. Eng. Technol. 3 (8), 1131–1137.

Kummer, M.P., Abbott, J.J., Kratochvil, B.E., Borer, R., Sengul, A., Nelson, B.J., 2010. OctoMag: an electromagnetic system for 5-DOF wireless micromanipulation. IEEE Trans. Robot. 26 (6), 1006–1017.

Lauga, E., DiLuzio, W.R., Whitesides, G.M., Stone, H.A., 2006. Swimming in circles: motion of bacteria near solid boundaries. Biophys. J. 90 (2), 400–412.

Li, Y., Moon, K., Wong, C., 2010. Nano-conductive adhesives for nano-electronics inter-connection. In: Wong, C., Moon, K.S., Li, Y. (Eds.), Nano-Bio-Electronic, Photonic and MEMS Packaging. Springer, Boston, MA.

Li, H., Du, Z., Dong, W., 2016. Horizontal two-dimensional nano-positioner based on shear plate piezoelectric actuators. In: 2016 IEEE International Conference on Manipulation, Manufacturing and Measurement on the Nanoscale (3M-NANO), 18–22 July 2016, Chongqing, China, IEEE, pp. 95–100. https://doi.org/10.1109/3M-NANO.2016.7824916.

Loscri, V., Vegni, A.M., 2015. An acoustic communication technique of nanorobot swarms for nanomedicine applications. IEEE Trans. Nanobiosci. 14 (6), 598–607.

Mahoney, A.W., Sarrazin, J.C., Bamberg, E., Abbott, J.J., 2011. Velocity control with gravity compensation for magnetic helical microswimmers. Adv. Robot. 25 (8), 1007–1028.

Mali, S., 2013. Nanotechnology for surgeons. Ind. J. Surg. 75 (6), 485–492.

Manjunath, A., Kishore, V., 2014. The promising future in medicine: nanorobots. Biomed. Sci. Eng. 2 (2), 42–47.

Martel, S., 2008. Nanorobots for microfactories to operations in the human body and robots propelled by bacteria. *Facta universitatis*. Ser. Mech. Automat. Control Robot. 7 (1), 1–8.

Martel, S., Felfoul, O., Mohammadi, M., Mathieu, J.B., 2008. Interventional procedure based on nanorobots propelled and steered by flagellated magnetotactic bacteria for direct targeting of tumors in the human body. In: 2008 30th Annual International Conference of the IEEE Engineering in Medicine and Biology Society, 20–25 Aug. 2008, Vancouver, BC, Canada, IEEE, pp. 2497–2500. https://doi.org/10.1109/IEMBS.2008.4649707.

Martel, S., Mohammadi, M., Felfoul, O., Lu, Z., Pouponneau, P., 2009. Flagellated magnetotactic bacteria as controlled MRI-trackable propulsion and steering systems for medical nanorobots operating in the human microvasculature. Int. J. Robot. Res. 28 (4), 571–582.

Martel, S., 2011. Flagellated bacterial nanorobots for medical interventions in the human body. In: Surgical Robotics. Springer, Boston, MA, pp. 397–416.

Méallet-Renault, R., Denjean, P., Pansu, R.B., 1999. Polymer beads as nano-sensors. Sens. Actuators B Chem. 59 (2–3), 108–112.

Merina, R.M., 2010. Use of nanorobots in heart transplantation. In: 2010 International Conference on Emerging Trends in Robotics and Communication Technologies (INTER-ACT), 3–5 Dec. 2010, Chennai, India, IEEE, pp. 265–268. https://doi.org/10.1109/INTERACT.2010.5706155.

Miraz, M., Ali, M., Excell, P., Picking, R., 2018. Internet of nano-things, things and everything: future growth trends. Future Internet 10 (8), 68.

Petrina, A., 2012. Nanorobotics: simulation and experiments. Automat. Doc. Math. Linguist. 46 (4), 159–169.

Peyer, K.E., Siringil, E., Zhang, L., Nelson, B.J., 2014. Magnetic polymer composite artificial bacterial flagella. Bioinspirat. Biomimetics 9 (4), 046014.

Quader, N., Al-Arif, S.M.R., Shaon, M.A.M., Islam, K.K., Ridwan, A.R., 2011. Control of autonomous nanorobots in neural network. In: 2011 4th International Conference on Biomedical Engineering and Informatics (BMEI), 15–17 Oct. 2011, Shanghai, China, IEEE, vol. 3, pp. 1399–1402.

Ramsden, J., 2018. Applied Nanotechnology: The Conversion of Research Results to Products (Micro and Nano Technologies). William Andrew, UK, ISBN: 9780128133439.

Rasooly, A., Herold, K., 2009. Biosensors and Biodetection. Humana Press, New York.

Roszek, B., De Jong, W.H., Geertsma, R.E., 2005. Nanotechnology in Medical Applications: State-of-the-Art in Materials and Devices. RIVM Report 265001001/2005.

Schuerle, S., Erni, S., Flink, M., Kratochvil, B.E., Nelson, B.J., 2013. Three-dimensional magnetic manipulation of micro-and nanostructures for applications in life sciences. IEEE Trans. Magn. 49 (1), 321–330.

Shang, W., Lu, H., Wan, W., Fukuda, T., Shen, Y., 2016. Vision-based nano robotic system for high-throughput non-embedded cell cutting. Sci. Rep. 6, 22534.

Shen, Y., Fukuda, T., 2014. Micro-nanorobotic manipulation in single cell analysis. Robot. Biomimetic 1 (1), 21. https://doi.org/10.1186/s40638-014-0021-4.

Shen, Y., Nakajima, M., Kojima, S., Homma, M., Fukuda, T., 2010. Nano knife fabrication and calibration for single cell cutting inside environmental SEM. In: 2010 International Symposium on Micro-NanoMechatronics and Human Science, 7–10 Nov. 2010, Nagoya, Japan, IEEE, pp. 316–320. https://doi.org/10.1109/MHS.2010.5669527.

Sinha, N., Wabiszewski, G.E., Mahameed, R., Felmetsger, V.V., Tanner, S.M., Carpick, R.W., Piazza, G., 2009, June. Ultra thin AlN piezoelectric nano-actuators. In: TRANSDUCERS 2009–2009 International Solid-State Sensors, Actuators and Microsystems Conference, 21–25 June 2009, Denver, CO, USA, IEEE, pp. 469–472. https://doi.org/10.1109/SENSOR.2009.5285460.

Sitti, M., 2007. Microscale and nanoscale robotics systems [Grand Challenges of Robotics]. IEEE Robot. Automat. Mag. 14 (1), 53–60.

Song, B., Yang, R., Xi, N., Patterson, K.C., Qu, C., Lai, K.W.C., 2012. Cellular-level surgery using nano robots. J. Lab. Automat. 17 (6), 425–434.

Stewart, D., 1965. A platform with six degrees of freedom. J. Power Energy 180 (15), 371–386.

Subramanian, S., Rathore, J.S., Sharma, N.N., 2009. Design and analysis of helical flagella propelled nanorobots. In: 2009 4th IEEE International Conference on Nano/Micro Engineered and Molecular Systems, 5–8 Jan. 2009, Shenzhen, China, IEEE, pp. 950–953. https://doi.org/10.1109/NEMS.2009.5068731.

Thammawongsa, N., Zainol, F.D., Mitatha, S., Ali, J., Yupapin, P.P., 2013. Nanorobot controlled by optical tweezer spin for microsurgical use. IEEE Trans. Nanotechnol. 12 (1), 29–34.

Tiemerding, T., Fatikow, S., 2018. Software for small-scale robotics: a review. Int. J. Autom. Comput. 15 (5), 515–524.

Trihirun, S., Achalakul, T., Kaewkamnerdpong, B., 2013. Modeling nanorobot control for blood vessel repair: a non–Newtonian blood model. In: The 6th 2013 Biomedical Engineering International Conference (BMEiCON), 23–25 Oct. 2013, Amphur Muang, Thailand, IEEE, pp. 1–5. https://doi.org/10.1109/BMEiCon.2013.6687727.

Tsitkov, S., Hess, H., 2017. Rise of the nanorobots: advances in control, molecular detection, and nanoscale actuation are bringing us closer to a new era of technology enhanced by nanorobots. IEEE Pulse 8 (4), 23–25.

Wamala, I., Roche, E.T., Pigula, F.A., 2017. The use of soft robotics in cardiovascular therapy. Expert. Rev. Cardiovasc. Ther. 15 (10), 767–774.

Wang, C., Wu, J., Jiang, K., Humphrey, M.G., Zhang, C., 2017. Stable Ag nanoclusters-based nano-sensors: rapid sonochemical synthesis and detecting Pb^{2+} in living cells. Sens. Actuators B Chem 238 (2017), 1136–1143.

Wautelet, M., 2009. Nanotechnologies. Institution of Engineering and Technology, London, UK.

Wong, C.P., Moon, K.S., Li, Y., 2010. Nano-Bio-Electronic, Photonic and MEMS Packaging. Springer, New York, NY, USA, ISBN: 978-1-4419-0039-5.

Xu, J., 2013. Motion simulation of an artificial flagellum nanorobot. In: 2013 13th IEEE International Conference on Nanotechnology (IEEE-NANO 2013), 5–8 Aug. 2013, Beijing, China, IEEE, pp. 1208–1211. https://doi.org/10.1109/NANO.2013.6720954.

Yang, W.D., Wang, X., Fang, C.Q., Lu, G., 2014. Electromechanical coupling characteristics of carbon nanotube reinforced cantilever nano-actuator. Sens. Actuators A Phys. 220 (2014), 178–187.

Zhang, L., Petit, T., Peyer, K.E., Nelson, B.J., 2012. Targeted cargo delivery using a rotating nickel nanowire. Nanomed.: Nanotechnol. Biol. Med. 8 (7), 1074–1080.

Zhang, L., Abbott, J.J., Dong, L., Kratochvil, B.E., Bell, D., Nelson, B.J., 2009. Artificial bacterial flagella: fabrication and magnetic control. Appl. Phys. Lett. 94 (6), 064107.

Zhang, L., Peyer, K.E., Nelson, B.J., 2010. Artificial bacterial flagella for micromanipulation. Lab Chip 10 (17), 2203–2215.

Zhao, Q., Li, M., Luo, J., Dou, L., Li, Y., 2015. A nanorobot control algorithm using acoustic signals to identify cancer cells in non–Newtonian blood fluid. In: 2015 IEEE International Conference on Mechatronics and Automation (ICMA), 2–5 Aug. 2015, Beijing, China, IEEE, pp. 912–917. https://doi.org/10.1109/ICMA.2015.7237607.

Further reading

Soliman, M., Azar, A.T., Saleh, M.A., Ammar, H.H., 2020. Path planning control for 3-omni fighting robot using PID and fuzzy logic controller. In: The International Conference on Advanced Machine Learning Technologies and Applications (AMLTA2019). AMLTA 2019. Advances in Intelligent Systems and Computing, vol. 921. Springer, Cham, pp. 442–452.

CHAPTER TWELVE

Impedance control applications in therapeutic exercise robots

Erhan Akdogan[a] and Mehmet Emin Aktan[b]

[a]Department of Mechatronics Engineering, Mechanical Engineering Faculty, Yıldız Technical University, Istanbul, Turkey
[b]Department of Mechatronics Engineering, Engineering Faculty, Bartın University, Bartın, Turkey

1 Introduction

Rehabilitation is a treatment process performed to increase the quality of life of an individual with physical disability, which might be congenital or happen due to an accident, aging, or an illness and to reach the optimum functional level (İnal, 2000; Akdemir and Akkus, 2006). In parallel with the increase in the world population, the need for rehabilitation is increasing. Various medical methods have been developed to treat limb injuries as much as possible. Accordingly, therapeutic exercises have a crucial role in physical medicine and rehabilitation area. They include specific exercise movements that are performed by a physiotherapist for the target limb. The purpose of these exercises is to regain range of motion (ROM) and rebuild muscle strength, endurance, elasticity of tissues, and patient's motor skills. The physiotherapist performs therapeutic exercises either manually or by various exercise devices. Scientists have done many studies on rehabilitation robots by stating that performing these exercises with robotic systems will provide many advantages and convenience both for patients and physiotherapists. These studies have gained momentum in the last 15 years. Some of these studies have been turned into commercial products (Sobh and Xiong, 2012). The functional recovery is a long and laborious process and usually involves complex movements, dedicated work, and compliance to the treatment schedule. Robotic rehabilitation contributes to this process and shortens the duration of treatment. In particular, patients' access to the hospitals is a serious challenge. Rehabilitation robots allow home care of patients through internet-based remote control and programmable features. In the rehabilitation process, the measurement and recording of biomechanical and biological parameters of the patient with high accuracy and precision is extremely important in terms of monitoring the treatment process and contributing to the treatments to be applied in other patients. Rehabilitation robots offer effective solutions to this problem with their sensors with high precision, accuracy, and resolution and databases that can store and analyze a large number of different data. In addition, they prevent the problems caused by subjective measurement and evaluation in manual treatment performed by a pyhsiotherapist. Therapeutic exercises consist of movements that require repetitive and same exercise conditions. A rehabilitation robot can apply the same force and ROM to the patient in the desired number and duration. The use of rehabilitation robots in rehabilitation centers and hospitals is increasing day by day due to the proven successes. Therapeutic exercises include ROM, strengthening, and active-assistive exercise movements. These movements correspond to position and force control in robot control. There is human (patient or physiotherapist)–robot interaction in a rehabilitation robot. Therefore, effective control, sensitive-fast feedback, and safety are very important factors in this interaction. The most known technique for position control of robots is proportional-derivative-integral (PID) control. Impedance control is used as the most effective control method for the systems that are needed for human-robot interaction. Impedance control method was developed by Hogan (1985). It performs force and position control by adjusting the mechanical impedance of the end effector of robot. There are several versions of impedance control: position-based impedance control, force-based impedance control, hybrid impedance control, and variable impedance control. In position-based impedance control, position tracking can be performed by providing the target resistance level. In the force-based impedance control, the resistance of the robot end effector against the patient can be adjusted. Hybrid impedance control method was developed by Anderson and Spong (1988). In this technique, "impedance control" and "hybrid force-position control" strategies are combined under the same control structure. Thus, both position- and force-based impedance controls are combined within a single control structure. This chapter is organized as follows: related work, theory of therapeutic exercises, impedance control techniques, modeling therapeutic exercises with impedance control methods, and impedance control-based robotic systems developed by the authors.

2 Related work

There are many studies in the literature on the use of impedance control methods in therapeutic exercise robots. Krebs et al. (1998, 2004) and

Hogan et al. (1995) developed a three degree-of-freedom (DOF) rehabilitation system named MIT-MANUS (Massachusetts Institute of Technology-MANUS) for shoulder and elbow rehabilitation. The impedance control method is used in the control of the system. The robot can perform passive, active-assistive, and resistive exercises. Okada et al. (2001) have developed an impedance-controlled robotic system that can perform passive exercises for lower limbs in spastic patients. Nef et al. (2007) developed an exoskeleton robot for the shoulder and elbow rehabilitation which they called ARMin. This system has six DOF, including four active and two passive, and can do passive and active-assistive exercises. It has the ability to perform trajectory tracking applications, to give auditory and visual feedback, and to make gravity compensation. Impedance and admittance control techniques are used in the control of the system. Denève et al. (2008) developed a three-DOF exoskeleton robot for shoulder and elbow rehabilitation. The robot can perform passive, active-assistive, and resistive exercises. PI and PID position control and impedance control methods were used in the control of the system. Oblak et al. (2009) developed a two-DOF robot system for the rehabilitation of the wrist and forearm called Universal Haptic Device (UHD). UHD can perform daily activities (reaching, dropping, etc.). The system has two operating modes: wrist and forearm. The impedance controller was used as controller. Kiguchi and Hayashi (2012) developed a seven-DOF exoskeleton robot system. The system is designed for wrist, forearm, elbow, and shoulder rehabilitation and can do active-assistive exercises. Impedance control method is used in the robot control. Ren et al. (2013) developed a six DOF, impedance-controlled exoskeleton robot system called as IntelliArm. The system is designed for wrist, forearm, elbow, and shoulder rehabilitation and can perform passive, active, active-assistive, and resistive exercises. Yeong et al. (2009) have developed a three-DOF wrist and forearm rehabilitation robot called ReachMAN (reach and manipulation). Admittance and impedance control methods were used in robot control. There are modes of pick and place, eating-drinking, and therapeutic exercise. Fraile et al. (2016) developed a robot for the rehabilitation of shoulder and elbow with two DOF, called E2Rebot. There is a handle on the robot which can be moved in X- and Y-axes. Impedance control method is used in the robot control. Kim and Deshpande (2017) developed a 5 + 5-DOF exoskeleton robot for the rehabilitation of shoulders, elbows, and forearms, called Harmony, of the right and left arms. The impedance control method is used. The authors also modeled therapeutic exercises using the impedance control method and developed robotic systems for lower- and upper-limb rehabilitation called as PHYSIOTHERABOT© (Akdogan and Adli, 2011), PHYSIOTHERABOT/w1© (Akdogan et al., 2018), and DIAGNOBOT© (Aktan and Akdogan, 2018). The detailed information about these systems and explanation of the control methods are given in this chapter.

3 Background of therapeutic exercises

3.1 Movement types

There are five fundamental exercise movements in rehabilitation (Fig. 1) (Griffith, 2000).

Flexion: Bending limb or joint

Extension: Extending limb or joint

Abduction: Movement of a limb away from the midline of the body

Adduction: Movement of a limb closer to the midline of the body

Rotation: Rotating limb or joint

3.2 Exercise types

Therapeutic exercises are one of the most important applications of rehabilitation. Unlike drug therapy, exercise procedures are not fully defined. This poses negativity for patients. Physical exercises improve strength, endurance, body, and limb motion capability.

Therapeutic exercises can be classified as passive, active–assistive, and resistive exercises. The classification of therapeutic exercises is given in Fig. 2. Passive exercises are especially applied in patients without muscular contraction. They can be done manually or with the help of an exercise device. The active exercises include voluntary muscular contractions. These exercises can include active ROM or general stroke rehabilitation exercises and muscles move through therapeutic movements. The purpose of resistive exercises is to increase muscular strength. Resistive exercises can also be performed by hand or with the aid of a therapeutic device. Isotonic exercises are increase muscular strength, power, and endurance based on lifting a constant amount of weight at variable speeds through a ROM. In contrast, isometric exercises are a type of strength exercise in which the joint angle and muscle length do not change during muscular contraction. In the isokinetic exercises, the joint velocity is constant and the system applies force against the person. Generally, athletes do this exercise.

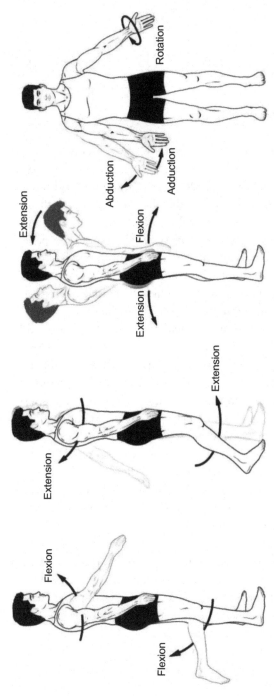

Fig. 1 Fundamental exercise movements (Openstax, 2013).

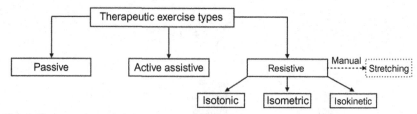

Fig. 2 Therapeutic exercise types.

Passive exercise: In passive exercise, movements to the patient are performed by a physiotherapist or by a device such as a continuous passive motion. Various robotic devices are also used for this purpose. The purpose of this exercise is to enable the patient to reach the normal ROM. It is usually performed in stroke patients.

Active-assistive exercise: In this type of exercise, the patient is assisted to complete the movement when he cannot complete. The patient moves his limb to the point where he can move. The movement is then completed by an external force. Unlike the passive exercise, it provides some increase in muscular strength and coordination.

Resistive exercises: The resistive exercises are performed against a force. Muscle strength increases via resistive exercises. They are performed by a force against dynamic or static muscular contraction.

  Isometric (static) exercise: It is a static exercise with muscular contraction without any elongation of muscle length. Exercises such as applying force to a standing object, keeping a certain weight without moving the joint are isometric exercises.

  Isotonic (kinetic) exercise: In this exercise, resistive movement is made within a certain ROM limits. There is more strength increase than isometric movements. Isotonic exercise is done against gravity. They are performed with dynamic muscular contractions against a constant resistance over the ROM.

  Isokinetic exercise: It was developed in the early 1960s by American biomechanics expert James Perrine. With an isokinetic machine, the person works at a constant speed. Isokinetic exercise is a constant-speed resistive exercise. The speed remains constant, with the resistance set according to the muscle torque. Constant resistance increases as muscle effort increases. The basic theoretical advantage of this technique is to provide maximum muscular tension at full ROM. These exercises can be performed with an isokinetic exercise device or robotic systems.

Table 1 Control methods according to exercise types.

Exercise type	Control method
Passive	Position control
Active assistive	Position and force control
Isometric exercise	Force control
Isotonic exercise	Force control
Isokinetic	Force and velocity control
Manual	Position and force control

Manual exercises: They are performed by the physiotherapist, depending on his expertise. In particular, stretching exercise is performed by physiotherapists. The condition of the muscle is observed by the eye and hand. According to the psychology of the patient and the physical condition of the muscle during the treatment, the physiotherapist changes the exercise process. On the other hand, Akdogan et al. (2009) achieved modeling of stretching exercise using robot control methods.

These exercises can be modeled by position and force control in terms of robot control. Accordingly, the relationship between exercise types and control methods is given in Table 1.

4 Impedance control techniques

The impedance control is based on the performing of force and position control by adjusting the mechanical impedance of the end effector of the robot. This mechanical impedance is caused by the relationship between the external forces and the end effector of the robot that is caused by the contact of the robot manipulator with the environment. Mechanical impedance is the behavior of mechanism elasticity against external force applied. Various impedance models have been developed. These models are mainly based on position and force. Hybrid control-based hybrid impedance control method has also been developed. Researchers have used this control method in robot control with some small differences. Impedance control is known as the most effective control method in human-robot interactive systems. Therefore, it is frequently used in rehabilitation studies. Therapeutic exercise movements require either position or force control. However, in some cases they require both position and force control. The use of impedance control modes according to exercise types is shown in Table 2. The types of impedance control techniques are explained in the following.

Table 2 Exercise type and impedance control mode relationship.

Exercise type	Control method
Passive	Position-based impedance control
Active assistive	Hybrid impedance control
Isometric exercise	Force-based impedance control
Isotonic exercise	Force-based impedance control
Isokinetic	Force-based impedance control
Manual	Hybrid impedance control

4.1 Position-based impedance control

The position-based impedance control model is frequently used in passive exercises requiring position control and active-assistive exercises that require position and force control. This control method was used in PHYSIOTHERABOT and DIAGNOBOT. In this section, the general model of this method and another model developed by Yoshikawa (1990) which has small differences from the general model were also introduced.

4.1.1 General model

The dynamic equation of a robot is given by the following nonlinear equation:

$$\tau = M(q)\ddot{q} + C(q,\dot{q}) + G(q) + f(\dot{q}) \tag{1}$$

where $M(q)$, $C(q,\dot{q})$, $G(q)$, and $f(\dot{q})$ are the inertia, the Coriolis and centrifugal forces, the gravitational force, and friction forces, respectively. The vectors q, $\dot{q}$, and $\ddot{q}$ are the angular position, velocity, and acceleration of the robot joints, respectively.

The main equation of position-based impedance control is given in the following equation:

$$M_d(\ddot{x} - \ddot{x}_d) + B_d(\dot{x} - \dot{x}_d) + K_d(x - x_d) = -F_{ext} \tag{2}$$

where M_d, B_d, and K_d are symmetrical matrices that denote the desired inertia, damping, and stiffness matrices, respectively. The vector x denotes the actual end effector position and x_d denotes the desired end effector position. The vector F_{ext} is the force applied to the manipulator, which is the output of the force sensor. Eq. (2) is arranged as

$$\ddot{x} = \ddot{x}_d + M_d^{-1}\left[-B_d(\dot{x} - \dot{x}_d) - K_d(x - x_d) - F_{ext}\right] \tag{3}$$

The relationship between linear velocity $\dot{x}$ and angular velocity $\dot{\theta}$ is given by following equation:

$$\dot{x} = J(q)\dot{q} \tag{4}$$

where $J(q)$ is the Jacobian matrix and the linear acceleration is

$$\ddot{x} = \dot{J}(q)\dot{q} + J(q)\ddot{q} \tag{5}$$

In the robot dynamic equation given by Eq. (1), let us choose the control input as follows:

$$\tau = M(q)u + C(q,\dot{q}) + G(q) + f(\dot{q}) + J(q)^T F_{ext} \tag{6}$$

where $u = \ddot{q}$. Accordingly, Eq. (5) can be written as

$$u = J(q)^{\dagger}(\ddot{x} - \dot{J}(q)\dot{q}) \tag{7}$$

where $(.)^{\dagger}$ denotes Pseudo inverse of matrix. If we replace Eq. (3) in Eq. (7)

$$u = J(q)^{\dagger}(\ddot{x}_d + M_d^{-1}[-B_d(\dot{x} - \dot{x}_d) - K_d(x - x_d) - F_{ext}] - \dot{J}(q)\dot{q}) \tag{8}$$

If we replace Eq. (8) in Eq. (6), we obtain the control rule for position-based impedance control as follows:

$$\tau = M(q)J(q)^{\dagger}(\ddot{x}_d + M_d^{-1}[-B_d(\dot{x} - \dot{x}_d) - K_d(x - x_d) - F_{ext}] - \dot{J}(q)\dot{q}) + C(q,\dot{q})$$
$$+ G(q) + f(\dot{q}) - J(q)^T F_{ext} \tag{9}$$

4.1.2 Position-based impedance model of Yoshikawa

Yoshikawa (1990) developed a position-based impedance model. This model is used in PHYSIOTHERABOT. In this model, the contact force is expressed in terms of the desired impedance parameters:

$$M_d\ddot{x} + B_d\dot{x}_e + K_d x_e = F_{ext} \tag{10}$$

where x_d desired position vector and $x_e = x - x_d$. The dynamic equation defined in the joint space of the robot manipulator in contact with its environment is given by the following equation:

$$M(q)\ddot{q} + h_N(q,\dot{q}) = \tau + J^T(q)F_{ext} \tag{11}$$

where $M(q) \in \mathbb{R}^{n \times n}$ is inertia matrix at joint space, $h_N(q,\dot{q}) \in \mathbb{R}^{n \times 1}$ is Coriolis and centrifugal force vector at joint space, and $\tau \in \mathbb{R}^{n \times 1}$ is joint torque vector. The important point here is that since the robot has a relationship

with its environment, it defines the work space in Eq. (11), which is defined
in the joint space

$$M_x(q)\ddot{x} + h_x(q,\dot{q}) = J_x^{-T}(q)\tau + F_{ext} \qquad (12)$$

where $M_x(q) \in \mathbb{R}^{n \times n}$ is inertia matrix, $h_x(q,\dot{q}) \in \mathbb{R}^{n \times 1}$ is Coriolis and
centrifugal force vector, and $J_x \in \mathbb{R}^{n \times 1}$ is Jacobian vector. These vectors
and matrices are defined at workspace.

$$x = f_x(q) \qquad (13)$$
$$\dot{x} = J_x(q)\dot{q} \qquad (14)$$
$$\ddot{x} = \dot{J}_x\dot{q} + J_x\ddot{q} \qquad (15)$$

It is possible to express the $M_x(q)$ matrix in the work space and the vector
$h_x(q,\dot{q})$ consisting of nonlinear terms in terms of $M(q)$ and $h_N(q,\dot{q})$ in the
joint space.

$$M_x(q) = J_x^{-T}M(q)J_x^{-1}(q) \qquad (16)$$
$$h_x(q,\dot{q}) = J_x^{-T}h_N(q,\dot{q}) - M_x(q)\dot{J}_x(q)\dot{q} \qquad (17)$$

Using Eqs. (12), (16), (17), the required joint torque to obtain the desired
impedance parameters M_d, B_d, and K_d can be found by the following
equation:

$$\begin{aligned}
\tau = {} & h_N(q,\dot{q}) - M(q)J_x^{-1}(q)\dot{J}_x(q)\dot{q} - M(q)J_x^{-1}(q)M_d^{-1}(B_d\dot{y}_e + K_d x_e) \\
& + [M(q)J_x^{-1}(q)M_d^{-1} - J_x^T(q)]F_{ext}
\end{aligned} \qquad (18)$$

The block diagram of Yoshikawa' impedance model is given in Fig. 3.

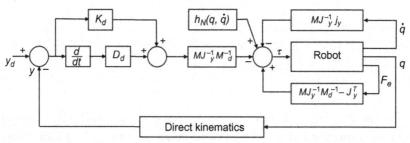

Fig. 3 Yoshikawa' model block diagram.

4.2 Force-based impedance control

For the force-based impedance control, the desired dynamics behavior of the system can be given as follows:

$$M_d\ddot{x} + B_d\dot{x} - F_d = -F_{ext} \tag{19}$$

which is equal to

$$\ddot{x} = M_d^{-1}(-B_d\dot{x} + F_d - F_{ext}) \tag{20}$$

where F_d is desired force. If the expression in Eq. (20) is written in Eq. (7):

$$u = \ddot{q} = J(q)^{\dagger}((M_d^{-1}(-B_d\dot{x} + F_d - F_{ext})) - \dot{J}(q)\dot{q} \tag{21}$$

Eq. (21) is written in Eq. (6), the general torque equation is obtained

$$\tau = M(q)J(q)^{\dagger}((M_d^{-1}(F_d - F_{ext} - B_d\dot{x})) - \dot{J}(q)\dot{q}) + C(q,\dot{q}) + G(q) + f(\dot{q})$$
$$+ J(q)^T F_{ext} \tag{22}$$

4.3 Hybrid impedance control

Hybrid impedance control method was developed by Anderson and Spong (1988). Hybrid impedance control consists of combining the position- and force-based impedance control under a single control rule. It is an extremely useful control method for exercises that require both position and force control. In hybrid impedance control, the controller can operate in force or position control mode. This selection is performed with a selection matrix (S). If the matrix "1," it works as position based, and if "0," it works as a force-based impedance controller. The desired dynamic behavior of the system is a combination of Eqs. (2), (19), which includes the switching matrix.

$$M_d(\ddot{x} - S\ddot{x}_d) + B_d(\dot{x} - S\dot{x}_d) + SK_d(x - x_d) + (I - S)F_d = -F_{ext} \tag{23}$$

Final control rule for the desired dynamics

$$\tau = M(q)J(q)^{\dagger}(S\ddot{x}_d + M_d^{-1}[(I-S)F_d - F_{ext} - B_d(\dot{x} - S\dot{x}_d)$$
$$- SK_d(x - x_d)] - \dot{J}(q)\dot{q}) + C(q,\dot{q}) + G(q) + F(q) - J(q)^T F_{ext} \tag{24}$$

4.4 Variable (angle-dependent) impedance control

In variable impedance control method, B_d varies according to the angle of the joint. When the ROM is 0 degree, B_d takes the maximum value (B_{dmax}).

B_d decreases as the joint moves to the maximum ROM, and B_d receives the smallest value (B_{dmin}) when it reaches the maximum θ value.

$$\Delta B_d = B_{dmax} - B_{dmin} \tag{25}$$

The change of B_d can be written depending on the actual position (θ) and maximum position (θ_{max})

$$B_d = \Delta B_d \frac{(\theta_{max} - |\theta|)}{\theta_{max}} \tag{26}$$

When θ reaches the θ_{max} value, B_d becomes zero. This leads to unstability of the system. For this reason, the B_{dmin} is added to Eq. (26) and the following equation is obtained.

$$B_d = [(\theta_{max} - |\theta|)(B_{dmax} - B_{dmin})(\theta_{max})^{-1}] + B_{dmin} \tag{27}$$

The resulting control law after combining Eqs. (22), (27) becomes

$$\tau = M(q)J(q)^\dagger ((M_d^{-1}(F_d - F_{ext} - [(\theta_{max} - |\theta|)(B_{dmax} - B_{dmin})(\theta_{max})^{-1}] \\ + B_{dmin})) - \dot{J}(q)\dot{q}) + C(q,\dot{q}) + G(q) + f(\dot{q}) + J(q)^T F_{ext} \tag{28}$$

5 Therapeutic exercise modeling via impedance control

The algorithms for modeling the therapeutic exercises with robot control are given in the following sections.

5.1 Passive exercise

This type of exercise is applied to patients with little or no muscle contraction, in particular. The patient's limb is moved in the ROM. There is no resistance during the motion. The therapist can describe exercise parameters. These are ROM, repetition number, and the velocity of movements via a graphical user interface (GUI). The passive exercise requires position control. Therefore, the robot manipulator can make its moves using the position-based impedance control. The desired trajectory is produced by the main controller of the robot manipulator in accordance with the specified ROM and velocity inputs. The algorithm of the passive exercise is shown in Fig. 4A.

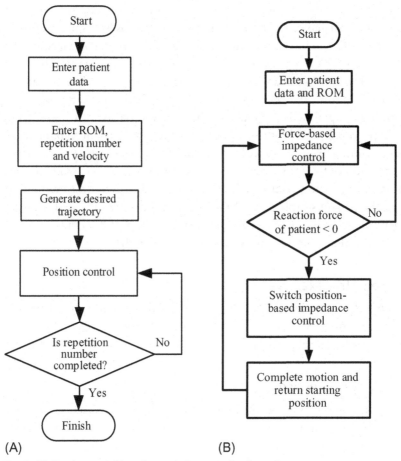

Fig. 4 (A) Passive and (B) active-assistive exercise flow chart.

5.2 Active-assistive exercise

The active-assistive exercise is a type of exercise performed by physiotherapist and patient together. In this exercise, the patient moves his limb to the extent he is capable of. The patient is assisted by the robot manipulator, which replaces the physiotherapist from the position where he can no longer move his limb. This exercise is applied to patients with a muscle degree of 2 and 3. (The degree of muscles are evaluated with a scale that has six levels. This test is done by physiotherapist manually.) The GUI input parameters are the ROM values. The patient's limb weight is eliminated by the robot via compensation of the gravity effects using impedance control. This way the patient is enabled to move his limb to this level easily despite his muscle weaknesses. In the selection of the impedance control parameters, criteria such as enabling the patient to move his limb at the lowest possible level of resistance and eliminating any possibility of causing vibration in the mechanism during the motion is considered. These parameter values are determined experimentally and are stored in the database. By evaluating the force sensor signals, the position at which the patient cannot move the limb during exercise is determined. This force value arising in the opposite direction of the movement is detected by the controller, which in turn activates the position control algorithm for the completion of the motion. Then, the patient's limb is moved to the limit of the ROM with constant velocity and then brought back to the starting position also at constant velocity. For the algorithm of this exercise refer to Fig. 4B.

5.3 Isometric exercise

Isometric exercise is performed with the application of constant resistance in a certain position. However, the ROM does not change whereas his muscle tone increases. The algorithm of the isometric exercise is shown in Fig. 5A.

5.4 Isotonic exercise

During an isotonic exercise, the limb is moved against a constant force. The ROM changes, but the counter-force is kept constant. In order to model this exercise, impedance controller is used in the force control mode. The input of the controller is the target force. The algorithm of the isotonic exercise is shown in Fig. 5B.

5.5 Isokinetic exercise

The purpose of this exercise is to maintain maximum muscle contraction by keeping the patient's limb at a constant level. It is applied athletes whose muscle degree is the highest level. When the exercise is started, the impedance control starts operating with low impedance parameters and the athlete moves his limb. The speed of limb movement is continually detected by the software. If the speed reaches the predetermined level, the generated limb force value is calculated and this force is applied instantaneously to the leg on the opposite direction of the movement. Thus, even if the patient tries to increase the limb speed, the force generated in the opposite direction to

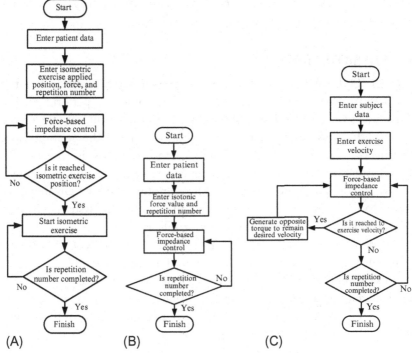

Fig. 5 (A) Isometric, (B) isotonic, and (C) isokinetic exercise flow chart.

the movement keeps the speed of the movement constant. The algorithm of the isokinetic exercise is shown in Fig. 5C.

6 Impedance-controlled rehabilitation robots

In this section, three therapeutic exercise robots that are controlled using impedance control are explained. First, three-DOF lower-limb robot that is called as PHYSIOTHERABOT© is introduced. Its controller is in the form of an intelligent controller supported by a position-based impedance controller. It models not only therapeutic exercises but also manual therapy that is performed by a physiotherapist. Second, three-DOF upper-limb robot that is called as PHYSIOTHERABOT/w1 is explained. It is controlled by a hybrid impedance controller to model upper-limb exercises. Finally, DIAGNOBOT© is an intelligent robotic rehabilitation system developed to assist physiotherapists and physicians in diagnosis and treatment during wrist and forearm rehabilitation. Its controller structure is developed via the force-based impedance control.

6.1 PHYSIOTHERABOT

PHYSIOTHERABOT is a three-DOF lower-limb rehabilitation robot. It can perform knee and hip rehabilitation. In addition to perform passive, active-assistive, and resistive exercises for these joints, it can learn the movements of the physiotherapist and apply it to the patient. The position-based impedance control was used to model the exercises in the system. Yoshikawa's model was used for this purpose. It has a unique patented mechanical design (Akdogan, 2008). All the motors of the system are placed on the base. Therefore, masses of motors do not affect the system dynamics. The knee joint of the robot has a pantograph structure.

For detailed information and experimental results of PHYSIOTHERABOT, please refer to Akdogan and Adli (2011) and Akdogan et al. (2009).

PHYSIOTHERABOT consists of four basic elements. These are physiotherapist, intelligent controller, robot manipulator, and patient. The general block diagram of the system is given in Fig. 6. In the system,

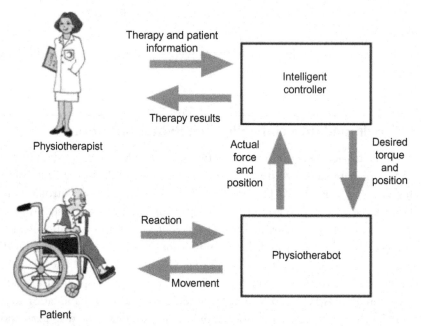

Fig. 6 PHYSIOTHERABOT general block diagram.

the physiotherapist first enters the patient information (age, weight, height, and foot length) into the system using the user interface. He also selects the movement type and exercise type and enters the exercise parameters (period, repetition number, ROM, velocity). The control of the system is carried out by the intelligent controller. The information required for the robot manipulator to move (position-torque) is calculated by the intelligent controller and sent to the manipulator. The response from the patient during the exercise can be evaluated by the intelligent controller and the process (applied position and force) can be changed. Detailed information about system elements is given later. The system is capable of performing passive, active-assistive, isotonic, isometric, isokinetic, and manual (called in the system: robotherapy) exercises. Manual exercises are performed in two stages: teaching mode and therapy mode. In the teaching mode, the physiotherapist performs the exercise with the robot manipulator. In the therapy mode, the robot manipulator applies exercises to the patient using the knowledge acquired in the teaching mode. The basic elements of the system are explained in following sections.

6.1.1 Intelligent controller

The intelligent controller is the management unit of the system. Communication between all the elements of the system is performed by this unit. The system has the ability to perform muscle testing, as well. As a result of the muscle test, the patient's muscular degree and ROM are detected and shown on the screen. Information such as posttherapy ROM, patient force, and patient response force can be observed graphically and numerically. The intelligent controller consists of central processing unit, conventional controller, database, rule base, and user interface. The intelligent controller block diagram is given in Fig. 7.

The system requires force and position control to perform exercises. In order to construct the controller structure, the desired force and position control have been carried out according to the exercise types with conventional control methods. For different types of exercises, a database and rule base have been created as it is necessary to select different conventional control techniques, keep control parameters in a database, and change the parameters. A user interface is available for the user to control the system. The communication between the conventional controller, database, rule base, and user interface is provided by the central processing unit. The functions of the units that constitute the intelligent controller according to the exercise types are given in Table 3. Position-based impedance control for

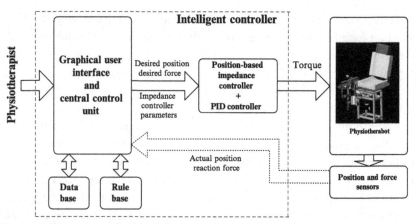

Fig. 7 Intelligent controller block diagram.

exercise types requiring force control and PID control method for exercise requiring position control were used. Some types of exercise require a combination of force and position control. In these exercise types, impedance and PID controls are used together. The reason for selecting impedance control is that the system can resist the desired softness or stiffness by changing the impedance parameters. The physiotherapist should be able to perform the movements in a very easy and gentle way while performing the exercise movements. Moreover, the resistance level is obtained by changing the values of the impedance parameters in the resistive exercises.

The central processing unit performs the following tasks as common to all types of exercise:
- Algorithm selection
- Send patient parameters that are stored in the database to the algorithms
- Send the data from the sensors to the rule base and database
- Send information from the rule base and the database to the controllers

6.1.2 Robot manipulator

The developed robotic system has three DOF and can perform flexion-extension for the knee and flexion-extension and abduction-adduction movements for the hip. Servo motors are used to actuate the mechanism and force sensors are used for force measurement. It can be adjusted according to the limb size and is suitable for both legs. The general structure of the robot manipulator is given in Fig. 8. The knee link (Link 2) of the PHYSIOTHERABOT was designed according to the parallelogram principle. The robot manipulator's motors are placed on the base to reduce the effects of motor weights on the robot manipulator dynamics.

Table 3 The functions of the intelligent control unit

Mode	Central control unit	Database	Rule base
Muscle test	– Sends force and position data to rule base. – Sends maximal torque data to controller.	Stores maximal torque data and patient muscle level.	Determines muscle degree.
Passive	Determines the motion trajectory according to ROM and velocity and sends it to the database and PID controller.	Stores motion trajectory.	–
Active assistive	According to information from the force sensor, "patient cannot move the limb" information is sent to the rule base.	–	Activates PID controller if force is detected in reverse direction on force sensor.
Isotonic	– According to the resistance level, takes the impedance parameters from the database. – Evaluates the position data and calculates the number of repetitions and sends it to the rule base. – Sends the impedance parameters according to the selected technique from the rule base to the controller.	–	– Stop therapy if the number of repeats is complete. – Determines the appropriate impedance parameters according to the resistance level.
Isometric	Sends the weight, duration, and length of the limb from the database to the controller.	Stores length of the limb.	Detects impedance parameters according to the resistance level.

Continued

Table 3 The functions of the intelligent control unit—cont'd

Mode	Central control unit	Database	Rule base
Isokinetic	Sends the velocity information to the controller.	–	–
Teaching	Determines the maximum force, and position data during teaching and sends these values to the database	Stores the force and position data during the therapy applied by the physiotherapist.	–
Intelligent therapy	– Updates the rule base with maximum force and position information in the database and send them to the controller. – It takes the maximum position data in the database and the force-position data required for the healthy people to reach that position and sends it to the controller	Holds the duration, maximum force, and position data during teaching and exercise parameters.	Identifies the relevant data file according to the patient's weight and the maximum ROM obtained during learning. When the force and position limit values are exceeded according to the real-time data from the central processing unit, the PD controller activates.

6.1.3 Electronics hardware

The electronics hardware block diagram of the PHYSIOTHERABOT is given in Fig. 9. System hardware consists of servo motors, reductors, motor drivers, data acquisition cards, encoders, force sensors, limit switches, and emergency stop buttons.

Digital force and position data from the intelligent controller are converted to analog current (torque) data via DAQs and sent to the motor drivers. Data from the force and position sensors on the robot manipulator are sent to the intelligent controller via the DAQs. The limit switches prevent the mechanism from moving beyond the required limits.

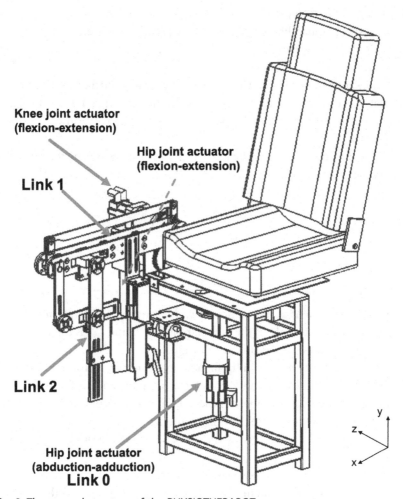

Fig. 8 The general structure of the PHYSIOTHERABOT.

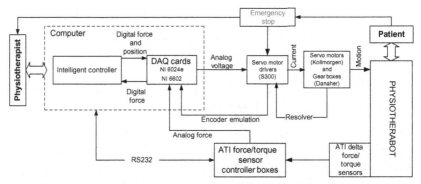

Fig. 9 The electronics hardware diagram of the PHYSIOTHERABOT.

6.1.4 Dynamic analysis

Knee and hip rehabilitation were independently modeled and controlled in PHYSIOTHERABOT. The dynamic analysis and control method is explained here.

Knee joint dynamic analysis: The general dynamic model of a robot manipulator was given in Eq. (11). The knee joint is a single DOF and can be modeled like a pendulum driven by the axis of rotation as shown in Fig. 10.

The dynamic model for the knee joint can be expressed by the following equation:

$$I_2\ddot{\theta}_2 + \tau_{grav2} = \tau_2 + J_y^T F_{ext} \tag{29}$$

$$J_{y2} = J_{y2}^T = L_{g2} \tag{30}$$

$$J_y^{-1} = 1/L_{g2} \tag{31}$$

$$h_{N2} = \tau_{grav2} = mg\sin\theta_2 L_{g2} \tag{32}$$

where I_2 is moment of inertia of link 2, $\ddot{\theta}_2$ is angular acceleration of link 2, τ_{grav2} is gravity effect, J_{y2} is the Jacobian of link 2, and L_{g2} is the distance of the center of mass of the link 2 to origin.

Hip joint dynamic analysis: Hip flexion-extension movement is a motion with two DOF. Links 1 and 2 work together to perform this movement in the system. Fig. 11 indicates the center of masses of links 1 and 2 that perform hip flexion-extension movement.

The physical parameters of the PHYSIOTHERABOT are given in Table 4.

Transforming force sensor coordinates into workspace coordinates: There are two force sensors in the system. These sensors are positioned to the ankle in link 2 to measure the knee forces, and to the thigh in link 1 to measure

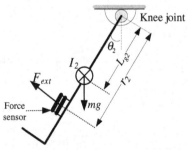

Fig. 10 The model of link 2.

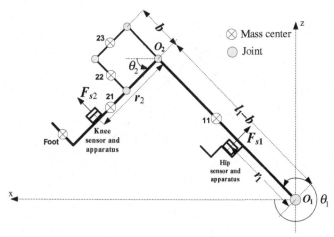

Fig. 11 The center of mass of links 1 and 2.

Table 4 Physical parameter values of the PHYSIOTHERABOT.

Part number	Mass (kg)	l_g (m)
11	1.955	0.3
21	3.25	Variable
22	0.75	–
23	0.18	0.09
Hip apparatus (ha)	4.25	0.25

Note: l_g is the distance of center of mass to the origin.

the hip forces (Fig. 12). Since the axes of these sensors and the workspace are different from each other, necessary coordinate transformations should be made. The axes of the sensors and the workspace and the angles between them are given in Fig. 12. In the figure, "r" subscript represents the work space of the robot and "s" subscript represents the axes of the sensors.

The equivalent of the forces measured from the hip sensor (f_{s1x}, f_{s1z}) in the workspace coordinates (X_r, Z_r) is given in the following equation:

$$F_{x1} = F_{s1x} \cos\theta_1 - F_{s1z} \sin\theta_1$$
$$F_{z1} = -F_{s1x} \sin\theta_1 - F_{s1z} \cos\theta_1 \tag{33}$$

The equation of the forces measured from the knee sensor (f_{s2x}, f_{s2z}) in the workspace coordinates (x_r, z_r) is given in the following equation:

$$F_{x2} = F_{s2x} \cos\theta_2 - F_{s2z} \sin\theta_2$$
$$F_{z2} = -F_{s2x} \sin\theta_2 - F_{s2z} \cos\theta_2 \tag{34}$$

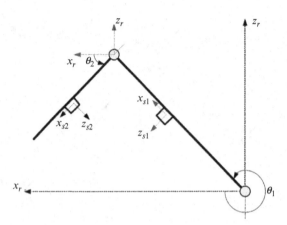

Fig. 12 Sensor and workspace axes.

Coordinates of sensors mounted to the links and Jacobian matrix: The coordinates of sensors connected to links 1 and 2 and the calculation of their velocity are required for the calculation of the Jacobian matrix. The coordinates of the sensors connected to the links and the calculations of the Jacobian matrix are given here.

The coordinate of the hip sensor connected to the link 1 and the Jacobian matrix: The Jacobian matrix is obtained using the following equation:

$$J = \frac{\partial p}{\partial q} \tag{35}$$

where p is the position of the sensor.

The coordinates in the x- and z-axes and partial derivatives of the sensor are given in the following equation:

$$
\begin{aligned}
x_1 &= l_{g1}\cos(2\pi - \theta_1) = l_{g1}\cos\theta_1 \rightarrow \delta x_1 = -l_{g1}\delta\theta_1\sin\theta_1 \\
z_1 &= l_{g1}\cos(2\pi - \theta_1) = l_{g1}\cos\theta_1 \rightarrow \delta z_1 = -l_{g1}\delta\theta_1\cos\theta_1
\end{aligned} \tag{36}
$$

$$
\begin{bmatrix} \delta x_1 \\ \delta z_1 \end{bmatrix} = \begin{bmatrix} -l_{g1}\sin\theta_1 \\ -l_{g1}\cos\theta_1 \end{bmatrix}\delta\theta_1
$$

The Jacobian matrix for the link 1 is given in the following equation:

$$
J_{x1} = \begin{bmatrix} -l_{g1}\sin\theta_1 \\ -l_{g1}\cos\theta_1 \end{bmatrix} \tag{37}
$$

The coordinate of the knee sensor connected to the link 2 and the Jacobian matrix: The coordinates in the x- and z-axes and partial derivatives of the sensor are given in the following equation:

$$
\begin{aligned}
x_2 &= l_1 \cos\theta_1 + r_1 \cos\theta_2 \rightarrow \delta x_2 = -l_1 \delta\theta_1 \sin\theta_1 - r_1 \delta\theta_2 \sin\theta_2 \\
z_2 &= -l_1 \sin\theta_1 - r_1 \sin\theta_2 \rightarrow \delta z_2 = -l_1 \delta\theta_1 \cos\theta_1 - r_1 \delta\theta_2 \cos\theta_2
\end{aligned}
\tag{38}
$$

$$
\begin{bmatrix} \delta x_2 \\ \delta z_2 \end{bmatrix} = \begin{bmatrix} -l_1 \sin\theta_1 & -r_1 \sin\theta_2 \\ -l_1 \cos\theta_1 & -r_1 \cos\theta_2 \end{bmatrix} \begin{bmatrix} \delta\theta_1 \\ \delta\theta_2 \end{bmatrix}
$$

Jacobian matrix for link 2 is given in Eq. (39). The transpose, inverse, and derivative of the Jacobian are given in Eqs. (40)–(42), respectively.

$$
J_2 = \begin{bmatrix} -l_1 \sin\theta_1 & -r_1 \sin\theta_2 \\ -l_1 \cos\theta_1 & -r_1 \cos\theta_2 \end{bmatrix}
\tag{39}
$$

$$
J_2^T = \begin{bmatrix} -l_1 \sin\theta_1 & -l_1 \cos\theta_1 \\ -r_1 \sin\theta_2 & -r_1 \cos\theta_2 \end{bmatrix}
\tag{40}
$$

$$
J_2^{-1} = \begin{bmatrix} -r_1 \cos\theta_2 & -r_1 \sin\theta_2 \\ l_1 \cos\theta_1 & -l_1 \sin\theta_1 \end{bmatrix} \frac{1}{l_1 r_1 \sin\theta_1 \cos\theta_2 + l_1 r_1 \cos\theta_1 \sin\theta_2}
\tag{41}
$$

$$
\dot{J}_2 = \begin{bmatrix} -l_1 \delta\theta_1 \cos\theta_1 & -r_1 \delta\theta_2 \cos\theta_2 \\ l_1 \delta\theta_1 \sin\theta_1 & r_1 \delta\theta_2 \sin\theta_2 \end{bmatrix}
\tag{42}
$$

Dynamic equations: For calculating the joint torques of the robot manipulator, the dynamic equations of the system must be calculated. In cases where no external force is not applied, the Lagrange equation given in the following equation is used in the joint torque calculation:

$$
\tau = \frac{d}{dt}\frac{\delta K}{\delta\dot{\theta}} - \frac{\delta K}{\delta\theta} + \frac{\delta P}{\delta\theta}
\tag{43}
$$

where P and K are the potential and kinetic energy, respectively.

The coordinates of the robot manipulator parts:

In order to calculate the kinetic and potential energies in the Lagrange equation, the positions and velocities of the robot manipulator and lower-limb parts (Figs. 11 and 13) must be found. These equations are given in the following:

Part 11:

$$
\begin{aligned}
x_{11} &= l_{g11} \cos\theta_1 \rightarrow \dot{x}_{11} = -l_{g11}\dot{\theta}_1 \sin\theta_1 \\
z_{11} &= -l_{g11} \sin\theta_1 \rightarrow \dot{z}_{11} = -l_{g11}\dot{\theta}_1 \cos\theta_1
\end{aligned}
\tag{44}
$$

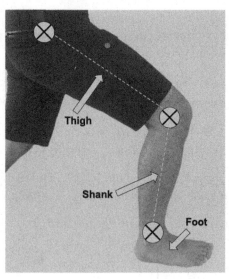

Fig. 13 Lower-limb parts.

Hip apparatus (ha) part:

$$x_{ha} = 0.25\cos\theta_1 \rightarrow \dot{x}_{ha} = -0.25\dot{\theta}_1\sin\theta_1$$
$$z_{ha} = -0.25\sin\theta_1 \rightarrow \dot{z}_{ha} = -0.25\dot{\theta}_1\cos\theta_1 \qquad (45)$$

Thigh part:

$$x_{thigh} = l_{gthigh}\cos\theta_1 \rightarrow \dot{x}_{thigh} = -0.144l_{human}\sin\theta_1$$
$$z_{11} = -l_{gthigh}\sin\theta_1 \rightarrow \dot{z}_{thigh} = -0.144l_{human}\cos\theta_1 \qquad (46)$$
$$l_{gthigh} = 0.144l_{human}$$

Part 22:

$$x_{22} = (l_1 - b + l_{g22})\cos\theta_1 + \dot{l}_{21}\cos\theta_2$$
$$\dot{x}_{22} = -(l_1 - b + 0.2)\dot{\theta}_1\sin\theta_1 - 0.2\dot{\theta}_2\sin\theta_2$$
$$z_{22} = (l_1 - b + l_{g22})\sin(2\pi - \theta_1) - \dot{l}_{21}\sin\theta_2 \qquad (47)$$
$$\dot{z}_{22} = -(l_1 - b + 0.2)\dot{\theta}_1\cos\theta_1 - 0.2\dot{\theta}_2\cos\theta_2$$
$$\dot{l}_{21} = 0.2\,[\mathrm{m}]$$

Shank part:

$$x_{shank} = l_1 \cos\theta_1 + l_{gshank} \cos\theta_2 \rightarrow \dot{x}_{shank} = -l_1\dot{\theta}_1 \sin\theta_1 - l_{gshank}\dot{\theta}_2 \sin\theta_2$$
$$z_{shank} = -l_1 \sin\theta_1 - l_{gshank} \sin\theta_2 \rightarrow \dot{Z}_{shank} = -l_1\dot{\theta}_1 \cos\theta_1 - l_{gshank}\dot{\theta}_2 \cos\theta_2$$
$$l_{gshank} = 0.144 l_{human}$$

(48)

Foot part:

$$x_{foot} = l_1 \cos\theta_1 + l_2 \cos\theta_2 + l_{gfoot} \cos\left(\frac{\pi}{2} - \theta_2\right) \rightarrow \dot{x}_{foot}$$
$$= -l_1\dot{\theta}_1 \sin\theta_1 - l_{23}\dot{\theta}_2 \sin\theta_2 - l_{gfoot}\dot{\theta}_2 \sin\left(\frac{\pi}{2} - \theta_2\right)$$
$$z_{foot} = -l_1 \sin\theta_1 - l_{g2} \sin\theta_2 + l_{gfoot} \sin\left(\frac{\pi}{2} - \theta_2\right) \rightarrow \dot{z}_{foot}$$

(49)

$$= -l_1\dot{\theta}_1 \cos\theta_1 - l_{g23}\dot{\theta}_2 \cos\theta_2 - l_{gfoot}\dot{\theta}_2 \sin\left(\frac{\pi}{2} - \theta_2\right)$$
$$l_{gfoot} = 0.429 l_{foot}$$

where l_{human} is height of patient, l_{gthigh}, l_{gfoot}, and l_{gshank} are mass of centers of thigh, foot, and shank, respectively.

The kinetic energy of the robot manipulator parts: In order to calculate the kinetic energy used in the Lagrange equation, the kinetic energies of the robot manipulator components are calculated separately and given in the following:

Part 11:

$$K_{11} = \frac{1}{2}m_{11}l_{g11}^2\dot{\theta}_1^2 + \frac{1}{2}I_{G11}\dot{\theta}_1^2 = 0.122\dot{\theta}_1^2$$

(50)

Hip apparatus(ha) part:

$$K_{ha} = \frac{1}{2}I_{oha}\dot{\theta}_1^2 = 0.132\dot{\theta}_1^2$$

(51)

Thigh part:

$$K_{thigh} = \frac{1}{2}I_{thigh}\dot{\theta}_1^2 = 10^{-7}[(21,186BW - 222,796) + 20.736m_{thigh}l_{human}^2]$$

(52)

where BW represents body weight and m is mass of part.

Part 21:

$$K_{21} = \frac{1}{2}m_{21}(\dot{x}_{21}^2 + \dot{z}_{21}^2)^2 + \frac{1}{2}I_{G21}\dot{\theta}_2^2 = 1.625(\dot{x}_{21}^2 + \dot{z}_{21}^2) + 0.04\dot{\theta}_2^2$$

(53)

Part 22:

$$K_{22} = \frac{1}{2}m_{22}(\dot{x}_{22}^2 + \dot{z}_{22}^2)^2 + \frac{1}{2}I_{G22}\dot{\theta}_2^2 = 0.375(\dot{x}_{22}^2 + \dot{z}_{22}^2) + 2.83\dot{\theta}_1^2 \quad (54)$$

Part 23:

$$K_{23} = \frac{1}{2}m_{23}(\dot{x}_{23}^2 + \dot{z}_{23}^2)^2 + \frac{1}{2}I_{G23}\dot{\theta}_2^2 = 0.9(\dot{x}_{23}^2 + \dot{z}_{23}^2) + 4.6587 \times 10^{-4}\dot{\theta}_2^2$$
$$(55)$$

Shank part:

$$K_{shank} = \frac{1}{2}m_{shank}(\dot{x}_{shank}^2 + \dot{z}_{shank}^2)^2 + \frac{1}{2}I_{Gshank}\dot{\theta}_2^2$$
$$= 10^{-7}(5341\,BW + 44{,}749) \quad (56)$$

Foot part:

$$K_{foot} = \frac{1}{2}m_{foot}(\dot{x}_{foot}^2 + \dot{z}_{foot}^2)^2 + \frac{1}{2}I_{Gfoot}\dot{\theta}_1^2 = 10^{-7}(355BW + 7296) \quad (57)$$

The potential energy of the robot manipulator parts: In order to calculate the potential energy used in the Lagrange equation, the potential energies of the robot manipulator components are calculated separately and given the following:

Part 11:

$$P_{11} = m_{11}gz_{11} = -5.75\sin\theta_1 \quad (58)$$

Part hip apparatus:

$$P_{ha} = m_{ha}gz_{ha} = -10.423\sin\theta_1 \quad (59)$$

Part thigh:

$$P_{thigh} = m_{thigh}gz_{thigh} = -1.412(m_{thigh}l_{human})\sin\theta_1 \quad (60)$$

Part 21:

$$P_{21} = m_{21}gz_{21} = -31.88(0.68\sin\theta_1 + l_{g21}\sin\theta_2) \quad (61)$$

Part 22:

$$P_{22} = m_{22}gz_{22} = 7.3575[(0.68 - b + l_{g22})\sin(2\pi - \theta_1) - 0.2\sin\theta_2] \quad (62)$$

Part 23:

$$P_{23} = m_{23}gz_{23} = -1.7658(0.68\sin\theta_1 + 0.09\sin\theta_2) \quad (63)$$

Shank part:

$$P_{\text{shank}} = m_{\text{shank}}gz_{\text{shank}} = -m_{\text{shank}}9.81(0.68\sin\theta_1 + 0.144l_{\text{human}}\sin\theta_2) \quad (64)$$

Foot part:

$$
\begin{aligned}
P_{\text{foot}} &= m_{\text{foot}}gz_{\text{foot}} \\
&= -m_{\text{foot}}9.81\left(0.68\sin\theta_1 + 0.5\sin\theta_2 - l_{\text{foot}}0.429\sin\left(\frac{\pi}{2} - \theta_2\right)\right)
\end{aligned} \quad (65)
$$

The contribution of the parts to joint torques: The contributions of the parts to the joint torques calculated using the Lagrange equation:
For Joint 1:

$$\tau_{111} = 0.244\ddot{\theta}_1 - 5.75\cos\theta_1 \quad (66)$$

$$\tau_{1\text{ha}} = 0.265\ddot{\theta}_1 - 10.243\cos\theta_1 \quad (67)$$

$$\tau_{1\text{thigh}} = I_{\text{othigh}}\ddot{\theta}_1 - 1.412m_{\text{thigh}}l_{\text{human}}\cos\theta_1 \quad (68)$$

$$
\begin{aligned}
\tau_{121} = {}& 3.25l_1^2\ddot{\theta}_1 + 3.25l_1 l_{g1}\ddot{\theta}_2\cos(\theta_1 - \theta_2) \\
&+ 3.25l_1 l_{g1}\dot{\theta}_2(\dot{\theta}_2 - \dot{\theta}_1)\sin(\theta_1 - \theta_2) \\
&+ 3.25l_1 l_{g1}\dot{\theta}_1\dot{\theta}_2\sin(\theta_1 - \theta_2) - 21.67\cos\theta_1
\end{aligned} \quad (69)
$$

$$
\begin{aligned}
\tau_{122} = {}& (0.75(l_1 - b + 0.2)^2 + 5.66)\ddot{\theta}_1 + 0.15\ddot{\theta}_2\cos(\theta_1 - \theta_2) \\
&+ 0.15\dot{\theta}_2(\dot{\theta}_2 - \dot{\theta}_1)\sin(\theta_1 - \theta_2) - 0.15\dot{\theta}_1\dot{\theta}_2(l_1 - b + 0.2)\cos(\theta_2 - \theta_1) \\
&+ 7.3575(0.6 - b + l_{g22})\cos\theta_1
\end{aligned} \quad (70)
$$

$$
\begin{aligned}
\tau_{123} = {}& 1.8l_1^2\ddot{\theta}_1 + 1.8l_1 l_{g23}\ddot{\theta}_2\cos(\theta_1 - \theta_2) + 1.8l_1 l_{g23}\dot{\theta}_2(\dot{\theta}_2 - \dot{\theta}_1)\sin(\theta_1 - \theta_2) \\
&+ 1.8l_1 l_{g23}\dot{\theta}_1\dot{\theta}_2\sin(\theta_1 - \theta_2) - 1.2\cos\theta_1
\end{aligned} \quad (71)
$$

$$
\begin{aligned}
\tau_{1\text{shank}} = {}& m_{\text{shank}}l_1^2\ddot{\theta}_1 + l_{g\text{shank}}\ddot{\theta}_2\cos(\theta_1 - \theta_2) + l_1 l_{g\text{shank}}\dot{\theta}_2(\dot{\theta}_2 - \dot{\theta}_1)\sin(\theta_1 - \theta_2) \\
&+ m_{\text{shank}}l_1 l_{g\text{shank}}\dot{\theta}_1\dot{\theta}_2\sin(\theta_1 - \theta_2) - m_{\text{shank}}6.6708\cos\theta_1
\end{aligned} \quad (72)
$$

$$
\begin{aligned}
\tau_{1\text{foot}} = {}& l_1^2\ddot{\theta}_1 + I_{\text{foot}}\ddot{\theta}_1 + m_{\text{foot}}l_1\ddot{\theta}_2(Bc\theta_1 - A\sin\theta_1) + m_{\text{foot}}[2l_1 l_2\dot{\theta}_2^2\cos\theta_1\sin\theta_2 \\
&+ 2\dot{\theta}_2^2 l_1 l_{g\text{foot}}\sin\theta_2\cos\theta_1 - 2\dot{\theta}_2 l_1\dot{\theta}_1 B\sin\theta_1 \\
&+ 2\dot{\theta}_2^2\sin\theta_1 l_1 l_{g\text{foot}}\sin\theta_2 + 2\dot{\theta}_2^2 l_1 l_2\sin\theta_2\sin\theta_1 \\
&- 2\dot{\theta}_2\dot{\theta}_1 l_1 Ac\theta_1] - m_{\text{foot}}\dot{\theta}_2\dot{\theta}_2(l_2\sin\theta_2\sin\theta_1 \\
&+ l_{g\text{foot}}\sin\theta_2\sin\theta_1) + 2\dot{\theta}_2\dot{\theta}_1(\sin\theta_1 l_{g\text{foot}}\cos\theta_2 \\
&+ \sin\theta_1 l_2\sin\theta_2) - 6.67\cos\theta_1
\end{aligned} \quad (73)
$$

$$A = l_{gfoot}\sin\left(\frac{\pi}{2}-\theta_2\right) - l_2\sin\theta_2 \tag{74}$$

$$B = l_2\cos\theta_2 - l_{gfoot}\sin\left(\frac{\pi}{2}-\theta_2\right) \tag{75}$$

For Joint 2:

$$\tau_{211} = \tau_{2ha} = \tau_{2thigh} = 0 \tag{76}$$

$$\tau_{221} = 3.25 l_1 l_{g1}\ddot{\theta}_1\cos(\theta_1-\theta_2) + 3.25 l_{g21}\ddot{\theta}_2 + 0.08\ddot{\theta}_2 + 3.25 l_1 l_{g1}\dot{\theta}_1(\dot{\theta}_2-\dot{\theta}_1)\sin(\theta_1-\theta_2)$$
$$- 3.25\dot{\theta}_1\dot{\theta}_2\sin(\theta_1-\theta_2) - 31.88 l_{g21}\cos\theta_2 \tag{77}$$

$$\tau_{222} = 0.15(l_1-b+0.2)\ddot{\theta}_1\cos(\theta_1-\theta_2) + 0.03\ddot{\theta}_2 + 0.15\dot{\theta}_1(l_1-b+0.2)(\dot{\theta}_2-\dot{\theta}_1)\sin(\theta_1-\theta_2)$$
$$- 0.15\dot{\theta}_1\dot{\theta}_2(l_1-b+0.2)\sin(\theta_1-\theta_2) - 1.4715\cos\theta_2 \tag{78}$$

$$\tau_{223} = 1.8 l_{g23} l_1\ddot{\theta}_1\cos(\theta_1-\theta_2) + 1.8 l_{g23}\ddot{\theta}_2 + 4.6587\times10^{-4}\ddot{\theta}_2 + 1.8 l_1 l_{g23}\dot{\theta}_1(\dot{\theta}_2-\dot{\theta}_1)\sin(\theta_1-\theta_2)$$
$$- 1.8 l_1 l_{g23}\dot{\theta}_1\dot{\theta}_2\sin(\theta_1-\theta_2) - 0.16\cos\theta_2 \tag{79}$$

$$\tau_{2shank} = m_{shank} l_1 l_{gshank}\ddot{\theta}_1\cos(\theta_1-\theta_2) + \ddot{\theta}_2(m_{shank} l^2_{gshank} + I_{gshank})$$
$$+ m_{shank} l_1 l_{gshank}\dot{\theta}_1(\dot{\theta}_2-\dot{\theta}_1)\sin(\theta_1-\theta_2) \tag{80}$$
$$- m_{shank} l_1 l_{gshank}\dot{\theta}_1\dot{\theta}_2\sin(\theta_1-\theta_2) - m_{shank}1.41264\cos\theta_2$$

$$\tau_{2foot} = m_{foot} l_1\ddot{\theta}_1\cos\theta_1 B + m_{foot}\ddot{\theta}_2 B^2 + m_{foot} l_1\dot{\theta}_1\left(B\cos\theta_1 - \dot{\theta}_1 B\sin\theta_1\right)$$
$$- m_{foot}\dot{\theta}_1\dot{\theta}_2 l_1\left(-l_2\sin\theta_2\cos\theta_1 - l_{gfoot}\cos\left(\frac{\pi}{2}-\theta_2\right)\cos\theta_1\right.$$
$$\left. - l_{gfoot}\cos\left(\frac{\pi}{2}-\theta_2\right)\sin\theta_1 + l_2\cos\theta_2\sin\theta_1\right)$$
$$- m_{foot}9.81\left(0.5\cos\theta_2 - l_{foot}0.429\sin\left(\frac{\pi}{2}-\theta_2\right)\right) \tag{81}$$

Joint torques: The joint torques can be calculated by using the sum of the contributions of the parts to the joint torque and using the following equations:

$$\tau_1 = \tau_{111} + \tau_{1ha} + \tau_{1thigh} + \tau_{121} + \tau_{122} + \tau_{123} + \tau_{1shank} + \tau_{1foot} \tag{82}$$

$$\tau_2 = \tau_{211} + \tau_{2ha} + \tau_{2thigh} + \tau_{221} + \tau_{222} + \tau_{223} + \tau_{2shank} + \tau_{2foot} \tag{83}$$

Robot dynamic equation for hip exercises: In the case of external force acting, the dynamic equation of a robot manipulator is as in Eq. (11). Coriolis, gravity, and other effects and inertial matrix M can be calculated using Eqs. (82), (83), (11). The inertia matrix in Eq. (11) and all the information in the h_N vector are given in the equations between Eqs. (66) and (81). Accordingly, the $M(\theta)$ matrix

$$M = \begin{bmatrix} M_{11} & M_{12} \\ M_{21} & M_{22} \end{bmatrix} \tag{84}$$

$$M_{11} = 0.244 + 0.265 + I_{othigh} + 3.25l_1^2 + (0.75(l_1 - b + 0.2)^2 + 5.66) + 1.8l_1^2 + l_1^2 + I_{foot} \tag{85}$$

$$
\begin{aligned}
M_{12} &= 3.25l_1 l_{g1}\cos(\theta_1 - \theta_2) + 0.15\cos(\theta_1 - \theta_2) + 1.8l_1 l_{g23}\cos(\theta_1 - \theta_2) \\
&\quad + l_{gshank}\cos(\theta_1 - \theta_2) + m_{foot}l_1(B\cos\theta_1 - A\sin\theta_1)
\end{aligned} \tag{86}
$$

$$
\begin{aligned}
M_{21} &= 3.25l_1 l_{g1}\cos(\theta_1 - \theta_2) + 0.15\cos(l_1 - b + 0.2)\cos(\theta_1 - \theta_2) + 1.8l_{g23}l_1\cos(\theta_1 - \theta_2) \\
&\quad + m_{shank}l_1 l_{gshank}\cos(\theta_1 - \theta_2) + m_{foot}Bl_1\cos\theta_1
\end{aligned} \tag{87}
$$

$$M_{22} = 3.25l_{g21} + 0.11 + 1.8l_{g23} + 4.6587 \times 10^{-4} + m_{shank}l_{gshank}^2 + I_{gshank} + m_{foot}B^2 \tag{88}$$

$h_N(\theta, \dot\theta)$ vector:

$$h_N(\theta, \dot\theta) = \begin{bmatrix} h_{N1}(\theta, \dot\theta) \\ h_{N2}(\theta, \dot\theta) \end{bmatrix} \tag{89}$$

$$
\begin{aligned}
h_{N1}(\theta, \dot\theta) =\ & -5.75\cos\theta_1 - 10.243\cos\theta_1 - 1.412 m_{thigh}l_{human}\cos\theta_1 + 3.25l_1 l_{g1}\dot\theta_2(\dot\theta_2 - \dot\theta_1)\sin(\theta_1 - \theta_2) \\
& + 3.25l_1 l_{g1}\dot\theta_1\dot\theta_2\sin(\theta_1 - \theta_2) - 21.67\cos\theta_1 + 0.15\dot\theta_2(\dot\theta_2 - \dot\theta_1)\sin(\theta_1 - \theta_2) \\
& - 0.15(l_1 - b + 0.2)\dot\theta_2\dot\theta_1\cos(\theta_2 - \theta_1) + 7.3575(0.6 - b + l_{g22})\cos\theta_1 + 1.8l_1 l_{g23}\dot\theta_2(\dot\theta_2 \\
& - \dot\theta_1)\sin(\theta_1 - \theta_2) + 1.8l_1 l_{g23}\dot\theta_2\sin(\theta_1 - \theta_2) - 1.2\cos\theta_1 + l_1 l_{gshank}\dot\theta_2(\dot\theta_2 \\
& - \dot\theta_1)\sin(\theta_1 - \theta_2) + m_{shank}l_1 l_{gshank}\dot\theta_1\dot\theta_2\sin(\theta_1 - \theta_2) - m_{shank}6.6708\cos\theta_1 \\
& + m_{foot}[2\dot\theta_2 c_1 l_1 l_2\sin\theta_2 + 2\dot\theta_2^2 l_1 l_{foot}\theta_1\sin\theta_2\cos\theta_1 - 2\dot\theta_2 l_1\dot\theta_1 B\sin\theta_1 \\
& + 2\dot\theta^2\sin\theta_1 l_1 l_{gfoot}\sin\theta_2 + 2\dot\theta_2^2 l_1 l_2\sin\theta_1\sin\theta_2 - 2\dot\theta_2\dot\theta_1 l_1 A\cos\theta_1] \\
& - m_{foot}\dot\theta_1\dot\theta_2(l_2\sin\theta_2\sin\theta_1 + 2\dot\theta_1\dot\theta_2(\sin\theta_1 l_{gfoot} + \sin\theta_1 l_2\theta_2)) - 6.67\cos\theta_1
\end{aligned}
$$

$$
\begin{aligned}
h_{N2}(\theta, \dot\theta) =\ & -3.25l_1 l_{g1}\dot\theta_1(\dot\theta_2 - \dot\theta_1)\sin(\theta_1 - \theta_2) - 3.25\dot\theta_1\dot\theta_2\sin(\theta_1 - \theta_2) - 31.88l_{g21}\cos\theta_2 \\
& + 0.15\dot\theta_1(\dot\theta_2 - \dot\theta_1)\sin(\theta_1 - \theta_2)(l_1 - b + 0.2) - 0.15(l_1 - b + 0.2)(\dot\theta_2\dot\theta_1)\sin(\theta_1 - \theta_2) \\
& - 1.4715\cos\theta_2 + 1.8l_1 l_{g23}\dot\theta_1(\dot\theta_2 - \dot\theta_1)\sin(\theta_1 - \theta_2) - 1.8l_1 l_{g23}\dot\theta_1\dot\theta_2\sin(\theta_1 - \theta_2) \\
& - 0.16\cos\theta_2 + m_{shank}l_1 l_{gshank}\dot\theta_1(\dot\theta_2 - \dot\theta_1)\sin(\theta_1 - \theta_2) - m_{shank}l_1 l_{gshank}\dot\theta_1\dot\theta_2\sin(\theta_1 - \theta_2) \\
& - m_{shank}1.41264\cos\theta_2 + m_{foot}\dot\theta_1 l_1(B\cos\theta_1 - \dot\theta_1 B\sin\theta_1) + m_{foot}\dot\theta_1\dot\theta_2 l_1(-l_2\sin\theta_2\cos\theta_1 \\
& - l_{gfoot}\cos\left(\frac{\pi}{2} - \theta_2\right)\cos\theta_1 - l_{gfoot}\cos\left(\frac{\pi}{2} - \theta_2\right)\sin\theta_1 + l_2\cos\theta_2\sin\theta_1) \\
& - m_{foot}9.81\left(0.5\cos\theta_2 - l_{foot}\sin\left(\frac{\pi}{2} - \theta_2\right)0.429\right)
\end{aligned}
$$

Jacobian vector J_x:

$$J_x = \begin{bmatrix} J_{x1} \\ J_{x2} \end{bmatrix} \tag{90}$$

The equations related to the Jacobian matrix are given in Eqs. (37), (39), (40). The external forces F_{ext} acting on PHYSIOTHERABOT are given in Eqs. (33), (34).

6.1.5 Position-based impedance control of the PHYSIOTHERABOT

Impedance and PID control techniques, which are conventional control techniques, have been used in PHYSIOTHERABOT to perform knee and hip movements and to apply appropriate force and position values according to the exercise types.

For the exercises that require force control, impedance control and for the exercises requiring position control, PID control methods were used. Some types of exercise require a combination of force and position control. In these exercise types, impedance and PID control were used together. Table 5 shows the exercises that the system can perform and the conventional control methods used in these exercises.

The control of the system was performed in three stages, which could be performed for flexion-extension of the knee and flexion-extension and abduction-adduction movements of the hip. In the first stage, the control of the flexion-extension movement of the knee joint with single DOF was performed, and the hip flexion-extension movement control which required the joint movement of the knee and hip joint in the second stage, and the hip abduction-adduction movement which required a single DOF in the final stage was performed.

Control equations for knee joint: PHYSIOTHERABOT can perform knee flexion-extension movement and related exercises. One DOF is sufficient for this movement and exercises are performed by link 2. The reductor connected to the motor actuating link 2 has 100:1 reduction ratio. Therefore, the external torque values (gravity, inertia, and

Table 5 Control methods according to exercise types for PHYSIOTHERABOT.

Exercise type	Control method
Passive	PID control
Active assistive	PID control and impedance control
Isometric exercise	Impedance control and torque control
Isotonic exercise	Impedance control
Isokinetic	Impedance control and torque control
Robotherapy	Impedance control and PID control

external force) are divided into 100. Using Eq. (18), the impedance control rule for link 2 can be obtained as follows:

$$\tau_2 = \tau_{grav2} - \left[\frac{0.01 I_2}{L_{g2} M_d}(B_d \dot{\theta}_e + K_d \theta_e)\right] + \left[\frac{0.01 I_2}{L_{g2} M_d} - L_{g2}\right] 0.01 F_{ext} \qquad (91)$$

The PID position control rule for the one-DOF link 2 is given in the following:

$$\tau_2 = K_{p2}\theta_{e2} + K_{der2}\frac{d\theta_{e2}}{dt} + K_{int2}\int \theta_{e2} dt + \tau_{grav2} \qquad (92)$$

where K_{p2}, K_{der2}, and K_{int2} are proportional, derivative, and integral gains, respectively.

Control equations for hip joint: PHYSIOTHERABOT can perform hip flexion-extension and abduction-adduction movements and related exercises. Description of position-based impedance control and position control techniques used in the performing of these movements are given in this section.

$$\tau = h_N - MJ_x^{-1}\dot{J}_x\dot{q} - MJ_x^{-1}M_d^{-1}(B_d J_x \dot{y}_e + K_d y_e) + (MJ_x^{-1}M_d^{-1} - J_x^T)F_{ext} \qquad (93)$$

$$q = \begin{bmatrix} \theta_1 \\ \theta_2 \end{bmatrix}, \quad \tau = \begin{bmatrix} \tau_1 \\ \tau_2 \end{bmatrix}, \quad J_x = \begin{bmatrix} J_{x1} \\ J_{x2} \end{bmatrix}, \quad h_N = \begin{bmatrix} h_{N1} \\ h_{N2} \end{bmatrix}, \quad M = \begin{bmatrix} M_{11} & M_{12} \\ M_{21} & M_{22} \end{bmatrix}$$

$$M_d = \begin{bmatrix} M_{dx} & 0 \\ 0 & M_{dz} \end{bmatrix}, \quad B_d = \begin{bmatrix} B_{dx} & 0 \\ 0 & B_{dz} \end{bmatrix}, \quad K_d = \begin{bmatrix} K_{dx} & 0 \\ 0 & K_{dz} \end{bmatrix}$$

All parameters in Eq. (93) are calculated in the previous sections. When the system starts to move, the force and position data is constantly updated and the motor torques that drive the PHYSIOTHERABOT are calculated in real time and the system performs desired movement.

The PID position control rule for hip flexion-extension is given in the following equation:

$$\begin{bmatrix} \tau_1 \\ \tau_2 \end{bmatrix} = \begin{bmatrix} K_{p1} & 0 \\ 0 & K_{p2} \end{bmatrix}\begin{bmatrix} \theta_{1e} \\ \theta_{2e} \end{bmatrix} + \begin{bmatrix} K_{int1} & 0 \\ 0 & K_{int2} \end{bmatrix}\begin{bmatrix} \int \theta_{1e} dt \\ \int \theta_{2e} dt \end{bmatrix} + \begin{bmatrix} K_{der1} & 0 \\ 0 & K_{der2} \end{bmatrix}\begin{bmatrix} \dot{\theta}_{1e} \\ \dot{\theta}_{2e} \end{bmatrix} + \begin{bmatrix} \tau_{grav1} \\ \tau_{grav2} \end{bmatrix}$$

$$(94)$$

The hip abduction-adduction movement is a one-DOF movement performed by the motor located at the base. This movement is performed by link 0. The abduction-adduction movement in the system is modeled for passive exercises. Therefore, as the movement requires only position control, it is sufficient to use the PID control method given in the following equation:

$$\tau_0 = K_{p0}\theta_{e0} + K_{der0}\frac{d\theta_{e0}}{dt} + K_{int0}\int \theta_{e0}\, dt \tag{95}$$

6.2 PHYSIOTHERABOT/w1

Among the robots developed for rehabilitation purposes, those for the upper limb are much more than those for the lower limb. The reason for this is that people use their upper limbs more than their lower limbs to maintain their daily life activities. PHYSIOTHERABOT/w1, developed for the rehabilitation of upper limbs, has the following features:
- performs passive and active therapeutic exercises for wrist and forearm rehabilitation
- has three DOF
- portable and suitable for hospital and home use
- has user friendly interface
- obtains and stores patient data

For detailed information and experimental results of PHYSIOTHERABOT/w1, please refer to Akdogan et al. (2018).

PHYSIOTHERABOT/w1 consists of a physiotherapist, patient, robot manipulator, and human-machine interface units that enable the communication and control of the system. The physiotherapist enters the exercise data via the user interface. According to this data, the position and force trajectories are created according to the type of exercise chosen. This information is converted to torque data by the human-machine interface and sent to the robot manipulator. The human-machine interface includes a conventional hybrid impedance controller. The information from the position and force sensors on the robot manipulator are included in the control loop as feedback information. The robotic manipulator allows the patient to perform exercise movements in the appropriate position and forces.

6.2.1 Human-machine interface

Human-machine interface (HMI) has four different units. These are main controller and GUI, data base, rule base, and hybrid impedance controller.

Main controller manages whole system. It is the main control unit of the HMI. The database stores patient information and exercises trajectory data. The rule base consists impedance parameters values with respect to exercise types. Using GUI, user can enter patient individual information and exercise data such as repetition number, duration, movement type, and resistance level. Hybrid impedance controller generates related torque value to control robot manipulator.

6.2.2 Electronics hardware

The electronics hardware block diagram of the system is shown in Fig. 14. There are three actuators in the system. Position information is received via encoders mounted on the motors. The ATI Nano25 (six axes force/torque sensor) is used for the force measurements. National Instruments analog input (NI PCI-6225), analog output (NI PCI-6703), and encoder cards (NI PCI-6601) were used for data acquisition. EMG signals belonging to the patient are also recorded in the system. However, it is not used as a feedback element in the control cycle. At this point, the applications where the EMG participates in the control cycle are highly significant.

6.2.3 Robot manipulator

The robot manipulator made of 7000 series aluminum material has three rotational axes. With these axes, flexion-extension, ulnar-radial deviation, and pronation-supination movements are performed. The general structure of the robot manipulator can be seen from Fig. 15.

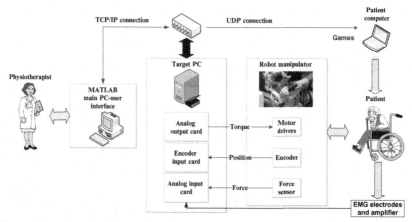

Fig. 14 Electronics hardware of the PHYSIOTHERABOT/w1.

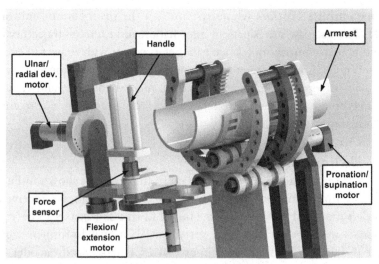

Fig. 15 The general structure of the PHYSIOTHERABOT/w1.

The patient's arm is placed on the armrest. The patient's hand is placed between the bars in the handle. With the force sensor located just below the handle, the force and torque values that are applied by the patient are measured. With this sensor which can measure in six axes (three for force and three for torque), the force is measured during flexion-extension and ulnar-radial deviation movements, and the torque values are measured during pronation-supination movements. As shown in Fig. 15, there are mechanical limitations for safety in each axis. By means of the pins placed in these limitations, the ROM of the joints can be limited to the desired values.

6.2.4 Kinematic and dynamic analysis
The kinematic and dynamic analysis of a robotic system is important for robot control. When performing kinematic and dynamic analysis of robots, classical manual calculation techniques can be used. However, as the DOF of the robot increases, these calculations become more complicated. For this reason, for three-DOF PHYSIOTHERABOT/w1, the analysis programs that can calculate the parameters related to the system model were used. Fig. 16 shows the axes of the robot manipulator for kinematic analysis. The Denavit-Hartenberg parameters according to the axes of the robot manipulator shown in Fig. 16 are given in Table 6.

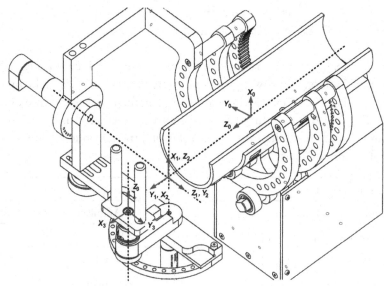

Fig. 16 The robot manipulator axes.

Table 6 Denavit-Hartenberg parameters.

Link	a_i	α_i	d_i	θ_i
1	0	90 degrees	0	q_1
2	0	90 degrees	0	90 degrees + q_2
3	l_1	0	$-l_2$	q_3

In Table 6, $l_1 = 75 \times 10^{-3}$ (m) and $l_2 = 30 \times 10^{-3}$ (m). In this case, the transformation matrices are as follows:

$$T_1^0 = \begin{bmatrix} \cos(q_1) & 0 & \sin(q_1) & 0 \\ \sin(q_1) & 0 & -\cos(q_1) & 0 \\ 0 & 1 & 0 & 0 \\ 0 & 0 & 0 & 1 \end{bmatrix} \tag{96}$$

$$T_2^1 = \begin{bmatrix} -\sin(q_2) & 0 & \cos(q_2) & 0 \\ \cos(q_2) & 0 & \sin(q_2) & 0 \\ 0 & 1 & 0 & 0 \\ 0 & 0 & 0 & 1 \end{bmatrix} \tag{97}$$

$$T_3^2 = \begin{bmatrix} \cos(q_1) & -\sin(q_3) & 0 & l_1\cos(q_3) \\ \sin(q_1) & \cos(q_3) & 0 & l_1\sin(q_3) \\ 0 & 0 & 1 & -l_2 \\ 0 & 0 & 0 & 1 \end{bmatrix} \tag{98}$$

The drawings of the manipulator were made by the solid modeling program. The link parameters obtained from these drawings are given in Table 7.

The dynamic equations of the system were obtained through the Robotics Toolbox which was programmed by Peter Corke. First, the information in Table 7 and Denavit-Hartenberg parameters (Table 6) is entered into the MATLAB program. Then the information entered in the program is combined to create the robot model. Finally, the dynamic parameters of the robot are obtained.

syms q1 q2 q3 real	% Angle variables
syms dq1 dq2 dq3 real	% Angular velocity variables
$M = robot.inertia[(q1\ \ q2\ \ q3)]$	% M matrix
$C = robot.coriolis[(q1\ \ q2\ \ q3)], [(dq1\ \ dq2\ \ dq3)]$	% C matrix
$G = robot.gravload[(q1\ \ q2\ \ q3)]$	% G matrix

6.2.5 Hybrid impedance control of the PHYSIOTHERABOT/w1

In the developed system, position control in the ROM of the robot manipulator is needed to model some exercises requiring position control such as passive and active-assistive exercises. However, due to contact with the patient and resistance to some exercises, the force control is needed. Because the system requires both position and force control, the hybrid impedance control method, which allows both position and force control within a single control rule, was used. Thus, with a single control rule, all therapeutic exercises could be modeled. The detailed information about hybrid impedance control is given in Section 4.1.4. The hybrid impedance control mode used in accordance with exercise types is given in Table 8.

Hybrid impedance parameters selection according to exercise types

There are four different levels for each type of exercises: low, medium, high, and very high. Thus, it becomes possible to perform exercises with patients having different muscular activation levels and to increase the efficiency of the exercises by incrementing the level as the patient progresses. The desired impedance parameters for these four levels are, respectively, given as follows:

Table 7 Mechanical parameters of links.

	Link 1	Link 2	Link 3
Mass (kg)	2.228 kg	1.101	0.374
Center of mass (mm)	[6.96 −113.55 −82.67]	[1.96 −83.72 −100.13]	[−11.72 0 7.93]
Inertia matrix (kg m²)	$\begin{bmatrix} 0.0732 & 0.0012 & -0.0060 \\ 0.0012 & 0.0414 & 0.0119 \\ -0.0060 & 0.0119 & 0.0462 \end{bmatrix}$	$\begin{bmatrix} 0.0170 & 0.00005 & -0.000250 \\ 0.00005 & 0.0091 & 0.0051 \\ -0.00025 & 0.0051 & 0.0083 \end{bmatrix}$	$\begin{bmatrix} 0.0006 & 0 & 0.0001 \\ 0 & 0.0008 & 0 \\ 0.0001 & 0 & 0.0003 \end{bmatrix}$

Table 8 Hybrid impedance control modes according to the exercise types for the PHYSIOTHERABOT/w1.

Exercise type	Hybrid impedance control mode
Passive	Position based ($s = 1$)
Active assistive	Hybrid (position ($s = 1$) + force ($s = 0$))
Isometric exercise	Force based ($s = 0$)
Isotonic exercise	Force based ($s = 0$)
Isokinetic	Force based ($s = 0$)

(Desired mass)	$m_d \in \{ 1.25, 2.50, 5.00, 7.50 \}$	(kg)
(Desired linear damping)	$b_d \in \{ 10, 20, 40, 80 \}$	(N s/m)
(Desired spring)	$k_d \in \{ 100, 200, 400, 800 \}$	(N/m)
(Desired inertia)	$\tilde{m}_d \in \{ 0.0125, 0.0250, 0.0500, 0.0750 \}$	(kg m^2)
(Desired rotational damping)	$\tilde{b}_d \in \{ 0.1, 0.2, 0.4, 0.8 \}$	(N m s/rad)
(Desired rotational spring)	$\tilde{k}_d \in \{ 1, 2, 4, 8 \}$	(N m/rad)

There are additional parameters to hold the robotic arm in a fixed position: $m_h = 22.5$, $b_h = 240$, $k_h = 2400$, $\tilde{m}_h = 0.225$, $\tilde{b}_h = 2.4$, and $\tilde{k}_h = 24$.

The impedance parameters M_d, B_d, K_d, and S are diagonal matrices for which the first three entries are related to the translational movement of the end effector on x_0, y_0, and z_0 axes and the last three entries are related to the rotational movement of the end effector around x_0, y_0, and z_0, respectively. For example, in order to prevent the end effector move on x_0 axis, the first entries of the impedance parameters M_d, B_d, and K_d are selected as m_h, b_h, and k_h. The first entry of the switching matrix S as position-based impedance control ($s = 1$), and the desired position for this axis is 0 degree.

Flexion-extension: During the flexion-extension movement, the origin of the coordinate system of the end effector moves in the y_0 and z_0 axes and rotates about the x_0 axis (see Fig. 16). As a result, impedance parameters with corresponding axes are selected as the desired impedance parameters whereas the others are selected as the hold impedance parameters to avoid possible radial-ulnar deviation and pronation-supination movements.

$$M_d = \text{diag}[100, m, m, \tilde{m}, \tilde{m}, 1]$$
$$B_d = \text{diag}[1000, b, b, \tilde{b}, \tilde{b}, 10]$$
$$K_d = \text{diag}[10,000, k, k, \tilde{k}, \tilde{k}, 100]$$
$$S = \text{diag}[1, s, s, s, 1, 1]$$

The operator diag[x_1, x_2, ..., x_n] denotes a block diagonal matrix whose elements on the main block diagonal are x_1, x_2, ..., x_n. The parameters s are chosen according to the exercise types which are listed in Table 8.

Radial-ulnar deviation: During the flexion-extension movement, the origin of the coordinate system of the end effector moves in the x_0 and z_0 axes and rotates about the y_0 axis.

$$M_d = \text{diag}[m, 100, m, 1, \tilde{m}, 1]$$
$$B_d = \text{diag}[b, 1000, b, 10, \tilde{b}, 10]$$
$$K_d = \text{diag}[k, 10{,}000, k, 100, \tilde{k}, 100]$$
$$S = \text{diag}[s, 1, s, 1, s, 1]$$

Pronation-supination: During pronation-supination movement, end effector only rotates about the z_0 axis.

$$M_d = \text{diag}[100, 100, 100, 1, 1, \tilde{m}]$$
$$B_d = \text{diag}[1000, 1000, 1000, 10, 10, \tilde{b}]$$
$$K_d = \text{diag}[10{,}000, 10{,}000, 10{,}000, 100, 100, \tilde{k}]$$
$$S = \text{diag}[1, 1, 1, 1, 1, s]$$

6.3 DIAGNOBOT

DIAGNOBOT is a robotic system developed for upper-limb rehabilitation. It has an intelligent control structure to perform diagnosis and therapy. It can perform flexion-extension and ulnar-radial deviation for the wrist and pronation-supination for the forearm movements. It is able to perform passive, active-assistive, stretching, isometric, isotonic, and resistive exercises. The PID and position-based impedance control were used for position-based exercises. The force-based impedance control and the angle-dependent impedance control were used for force-based exercises. For detailed information about the DIAGNOBOT, please refer to Aktan and Akdogan (2018) and Aktan (2018).

6.3.1 Robot manipulator

The general structure is shown in Fig. 17. The patient's arm is placed in the arm clamping mechanism actuated by the stepper motor. The manipulators were placed on the rotary table. The manipulators can be adjusted according to the length of the limbs. All manipulators are actuated by servo motor. Joint force and torque values are measured by the force and torque sensors. The patient's hand is placed between the bars of the handle. In the

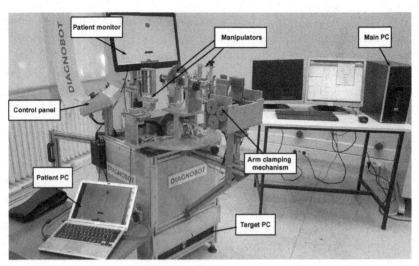

Fig. 17 General structure of the DIAGNOBOT.

movement of the flexion-extension and ulnar-radial deviation, the patient's arm has to be fastened by the arm clamping mechanism.

6.3.2 Electronics hardware

The block diagram of the electronics hardware is shown in Fig. 18. The physiotherapist is the main user of the system. He enters all the information relevant to the therapy. There are two PCs in the system: The Main PC is used to run the algorithms and the Target PC consists of DAQ cards. There is a Raspberry Pi ver.3 running the games for the isometric, isotonic, and resistive exercises. The algorithms were developed in MATLAB R2017a. The Simulink Real Time was used for the real-time prototyping. The UDP and TCP/IP protocol used for the communication between hardware. There are three servo motors (Maxon EC-Max 30) with the 103:1 reduction ratio and 500 pulse/rev encoders. There are also three servo motor drivers (Maxon EPOS 2 50/5) in the system. There are two force sensors (Burster 8523-200) and one torque sensor (Burster 8627-5710) to measure the torque and force applied by the patient. The force sensors were used in flexion-extension and ulnar-radial deviation units. The torque sensor was used in pronation-supination unit. The measurement ranges of each sensor are given in Table 9.

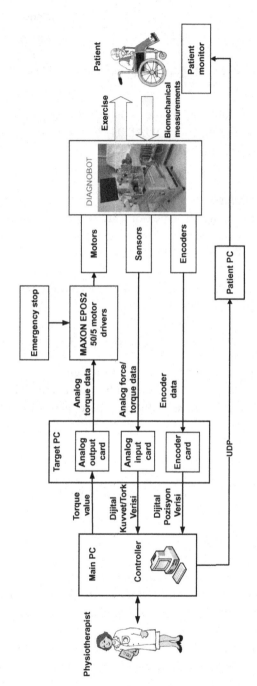

Fig. 18 Electronics hardware of the DIAGNOBOT.

Table 9 The measurement ranges of the sensors.

Sensors	Range
Torque sensor	± 10 Nm
Force sensors	± 200 N
Force sensor for measurement of the grasping force	50 kg

In the system, for the encoder input and analog inputs/outputs, Measurement Computing PCI QUAD04, NI PCI-6024E, and NI PCI-6040E DAQ cards are used, respectively.

6.3.3 Dynamic analysis

In this section, the dynamic equations are obtained and dynamic parameters of the system are calculated by using the experimental robot identification using optimized periodic trajectories method developed by Swevers et al. (1996).

The manipulator dynamic equation for a single robot link is expressed by the following equation:

$$\tau = M(q)\ddot{q} + gmr_y \sin(q) + gmr_x \cos(q) + f_v(\dot{q}) + f_c \, \text{sign}(\dot{q}) \qquad (99)$$

where r_y and r_x are the y and x position of the center of mass, respectively. The f_v and f_c are the viscous and Coulomb friction coefficients, respectively.

The inverse dynamics of Eq. (99) can be expressed by the following equation:

$$\tau = [\ddot{q} \quad g \cdot \sin(q) \quad g \cdot \cos(q) \quad \dot{q} \quad \text{sign}(\dot{q})] \cdot \begin{bmatrix} M \\ mr_y \\ mr_x \\ f_v \\ f_c \end{bmatrix}$$

$$= \phi(q, \dot{q}, \ddot{q}) \cdot p$$

In this equation, the robot position, velocity, and acceleration are known. The vector p contains unknown parameters. If at least six different vector ϕ and τ values corresponding to vector ϕ are known, vector p can be calculated. This is called the parameter estimation method (Swevers et al., 1996).

However, there are errors in the measurement of the velocity and the acceleration of the link and the motor torque. Therefore, more than six measurements were made and it is ensured that the robot manipulator follows a predetermined trajectory through a PID controller. Consider we have $i \in 1, \dots, M$ observations data, we can solve the equation of $Y = W \cdot p$:

$$W = \begin{bmatrix} \phi(q^1, \dot{q}^1, \ddot{q}^1) \\ \phi(q^2, \dot{q}^2, \ddot{q}^2) \\ \vdots \\ \phi(q^M, \dot{q}^M, \ddot{q}^M) \end{bmatrix}, \quad Y = \begin{bmatrix} \tau^1 \\ \tau^2 \\ \vdots \\ \tau^M \end{bmatrix} \quad (100)$$

The condition number of the matrix W represents how close the solution to the nonlinear differential equation is. To obtain the optimal trajectory, the optimal solution was found to minimize the condition number of the matrix W. The *fmincon* function of the MATLAB was used for the solution. Let us represent the trajectories as a finite Fourier series:

$$q(t) = \sum_{l=1}^{N} \frac{a_l}{\omega_f l} \sin(\omega_f l t) - \frac{b_l}{\omega_f l} \cos(\omega_f l t) + q_0 \quad (101)$$

$$\dot{q}(t) = \sum_{l=1}^{N} a_l \cos(\omega_f l t) + b_l \cos(\omega_f l t) \quad (102)$$

$$\ddot{q}(t) = \sum_{l=1}^{N} -a_l \omega_f l \sin(\omega_f l t) + b_l \omega_f l \cos(\omega_f l t) \quad (103)$$

where ω_f is the fundamental pulsation of the Fourier series, a_l and b_l are the coefficients, N is the number of harmonics, and q_0 is the robot configuration around which the robot excitation occurs. Boundary conditions for position, velocity, and acceleration can be given in obtaining the optimal trajectory. These boundary conditions are expressed as follows:

$$\delta^* = \arg \min \text{cond}(\delta, \omega_f)$$
$$q_{min} \leq q(t) \leq q_{max}$$
$$-\dot{q}_{max} \leq \dot{q}(t) \leq \dot{q}_{max}$$
$$-\ddot{q}_{max} \leq \ddot{q}(t) \leq \ddot{q}_{max}$$

where δ is the vector containing Fourier coefficients. δ^* is the optimal δ which minimizes the condition number of matrix W. In the optimization, $\omega_f = 0.1$, $N = 5$, $q_{min} = -90$ degrees, and $q_{max} = 90$ degrees were selected. With a PID controller, the robot manipulator for each unit is followed the optimal trajectory and the $q, \dot{q}, \ddot{q}$ values are saved and the vector W is obtained. The torque values applied by the servo motor during the following of the optimal trajectory are also saved and the Y vector is obtained. The

Table 10 Estimated parameters for each manipulator.

Parameters	Pro-Sup	Fle-Eks	Uln-Rad	
M	0.0277	0.0091	0.0177	kg m^2
mr_x	0.0082	0.2501	0.1878	kg m
mr_y	0.0182	0.4811	0.2733	kg m
f_v	0.0733	0.2433	0.3741	N m s/rad
f_c	0.1333	0.0766	0.0821	N m

Table 11 Control methods according to exercise types for the DIAGNOBOT.

Exercise type	Control method
Passive	PID control
Active assistive	PID and position-based impedance control
Isometric exercise	PID control
Isotonic exercise	Force-based and angle-dependent impedance control
Resistive	Force-based and angle-dependent impedance control

dynamic parameters of the system identified by using the least square estimation method in the following:

$$p = (W^T W)^{-1} W^T Y$$

The obtained dynamic parameters are given in Table 10.

6.3.4 Control of the DIAGNOBOT

Therapeutic exercises require force and position control. In the DIAGNOBOT, PID and position-based impedance control were used for position-based exercises. The force-based impedance control and the angle-dependent impedance control were used for force-based exercises. Exercise types and control methods are given in Table 11.

The control equations for DIAGNOBOT are explained in Section 4. For position-based exercises (Eq. 9) and general PID control equation are used. For force-based exercises (Eqs. 22, 28) are used.

7 Discussion

Impedance control is the most effective control method in human-robot interactive systems. Especially rehabilitation robots are very suitable for this control method in terms of patient-robot interaction. Therefore, various impedance control methods were used in the research studies in the field of rehabilitation robotics. There are four different fundamental impedance control structures: position based, force based, hybrid, and variable.

Therapeutic exercises consist of passive, active-assistive, and resistive exercises. These types of exercises require position and force control in terms of robotic. Therefore, these therapeutic exercises can be made by a robot to the patient by using the impedance control. Robotic rehabilitation has many advantages. These advantages have also been proven by clinical trials.

This chapter describes in detail how impedance control and therapeutic exercise modeling can be performed in three different therapeutic exercise robots previously developed by the authors. The PHYSIOTHERABOT is a three-DOF robot for lower-limb rehabilitation. The PHYSIOTHERABOT/w1 is a three-DOF robot for upper-limb rehabilitation. The DIAGNOBOT is an advanced version of the PHYSIOTHERABOT/w1 and contains differences in terms of mechanical and control. With these systems, therapeutic exercises were successfully modeled. However, modeling of manual exercises is still an important research topic. For this purpose, the use of artificial intelligence techniques combined with conventional control techniques can solve the problem. On the other hand, complex mechanical designs are needed to model rotational exercise movements, especially in lower limbs. For these movements, it is extremely important to ensure safety in robotic rehabilitation.

8 Conclusion

The studies on the use of impedance control in rehabilitation robots are ongoing. At this point, especially the selection of the impedance parameters and updating them according to the exercise performance of the patient are important research topics. The powerful control structures are needed to model physiotherapist movements precisely. In addition, studies are needed to determine the mechanical properties of manual exercises. This information can be used to design novel robotic systems which are able to perform more effective robotic rehabilitation.

Acknowledgments

These works were supported by the Scientific and Technological Research Council of Turkey (TUBITAK) under Grant Number 104M018 and 111M603 and Research Fund of the Yildiz Technical University under Grant Number 2015-06-04-DOP01 and 2012-06-04-YL01.

References

Akdemir, N., Akkus, Y., 2006. Rehabilitasyon ve hemşirelik. Hacettepe Üniversitesi Hemşirelik Yüksekokulu Dergisi 13 (1), 82–91.

Akdogan, E., 2008. The Rehabilitation Robot. Turkish Patent Institute, p. 2008/05687.

Akdogan, E., Adli, M.A., 2011. The design and control of a therapeutic exercise robot for lower limb rehabilitation: PHYSIOTHERABOT. Mechatronics 21 (3), 509–522. https://doi.org/10.1016/j.mechatronics.2011.01.005.

Akdogan, E., Tacgin, E., Adli, M.A., 2009. Knee rehabilitation using an intelligent robotic system. J. Intell. Manuf. 20 (2), 195. https://doi.org/10.1007/s10845-008-0225-y.

Akdogan, E., Aktan, M.E., Koru, A.T., Selçuk Arslan, M., Atlhan, M., Kuran, B., 2018. Hybrid impedance control of a robot manipulator for wrist and forearm rehabilitation: performance analysis and clinical results. Mechatronics 49, 77–91. https://doi.org/10.1016/j.mechatronics.2017.12.001.

Aktan, M.E., 2018. Teşhis ve Tedavi Amaçli Zeki Robotik Rehabilitasyon Sistemi (Ph.D. Thesis), Yildiz Technical University.

Aktan, M.E., Akdogan, E., 2018. Design and control of a diagnosis and treatment aimed robotic platform for wrist and forearm rehabilitation: DIAGNOBOT. Adv. Mech. Eng. 10 (1), 1–13. https://doi.org/10.1177/1687814017749705.

Anderson, R.J., Spong, M.W., 1988. Hybrid impedance control of robotic manipulators. IEEE J. Robot. Autom. 4 (5), 549–556.

Denève, A., Moughamir, S., Afilal, L., Zaytoon, J., 2008. Control system design of a 3-DOF upper limbs rehabilitation robot. Comput. Methods Prog. Biomed. 89 (2), 202–214. https://doi.org/10.1016/j.cmpb.2007.07.006.

Fraile, J.C., Pérez-Turiel, J., Baeyens, E., Viñas, P., Alonso, R., Cuadrado, A., Franco-Martín, M., Parra, E., Ayuso, L., García-Bravo, F., Nieto, F., Laurentiu, L., 2016. E2Rebot: a robotic platform for upper limb rehabilitation in patients with neuromotor disability. Adv. Mech. Eng. 8(8). https://doi.org/10.1177/1687814016659050.

Griffith, H.W., 2000. Spor Sakatliklari Rehberi. Bedray Publications. ISBN 6055989002.

Hogan, N., 1985. Impedance control: an approach to manipulation: part I—theory. J. Dyn. Syst. Meas. Control. 107 (1), 1–7.

Hogan, N., Krebs, H.I., Sharon, A., Charnnarong, J., 1995. Interactive robotic therapist. U. S. Patent # 5466213.

Kiguchi, K., Hayashi, Y., 2012. An EMG-based control for an upper-limb power-assist exoskeleton robot. IEEE Trans. Syst. Man Cybern. Part B (Cybernetics) 42 (4), 1064–1071. https://doi.org/10.1109/TSMCB.2012.2185843.

Kim, B., Deshpande, A.D., 2017. An upper-body rehabilitation exoskeleton harmony with an anatomical shoulder mechanism: design, modeling, control, and performance evaluation. Int. J. Robot. Res. 36 (4), 414–435. https://doi.org/10.1177/0278364917706743.

Krebs, H.I., Hogan, N., Aisen, M.L., Volpe, B.T., 1998. Robot-aided neurorehabilitation. IEEE Trans. Rehabil. Eng. 6 (1), 75–87.

Krebs, H.I., Ferraro, M., Buerger, S.P., Newbery, M.J., Makiyama, A., Sandmann, M., Lynch, D., Volpe, B.T., Hogan, N., 2004. Rehabilitation robotics: pilot trial of a spatial extension for MIT-Manus. J. Neuroeng. Rehabil. 1 (1), 5. https://doi.org/10.1186/1743-0003-1-5.

İnal, S., 2000. Kas Hastalıklarında Rehabilitasyon ve Ortezler. Çizge Press, Istanbul, Turkey.

Nef, T., Mihelj, M., Riener, R., 2007. ARMin: a robot for patient-cooperative arm therapy. Med. Biol. Eng. Comput. 45 (9), 887–900. https://doi.org/10.1007/s11517-007-0226-6.

Oblak, J., Cikajlo, I., Matjacic, Z., 2009. A universal haptic device for arm and wrist rehabilitation. In: 2009 IEEE International Conference on Rehabilitation Robotics, pp. 436–441.

Okada, S., Sakaki, T., Hirata, R., Okajima, Y., Uchida, S., Tomita, Y., 2001. TEM: a therapeutic exercise machine for the lower extremities of spastic patients. Adv. Robot. 14 (7), 597–606. https://doi.org/10.1163/156855301742030.

Openstax, 2013. Anatomy-Physiology. OpenStax CNX, Houston, TX.

Ren, Y., Kang, S.H., Park, H.S., Wu, Y.N., Zhang, L.Q., 2013. Developing a multi-joint upper limb exoskeleton robot for diagnosis, therapy, and outcome evaluation in neurorehabilitation. IEEE Trans. Neural Syst. Rehabil. Eng. 21 (3), 490–499. https://doi.org/10.1109/TNSRE.2012.2225073.

Sobh, T., Xiong, X., 2012. Prototyping of Robotic Systems: Applications of Design and Implementation. Idea Group Publishing. ISBN: 978-1466601765.

Swevers, J., Ganseman, C., Schutter, J.D., Brussel, H.V., 1996. Experimental robot identification using optimized periodic trajectories. Mech. Syst. Signal Process. 10 (5), 561–577.

Yeong, C.F., Melendez-Calderon, A., Gassert, R., Burdet, E., 2009. Reachman: a personal robot to train reaching and manipulation. In: 2009 IEEE/RSJ International Conference on Intelligent Robots and Systems, pp. 4080–4085.

Yoshikawa, T., 1990. Foundations of Robotics: Analysis and Control. MIT Press, Cambridge, MA, ISBN: 978-0262514583.

CHAPTER THIRTEEN

Architecture and application of nanorobots in medicine

Ramna Tripathi[a], Amit Kumar[b] and Akhilesh Kumar[c]
[a]Department of Physics, THDC-Institute of Hydropower Engineering & Technology, Tehri, India
[b]Department of ECE, THDC-Institute of Hydropower Engineering & Technology, Tehri, India
[c]Department of Physics, Govt. Girls P. G. College, Lucknow, India

1 Introduction

In a prescient talk in 1959, the late Nobel physicist Richard P. Feynman said, "There's Plenty of Room at the Bottom," and proposed to make miniature machine apparatuses and use these apparatuses to make still smaller machine apparatuses and so on, all the way down to the atomic level. Feynman was very much clear about the possible medical applications of the new technology. He added:

> "A friend of mine (Albert R. Hibbs) suggests a very interesting possibility for relatively small machines. He says that, although it is a very wild idea, it would be interesting in surgery if you could swallow the surgeon. You put the mechanical surgeon inside the blood vessel and it goes into the heart and looks around. It finds out which valve is the faulty one and takes a little knife and slices it out. Other small machines might be permanently incorporated in the body to assist some inadequately functioning organ."

Later, in his subsequent lecture in the same year, Feynman proposed the option of linking with biological cells and said, "We can manufacture an object that maneuvers at that level!" The vision of Feynman come to be a reality after the publishing of a technical paper by Eric Drexler in which he suggested that it might be possible to build nanodevices from biological parts that could examine and repair the cells of a human being. The exploration was followed a decade later by Drexler's technical book laying the foundations of molecular machines and molecular manufacturing systems (Drexler, 1986, 1992; Drexler et al., 1991) and subsequently supported by Freita's technical books on medical nanorobotics (Freitas, 1999, 2003, 2005).

Control Systems Design of Bio-Robotics and Bio-mechatronics
with advanced applications
https://doi.org/10.1016/B978-0-12-817463-0.00013-7

In the earlier 1990s, very little scientific work was done on nanorobotics, and was mostly based on concept generation, architecture, and modeling. Since then, many research papers have been published on systematic computational and experimental studies on nanorobotics. Nowadays, the field of nanorobotics keeps expanding and many scientists and researchers in their laboratories all over the world are focusing their activities on it. Nanorobotics became known to the public via science fiction movies, television, and books. In 1966, Isaac Asimov published a book entitled *Fantastic Voyage*, in which he described a minuscule submarine capable of moving through the human bloodstream (Asimov, 1966), while in 2002, in the very popular book, *Prey*, Michael Crichton introduced a swarm of intelligent nanorobots that threaten humankind (Crichton, 2002). Even though the concept of nanorobotics being described in this book is not at all related with the actual theory of nanorobots, it helped to generate public interest, which is important for the future growth of the field. Nanorobotics is a comparatively new field that grew out of the merging of robotics and nanotechnology during the late 1990s and early 2000s.

The term nanorobot was being used by the scientific community in the broadest possible way in the late 1990s, because this term included any type of active structure capable of anyone of the following, or any of them in combination: actuation, sensing, manipulation, propulsion, signaling, information processing, intelligence, and swarm behavior at the nanoscale. The term nanorobot includes large-scale manipulators with nanoscale precision accuracy and manipulation capabilities and micro-scale robotic devices with at least one nanoscale component (Weir et al., 2005). Nanorobots are basically theoretical microscopic devices built on nanometer dimensions (10^{-9} m). When nanorobotics has fully realized its potential from the current theoretical stage, nanorobots will work at atomic, molecular, and cellular level to perform tasks in both medical and industrial fields. This came as a natural evolution of the micro-robotics field that grew rapidly in the 1990s and of the nanotechnology field that exploded in the 2000s. Some of the most primitive appearances of the term occur in 1998 in the paper by Requicha et al. that focused on nanorobotic assembly (Requicha et al., 1998); Sitti and Hashimoto's paper on tele-nanorobotics (Sitti et al., 1998); and in 1999, Freitas' book on nanomedicine, where one can find a nice historical presentation of the nanorobotic concept for medical applications (Freitas, 1999). Prior to 1998, the term nanorobot had been clearly described by several researchers and was referred to as a "molecular machine," "nanomachine," or "cell repair machine" (Wowk, 1988; Dewdney, 1988).

This chapter emphasizes developments in the evolving domain of nanorobotics in medicine, especially on the design and application of cancer and cerebral aneurysm.

2 Design of nanorobotic systems for cancer therapy

Important parameters used for medical nanorobotic architecture and its control activation, along with essential technological background for advance manufacturing hardware for molecular machines are described below.

2.1 Mechanized technology

In manufacturing technology, complementary metal oxide semiconductor-very large scale integration (CMOS-VLSI) and verification hardware description language (VHDL) are playing an important role. The CMOS industry is guiding a pathway for assembly processes needed to manufacture components required to enable nanorobots, whereas to confirm the designs and to achieve a fruitful implementation, VHDL is being utilized in the integrated circuit manufacturing industry.

2.2 Chemical sensor

In the past decade, production of silicon-based chemical sensor and motion sensor arrays with the use of two-level system architecture hierarchy has been successfully achieved. These sensors are in widespread use from automotive to chemical industries for detection of air, water, element, and pattern recognition through embedded software programming and biomedical uses. Similarly the use of nanotechnology is also groom in this field as nanowires which decreased the estimated cost of energy demanded for data transfer and circuit operation by up to 60%. The CMOS-based biosensors using nanowires as material for circuit assembly can achieve superior efficiency for applications in detecting chemical changes, thus enabling medical treatment with increased precision and effectiveness. CMOS devices of 90 and 45 nm represent breakthrough technology devices that are being applied in manufacturing of nanorobots. Discovery or development of new materials such as strained channel with relaxed SiGe layer can reduce self-heating and improve the performance of nanorobots. In advance manufacturing techniques, silicon on insulator (SOI) technology has also been used to assemble high-performance logic sub 90 nm circuits. Circuit design methods to resolve bipolar effects and hysteretic variations based on SOI structures have been verified positively. In addition, chemical nanosensors can be implanted in nanorobots to monitor epithelial cadherin gradients. E-cadherin (or Cadherin-1, or CAM 120/80, or Uvomorulin) is a protein in the human body, encoded by the CDH1 (tumor suppressor) gene, which is designated as cluster of differentiation 324. Evolution of cancer and metastasis is directly proportional of loss of E-cadherin, which means E-cadherin down regulation decreases the strength of cellular adhesion within a tissue, causing an increase in cellular motility, which allows cancer cells to cross the basement membrane and invade surrounding tissues. E-cadherin is also used by pathologists to diagnose different kinds of breast cancer. When compared with invasive ductal carcinoma, E-cadherin expression is markedly reduced or absent in the great majority of invasive lobular carcinomas when studied by immunohistochemistry.

Nanorobots programmed for these tasks can perform detailed screening of the whole body of the patient. In this biomedical nanorobotic architecture, cellphones are used to retrieve patient information. A cellular application uses electromagnetic waves to command and detect the current status of nanorobots inside the patient's body. Contemporary advancement in FinFETs, double-gates, and 3D circuit technologies is capable of accomplishing surprising outcomes, which are expected to develop further.

2.3 Power supply

Active telemetry and power supply is the most effective and secure method of sustained energy supply for nanorobots in operation using CMOS. A similar procedure is also suitable for digital bit encoded data transfer from inside the human body.

It can be understood that nanocircuits with resonant electric properties can operate as a chip providing electromagnetic energy and supplying 1.7 mA at 3.3 V for power, allowing the operation of many tasks with very few or no power losses during the transmission. By utilizing a techniques that is widely used for commercial applications of radio frequency identification devices (RFID), with the use of inductive coupling, RF-based telemetry procedures have demonstrated encouraging results in patient monitoring and power transmission. In this procedure, the energy savings can be in the ranges of ~1 μW when the nanorobot is in inactive modes, and gets activated when signal patterns require it to do so. Some general nanorobotic tasks may require the device only to utilize low amounts of power once it has been strategically activated, for example, only ~1 mW RF signal required for communication. The easiest way to implement this architecture is by cellphone, which should be uploaded with the control software that includes the communication and energy transfer protocols, gaining both energy and data transfer capabilities.

2.4 Data transmission

The implanted devices and integrated sensors inside the human body to transmit health data of patients will provide exceptional advantage in constant medical monitoring. RFID chips have been developed as an integrated circuit device for medicine, and RFID for in vivo data collection and transmission has been successfully tested for electroencephalograms. Many researchers and scientists are working on single chip RFID CMOS-based sensors because CMOS with submicron SoC design may be used for very low power consumption in nanorobotic communication over longer distances through acoustic sensors. Nanorobots' active sonar communication frequencies may reach up to 20 μW@8 Hz at resonance rates with 3 V supply. In molecular machine architectures, to implant an embedded antenna for nanorobot RF communication with 200 nm size, a small loop planar device is adopted as an electromagnetic pick-up. It also has a good matching on low noise amplifier and is developed on gold nanocrystal with 1.4 nm^3 CMOS and nanoelectronics circuit technologies. Frequencies in the range of 1–20 MHz can be fruitfully used in biomedical applications without any damage to the device.

3 System implementation

Real-time 3D prototyping and simulation tools have significant benefits in nanotechnological developments because these tools have helped the semiconductor industry to attain faster VLSI developments. Simulation can anticipate the performance and provide support in device design, manufacturing, nanomechatronics control design, and hardware implementation. The simulation includes the nanorobot control design (NCD) software for nanorobot sensing and actuation, whereas nanorobot architectures include integrated nanoelectronics. The nanorobot architecture involves the use of cellphones for the early diagnosis of E-cadherin levels for smart chemotherapy drug delivery and in new tumor detection for cancer treatments. Nanorobots use an RFID CMOS transponder system for in vivo positioning using well-established communication protocols that allow tracking information about the nanorobot position and this information will help doctors in detecting tiny malignant tissues in the initial stages of development. The exterior of nanorobots is made up of diamondoid material, to which may be attached an artificial glycocalyx surface. The exterior material minimizes fibrinogen (and other blood proteins) adsorption and bioactivity, ensuring sufficient bio-compatibility to avoid immune system attack. Various types of molecules have been distinguished by a series of chemotactic biosensors whose binding sites have a different affinity for each kind of molecule. These sensors are also capa-ble of detecting obstacles that might require new trajectory planning for nanorobots. Nanorobots with sensory capabilities are able to detect and iden-tify changes of E-cadherin proteins gradients beyond permissible levels, which guide the nanorobots in detecting tumors even at early stages of cancer. There could be a variety of such sensors, for instance, chemical detection can be very selective for identifying various types of cells by their markers. In acoustic sensing, different frequencies are detected, which have different wavelengths depending on object sizes of attention.

4 Chemical signals inside the body

Depending on the requirement of communication in liquid workspaces, acoustic, light, RF and chemical signals are considered as prob-able alternatives for communication and data transmission. E-cadherin is a chemical signaling, which act as a transmission media between nanorobots. Chemical signals and their interaction with the bloodstream are a very important aspect to manage the application of nanorobots in cancer therapy. The signal sensing of nanorobots for simulated architecture in detecting gra-dient changes on E-cadherin signals are examined.

In order to improve response and bio sensing capabilities, nanorobots maintain positions near the vessel wall instead of floating throughout the vessel in the volume flow. The vein wall is modeled with a grid texture to enable better depth and distance perception in the 3D workspace. Another significant choice in chemical signaling is the measurement of time and detection of threshold at which the signal is considered to be received. Because of background concentration, some detection occurs even in the absence of the target signal. With threshold, diffusive capture rate (α) is used, for a sphere of radius (R) in a region with concentration as (the concentra-tion for other shapes such as cylinders is about the same):

$$\alpha = 4\pi DRC$$

With autonomous random motions for the molecules, detection over a time interval (Δt) is a Poisson process with mean value $\alpha \Delta t$. Table 1 shows the different parameters with chemical signals that one can use to find out the variables related with this concept. Scientists have discovered that, with the plaque on the vessel wall, fluid velocity near the target is lower than the average velocity.

When the nanorobot first detects a tumor for medical treatment, the nanorobot is programmed to attach on the tumor cell. The nanorobotic architecture is designed to send wireless communication about the accurate position of the tumor to the doctors. Then a predefined number of other nanorobots to support in perceptive chemotherapeutic action with precise drug delivery above the tumor are called for by sending signals. In a similar manner to quorum sensing in bacteria, it starts from monitoring the concentration of signals, chemical substances for near communication will attract or repeal nanorobots, and also estimates the number of nanorobots at the target. Due to this, nanorobots stop attracting other nanorobots when a sufficient number of nanorobots have responded to the initial signal. The amount of nanorobots can be changed depending on the stage of cancer and the tumor size, and may be defined by the oncologist based on the information received from the nanorobots through RF electromagnetic waves. The nanorobots at the plaque emit a different signal than others not already at the target, which is interpreted as an indication that others no longer need to respond. This mechanism allows them to be free to further search for other malignant tissues inside the body. Nanorobots will empower drug delivery and are also loaded with therapeutic

Table 1 Chemical signal and parameters.
Chemical signals

Production rate (Q)	10^4 molecules/s
Diffusion coefficient (D)	100 μm^2/s
Background concentration (C)	6×10^{-3} molecules/$(\mu m)^3$

Parameters	Values
Average fluid velocity (v)	1000 μm/s
Vessel diameter (d)	20 μm
Workspace length (L)	50 μm
Density of cells	2.5×10^{-3} cell/$(\mu m)^3$
Nanorobot	2 μm^3

chemicals, preventing the cancer from spreading further. The following control mechanisms are being considered:

- *Random*: Nanorobots moving inactively with the fluid reaching the target only if they bump into it due to Brownian motion.
- *Follow gradient*: Nanorobots monitor concentration intensity of E-cadherin signals when detected; they measure and follow the gradient until they reach the target. If the gradient estimate finds no additional signal in 50 ms, the nanorobot considers the signal to be a false positive signal and continues flowing with the fluid.
- *Follow gradient with attractant*: In addition to the previous mechanism, nanorobots arriving at the target release a different chemical signal, which is used by others to improve their ability to find the target. This mechanism involving peer-to-peer communication amongst the nanorobots is highly pertinent for improving nanorobotic performance.

5 Simulator results

Consider a fluid moving with uniform velocity (v) in the positive x axis direction. It contains a point source of chemical production rate (Q) in molecules per second. The diffusion coefficient is represented by D, and the diffusion equation is:

$$D\nabla^2 C = v\partial C/\partial x$$

With the boundary conditions of a steady point source at the origin and distance to the chemical signal source ($r = \sqrt{x^2 + y^2 + z^2}$) and no net flux across the boundary plane ($y = 0$), determines the steady-state concentration (C), i.e., time during which concentration (molecules/μm^3) remains stable or consistent, at point (x, y, z) is:

$$C(x, y, z) = (Q/2\pi Dr)e^{-v(r-x)/2D}$$

In nanorobotic behavior the fluid flow pushes the concentration of diffusing signal downstream. Subsequently, a nanorobot at more than a few microns away from the source won't detect the signal while it is still relatively near the source.

By taking the parameters from Table 1, one can detect an average higher signal concentration within about 10 ms when nanorobots are close enough. Thus, keeping their motion near the vessel wall, the signal detection happens after nanorobots have moved around 10 μm past the source. Thus, nearly five nanorobots per second arrive at the tumor cell in the small venule, which is one among many types of vessels present in the human body.

A design trade-off for chemical signals the nanorobots could release is designed by the equation $C(x, y, z) = (Q/2\pi Dr)e^{-v(r-x)/2D}$. Additional signals will use other molecules, which could, by design, have a different diffusion coefficient to use instead of the diffusion coefficient associated with the chemical from the target. From this equation, the effect of the fluid motion becomes significant at distances; this means that faster diffusion results in lower concentrations, demanding more time for other nanorobots to determine gradients. In future, if the signals are increasing in a steady, constant, and progressive manner, then chemical diffusion could be more efficient for nanorobotic communication. The nanorobots strike the target, if passing inside the human body, with a speed of nearly 0.1 μm. The nanorobots crossing within a few microns often detect the signal, which spreads a bit further upstream and away from the single tumor due to the slow fluid motion near the venule's wall and the cell's motion. Nanorobots flowing closer to the wall also benefit from slower fluid motion near the walls by having more time to detect signals. An "attractant" signal with the same value of D as the original signal has been used. Each nanorobot can release at one-tenth the rate of the target over the time. Individual performance has been observed throughout a set of analyses obtained from the NCD software, in which nanorobots use chemical sensors as the communication technique to interact dynamically in a 3D environment and to achieve a positive collective coordination. The virtual environment comprised of a small venule vessel that contains nanorobots, red blood cells (RBCs), and a single tumor cell, which is the target area on the vessel wall. Here, the target area is overlapped by the RBCs.

Table 2 provides a summary and comparison of the control techniques evaluated using the NCD simulator with the time required for 10 nanorobots and 20 nanorobots to identify and reach the target.

Each value is the mean of 30 repetitions in simulation, with standard deviation in parentheses. The error estimate for these mean values is

Table 2 Nanorobots: times in seconds to reach the target.

Control method	Nanorobots (10)	Nanorobots (20)
Random motion	0.73 (0.18)	1.47 (0.28)
Follow gradient	0.54 (0.17)	1.14 (0.24)
Gradient with "attractant"	0.46 (0.13)	0.79 (0.14)

$\sqrt{30}$ times smaller than the standard deviations listed here. For comparison, if every nanorobot passing through the vessel found the target, 20 nanorobots would arrive at the target in about 0.2 s, which enables nanorobots to detect and follow gradient concentration and thus increases the probability for nanorobots to find the target. In comparison, with random motion, here the nanorobots depict better performance by 23%. For gradient with an "attractant," the signals allow the nanorobots to find and reach the target in the 3D workspace 46% faster than that with random motion. The improvement in performance is remarkable in terms of response time, hence improving the chances of detecting and eliminating small tumors.

6 Design of nanorobotic systems for cerebral aneurysm
6.1 Nanorobot for intracranial therapy

Considering the properties of nanorobots to navigate as blood borne devices, they can aid significant treatment processes of complex diseases in early diagnosis and smart drug delivery (Freitas, 2005; Couvreur et al., 2006). Embedded technology plays an important role in nanorobotic application; through different embedded nanosensors, one can identify medical zones inside the human body. Various computational and numerical simulation techniques are being used that conclude various changes of chemical patterns for brain aneurysm. Various sensing methodologies are offered for nanorobots to identify harmful growth and level of difficulties in medicine, including specialized brain therapies (Leary et al., 2006; Gao et al., 2004). Nowadays, nanorobotic technology is used for treatment of patients from cerebral aneurysm, which solves all the problems.

6.2 Nanorobot hardware architecture

Usage of micro devices for medical treatments and instrumentation has investigated various important methods for aneurysm surgery (Ikeda et al., 2005; Roue, 2002). In a similar fashion to how the development of micro technology in the 1980s led to new tools for surgery, emerging nano-technologies will similarly facilitate further advancements in better diagnosis and new devices for medicine with the help of manufacturing of nanoelectronics (Rosner et al., 2002).

6.2.1 Manufacturing technology

The capacity to assemble nanorobots has resulted from new methodologies used in fabrication and different types of transducers used. Various changes on temperature, chemicals in the bloodstream, and electromagnetic signature effect are some of the key parameters in biomedical need. Nowadays, CMOS technology is used to manufacture components required to enable nanorobots. The combination of nanophotonics and nanotubes is enhancing the levels of resolution (Park et al., 2005a, b) to confirm the designs and to achieve a fruitful implementation; VHDL is being utilized in the integrated circuit manufacturing industry.

6.2.2 Chemical sensor

CMOS sensors using nanowires allow new medical applications, which conclude various chemical changes. Sensors with suspended arrays of nanowires assembled into silicon circuits resolve the problem of self-heating and thermal coupling (Fung et al., 2004). In addition, advancements in SOI technology are being used to assemble high-performance logic (Park et al., 2005a, b). Approaches of circuit design to solve bipolar effect and hysteretic variations problems, SOI structures have been confirmed (Bernstein et al., 2003). The best material to design CMOS IC nanosensors is carbon nanotube (Kishimoto et al., 1992). The protein nitric oxide synthase (NOS) offers positive or negative effects upon cells and tissues in cellular living processes. It has also been recognized that the correlations between higher levels of NOS and brain aneurysm have been established (Fukuda et al., 2000).

The antibody used for medical nanorobots helps in identifying higher concentrations of proteins that couple NOS forms in the intracellular bloodstream (NOS 2007). Nanobiosensors provide a well-organized technique for nanorobots to identify the exact locations with existence of NOS, which is represented by gradients in the brain enzymes.

6.2.3 Actuator

A set of fullerene structures has been presented for nanoactuators (Crowley, 2006). The use of CNTs as conductive structures permits electro-statically driven motions providing forces necessary for nano-manipulation. CNT self-assembly and SOI properties can collectively address CMOS high performance of design and manufacturing nanoelectronics and nanoactuators (Shi et al., 2006). In medical nanorobots, the use of CMOS as an actuator based on biological patterns and CNTs is adopted. In the similar fashion DNA can be used for coupling energy transfer (Schifferli et al., 2002a, b;

Ding et al., 2006) and proteins may serve as basis for ionic flux with electrical discharge ranges from 50 to 70 mV DC voltage gradients in cell membrane (Jenkner et al., 2004). An array format based on CNTs and CMOS techniques could be used to achieve nanomanipulators as an embedded system for integrating nanodevices of molecular machines (Zheng et al., 2005). Ion channels can be interfaced with electrochemical signals using sodium for the energy generation necessary for mechanical actuators' operation (Jenkner et al., 2004). Actuators are programmed to perform different manipulations, which assists the nanorobot in an active interaction with the bloodstream inside the body.

6.2.4 Power supply

In nanorobots, the use CMOS technology for active telemetry and power supply is the best and most secure way to guarantee power supply. The same technique can also be used for bit encoded data transfer from inside the human body (Mohseni et al., 2005). Nanocircuits with tuned electrical properties can work as a chip, providing energy in electromagnetic form to supply at 1.7 mA at 3.3 V. This also permits such tasks with few or no transmission losses (Sauer et al., 2005). RF-based telemetry has shown commendable results in patient monitoring and power transmission with the use of inductive coupling (Eggers et al., 2000).

6.2.5 Data transmission

Nanorobot architecture includes a single chip RFID CMOS-based sensor (Ricciardi et al., 2003) Using sensor data transfer as well as read and write data is feasible. Therefore, nanorobot active sonar communication frequencies may reach up to 20 μW@8 Hz at resonance rates with 3 V supply (Horiuchi et al., 2004). For brain aneurysm, chemical nanosensors are embedded in nanorobots to monitor NOS levels. After the last set of events recorded in a pattern array, information can be reflected back by wave resonance. The passive data transfer at $\sim$4.5 kHz frequency with approximate 22 μs delays are possible ranges for data communication. In molecular machine architecture, an antenna with 200 nm size for the nanorobot RF communication has been embedded. A small loop planar device is implemented as an electromagnetic pickup, having a good matching on low noise amplifier. The antenna is based on gold nanocrystal with 1.4 nm^3, CMOS, and nanoelectronics circuit technologies (Schifferli et al., 2002a, b; Sauer et al., 2005).

6.3 Implementation and simulation results

Nanorobots can be programmed to detect different levels of inducible nitric oxide synthase (iNOS) pattern signals using chemical sensors embedded nanoelectronics. The iNOS proteins serve as medical targets for detecting early stages of aneurysm development.

In the NOS subgroups, while eNOS acts as a positive protein, the nNOS is linked to neurodegenerative diseases like Alzheimer's and Parkinson's. nNOS plays a distinct role on endothelial cell degenerative changes (Kishimoto et al., 1992). In special cases, nNOS could result in negative effects with nitrosative stress, accelerating intracranial aneurysm rupture. Nanorobots injected in the bloodstream have been used as mobile medical devices. Fig. 1, showing the medical 3D environment, contains clinical data based on key morphological parameters in patients with cerebral aneurysm.

Integrated nanosensors, nanobioelectronics, and RF wireless communications (Cavalcanti et al., 2007) are incorporated into the nanorobot model in order to inform changes of gradients for iNOS signals (Fukuda et al., 2000), which assists the medical professional in deciding the treatment plan.

The nanorobots are designed at dimensions of 2 μm, which allows them to operate easily inside the body. The nanorobot model comprises of IC nanoelectronics and the platform architecture can instead use cellphones for data transmission and coupling energy. Computations are performed by embedded nanosensors. They are programmed for sensing and detecting NOS concentrations in the bloodstream. Due to background compounds, some detection occurs even without the NOS concentrations specified as the aneurysm target. High precision and a fast response are required for

Fig. 1 Aneurysm morphology—MCA, middle cerebral artery; BT, basilar trunk; BA, basilar artery.

biosensors. Additionally, false positives of NOS can often occur due to some positive functions of nitric oxide with semicarbazone (pNOS). Chemical detection in a complex and active environment is a vital aspect for nanorobots in the task of interacting within the human body.

In the nanorobot architecture, the integrated system contains engines for orientation, drive, and sensing and control mechanisms. Core morphologic features related to brain aneurysm are taken for modeling the study of nanorobots sensing and interaction with blood fluid patterns in the deformed vessel.

A critical issue on cerebral aneurysm is to detect and locate the vessel dilation. Nanorobots are required to track the aneurysm growth before a subarachnoid hemorrhage occurs.

If an electrochemical sensor detects NOS in low quantities or inside normal gradients, it generates a weak current lower than 50 nA.

In such cases, the nanorobot ignores the NOS concentration, assuming it as expected levels of intracranial NOS. However, if the NOS patterns reach concentration higher than 2 μL, it activates the embedded sensor, generating an electric current higher than 90 nA. Every time the activation of nanorobots takes place, an electromagnetic signal is back-propagated in the integrated system platform, which records the nanorobots' positions at the time of the signal generation. If large numbers of signals are generated, they indicate the early stages of brain aneurysm in the patient. It also informs the doctors about the location of the vessel bulb. The nanorobots provide their respective positions for the moment they detected a high concentration of NOS7.

To escape noise distortions and achieve a higher resolution, the system studies a strong evidence of intracranial aneurysm every time it receives back-propagated signals from a total of 100 nanorobots.

7 Medical application of nanorobots

Nanorobots are expected to provide novel treatments for patients suffering from various diseases. The advancement will result in an astonishing improvement in the medical arena. Existing advances in bimolecular computing are a promising step towards a future of nanoprocessors of increased intricacy and capabilities. Studies meant at developing biosensors and nanokinetic devices required for medical nanorobotics operation and locomotion is in progress. Application of nanorobots may boost biomedical involvement with slightly invasive surgeries. It will also help patients who need continuous monitoring of body functions. Monitoring diabetes and controlling glucose levels for patients will be a possible application of nanorobots. It is also expected to improve the competence of treatments through early diagnosis of probable severe diseases. For instance, nanorobots may be utilized to attach on transmigrating inflammatory cells or white blood cells, thus reaching inflamed tissues faster to help in their healing process. Nanorobots could be used to process specific chemical reactions in the human body as auxiliary devices for wounded organs. Nanorobots will be useful in chemotherapy to fight cancer through specific chemical dosage administration. Similar drug delivery methodology can be adopted to allow nanorobots to deliver anti-HIV or any other drug. Nanorobots could be used to locate and destroy kidney stones. One significant application of medical nanorobots could be the potential to locate atherosclerotic lesions in stenos blood vessels, primarily in coronary circulation, and treat them either mechanically, chemically, or pharmacologically.

Organic nanorobots that work on ATP and DNA-based molecular machines are also known as bio-nanorobots. The scheme is to develop ribonucleic acid and adenosine tri-phosphate devices. The usage of tailored microorganisms to achieve a bio-molecular computation, sensing, and actuation for nanorobots is also undergoing experiments.

Substitute methods for the development of molecular machines are the inorganic nanorobots. Development of inorganic nanorobots is based on customized nanoelectronics. In comparison to bio-nanorobots, inorganic nanorobots could achieve a much higher intricacy of incorporated nanoscale components. Using new diamondoid rigid materials are a likely advancement that could help in developing new materials for inorganic nanorobots.

The nano-build hardware integrated system is discussed here. It includes a combined set of modus operandi and new methodologies from nanotechnology targeted at mechanized manufacturing of nanorobots. It is used in 3D simulation and manufacturing design with integrated nanoelectronics. The challenge of manufacturing nanorobots perhaps will result from new methodologies in fabrication, computation, sensing, and manipulation. Real-time 3D prototyping apparatus are important in nanotechnological developments. This is expected to have direct impact on implementation of new approaches in manufacturing techniques. Simulation can forecast the performance of new nanodevices. Moreover, it will also help nanomechatronics designs and in the test of control and automation approaches. Here in this chapter, the focus is on the applications of nanorobots on cancer treatment and cerebral aneurysms.

8 Nanorobots in cancer treatment

Nanorobots are expected to provide substantial improvements in medicine through the miniaturization from microelectronics to nanoelectronics (Freitas, 2003). Cancer can be treated effectively with current levels of medical technologies and therapy tools. Thus far, a crucial factor in determining the likelihood of a patient with cancer surviving is early diagnosis. The treatment of cancer is reasonably successful when it is detected at least before the metastasis has begun. To achieve successful treatment for patients, professionally targeted drug delivery is also important to decrease the side effects from chemotherapy. Bearing in mind the capability of nanorobots to navigate as blood borne devices, nanorobots can help in targeted drug delivery. Nanorobots with implanted chemical biosensors could be used for detection of tumor cells at an early stage of tumor development in the patient's body. Nanosensors integrated on the nanorobots can be used to find intensity of E-cadherin signals. Hardware-based architecture for nanobioelectronics is in developmental stage for function of nanorobots in cancer therapy. Analyses and termination for the proposed model have been obtained through real-time 3D simulations.

9 Nanorobots in cerebral aneurysm

Endovascular treatment of brain aneurysms, arteriovenous malformations, and arteriovenous fistulas are expected to gain assistance from present research and developments in medical nanorobotics (Drexler, 1986). The first generation of nanotechnological prototypes of molecular machines are being examined and many encouraging device propulsion and sensing methodologies have been recognized (Asimov, 1966; Crichton, 2002). More complex molecular machines like nanorobots, having embedded nanoscopic tools for medical procedures (Sitti et al., 1998), are in the developmental phase. Sensors for biomedical applications are improving through teleoperated surgery and pervasive medicine (Drexler et al., 1991). The same technology provides the basis for manufacturing bimolecular actuators. These tools have pointedly helped the semiconductor industry to achieve faster VLSI developments (Ahuja et al., 2006). It may have a similar impact on the implementation of nanomanufacturing techniques, and on the progress of nanoelectronics. Simulation can foresee performance, help in device modeling, manufacturing analysis, nanomechatronics control investigation (Drexler et al., 1991), and hardware designs. For analysis, a real-time simulation based on clinical data is useful, demonstrating sensor and nanorobot behavior capabilities for detection of abnormal vessel dilatation in cases of cerebral aneurysm (Weir et al., 2005).

10 Conclusion

This chapter discussed the developments in new manufacturing technologies using nanotechnology because it is an investigative and treatment instrument for patients with cancer and cerebral aneurysm providing tailored treatments with better effectiveness and decreased side effects than those available today. Nano-medicine holds the promise to lead to an earlier diagnosis, better therapy, and improved follow up care, making the healthcare more effective and affordable. This chapter also provided a summary of nanodevices and nanorobotics in medicine. It was a small subset of the massive field of nanotechnology and nanobiotechnology. It is certainly possible that the use of nanorobotic technology will become ubiquitous in medicine within a generation.

11 Forthcoming nanomedicine

The arrival of molecular nanotechnology improves the effectiveness, comfort, and speed of forthcoming medical treatments. It will significantly reduce their risk, cost, and invasiveness. Nanotechnology can modify healthcare and human life more intensely than other developments. It is also contributing to shaping the modern industry, broadening the product development in pharma, biotech, diagnostic, and healthcare industries. Prospective healthcare will make use of delicate diagnostics for upgraded health risk assessment. The maximum influence can be anticipated in cardiovascular diseases, cancer, musculoskeletal conditions, neurodegenerative and psychiatric diseases, with the possibility for diabetes and viral infections to be cured with the application of nanotechnology. It will also lead to earlier diagnosis, better therapy, and improved follow-up care, making healthcare more effective and affordable. This technology will also work innovatively in constructing and employing nanorobots effectively for biomedical glitches. Applications of nanorobots in medicine holds a large number of promises from exterminating disease to retreating the ageing process (wrinkles, loss of bone mass, and age-related conditions are all treatable at the cellular level).

Nanomedicine will permit a more personalized treatment for countless ailments, by taking advantage of the in-depth understanding of diseases on a molecular level. Scientific researchers are on the verge of developing technologies on a scale an order of magnitude smaller than ever before. With the advancement of technology, we will be able to achieve improved control of the world around us and ourselves. Developing the ability to operate the world on a smaller scale has brought revolutionary changes in the scientific discoveries and the world at large. Whether it was the age of microscopes accompanying in the range of bacteriology or commencement of the atomic age with the learning of particle physics, nanotechnology is certain to change many of the patterns with which we think about disease diagnosis, treatment, prevention, and screening related to healthcare. Nanorobotics is evolving extensive possible uses across all fields of medicine and growing the number of therapeutic options available, it is also improving the effectiveness of existing treatments. Nanotechnology will touch our lives in uncountable ways through industries such as telecommunications and agriculture and more.

References

Ahuja, S., et al., 2006. A survey on wireless grid computing. J. Supercomput, 3–21.

Asimov, I., (1966). Fantastic Voyage. Houghton Mifflin, Boston.

Bernstein, K., et al., 2003. Design and CAD challenges in sub-90 nm CMOS technologies. In: ACM Proc. of the Int'l Conf. on Computer Aided Design (ICCAD'03), pp. 129–136.

Cavalcanti, A., et al., 2007. Medical nanorobot architecture based on nanobioelectronics. In: Recent Patents on Nanotechnology. Bentham Science, pp. 1–10.

Couvreur, P., et al., 2006. Nanotechnology: Intelligent Design to Treat Complex Disease. Pharmaceutical Research Springer, pp. 1417–1450.

Crichton, M. (2002). Avon. New York.

Crowley, R. (2006). Carbon Nanotube Actuator. 7099071US.

Dewdney, A.K., 1988. Nanotechnology—wherein molecular computers control tiny circulatory submarines. Sci. Am., 100–103.

Ding, B., et al., 2006. Operation of a DNA robot arm inserted into a 2D DNA crystalline substrate. Science, 1583–1585.

Drexler, K.E., 1986. Engines of Creation: The Coming Era of Nanotechnology. Anchor Books, New York.

Drexler, K.E., 1992. Nanosystems: Molecular Machinery, Manufacturing, and Computation. John Wiley & Sons (Association of American Publishers "Most Outstanding Computer Science Book"), New York.

Drexler, K.E., et al., 1991. Unbounding the Future: The Nanotechnology Revolution. William Morrow, New York.

Eggers, T., et al., 2000. Advanced hybrid integrated low-power telemetric pressure monitoring system for biomedical application. In: Proc. of Int'l Conf. on Micro Electro Mechanical Systems, pp. 23–37.

Freitas, R.A., 1999. Nanomedicine: Basic Capabilities. Landes Bioscience, Georgetown, TX. http://www.nanomedicine.com/NMI.html.

Freitas, R.A., 2003. Nanomedicine: Biocompatibility. Landes Bioscience, Georgetown. http://www.nanomedicine.com/NMIIA.htm.

Freitas, R.A., 2005. Current status of nanomedicine and medical nanorobotics. J. Comput. Theory Nanosci., 21–25.

Fukuda, S., et al., 2000. Prevention of rat cerebral aneurysm formation by inhibition of nitric oxide synthase. Circulation, 101 (21), 2532–2538.

Fung, C., et al., 2004. Ultra-low-power polymer thin film encapsulated carbon nanotube thermal sensors. In: IEEE Conf. on Nanotechnology, pp. 158–160.

Gao, Q., et al., 2004. Disruption of neural signal transducer and activator of transcription 3 causes obesity, diabetes, infertility, and thermal dysregulation. Proc. Natl. Acad. Sci. USA, 4661–4666.

Horiuchi, T., et al., 2004. A time-series novelty detection chip for sonar. Int. J. Robot. Automat. 19 (4), 171–177.

Ikeda, S., et al., 2005. In vitro patient-tailored anatomical model of cerebral artery for evaluating medical robots and systems for intravascular neurosurgery. In: IEEE Int'l Conf. on Intelligent Robots and Systems, pp. 1558–1563.

Jenkner, M., et al., 2004. Cell-based CMOS sensor and actuator arrays. IEEE J. Solid-State Circ., 2431–2437.

Kishimoto, J., et al., 1992. Localization of brain nitric oxide synthase (NOS) to human chromosome 12. Genomics, 802–804.

Leary, S., et al., 2006. Toward the emergence of nanoneurosurgery. Part III. Nanomedicine: Targeted nanotherapy, nanosurgery, and progress toward the realization of nanoneurosurgery. Neurosurgery, 1009–1025.

Mohseni, P., et al., 2005. Wireless multichannel biopotential recording using and integrated FM telemetry circuit. IEEE Trans. Neural Syst. Rehabil. Eng., 263–271.

Park, J. et al. (2005a). Method of Fabricating Nano SOI Wafer and Nano SOI Wafer Fabricated by the Same. 6884694US.

Park, J., et al. (2005b). Method of Fabricating Nano SOI Wafer and Nano SOI Wafer Fabricated by the Same. 6884694US.

Requicha, A.A.G., et al., 1998. Nanorobotic assembly of two-dimensional structures. In: IEEE International Conference on Robotics and Automation, pp. 3368–3374.

Ricciardi, L., et al., 2003. Investigation into the future of RFID in biomedical applications. In: Proc. of SPIE—The Int'l Society for Optical Engineering, pp. 199–209.

Rosner, W. et al. (2002). Circuit Configuration Having At Least One Nanoelectronic Component and a Method for Fabricating the Component. 6442042US.

Roue, C. (2002). Aneurysm Liner. 6350270US.

Sauer, C., et al., 2005. Power harvesting and telemetry in CMOS for implanted devices. IEEE Trans. Circ. Syst., 2605–2613.

Schifferli, K., et al., 2002a. Remote electronic control of DNA hybridization through inductive coupling to an attached metal nanocrystal antenna. Nature, 152–156.

Schifferli, K., et al., 2002b. Remote electronic control of DNA hybridization through inductive coupling to an attached metal nanocrystal antenna. Nature, 152–156.

Shi, J., et al., 2006. Self-assembly of gold nanoparticles onto the surface of multiwall carbon nanotubes functionalized with mercaptobenzene moieties. Springer J. Nanopart. Res., 743–747.

Sitti, M., et al., 1998. Tele-nanorobotics using atomic force microscope. In: IEEE/RSJ International Conference on Intelligent Robots and Systems, pp. 1739–1746.

Weir, N.A., et al., 2005. A Review of Research in the Field of Nanorobotics. Sandia National Laboratories.

Wowk, B., 1988. Cell Repair Technology. Cryonics Magazine, pp. 7–10.
Zheng, L., et al., 2005. A large-displacement CMOS micromachined thermal actuator with capacitive position sensing. Asian Solid-State Circ. Conf., 89–92.

Further reading

Fukuda, T., et al., 1995. Steering mechanism and swimming experiment of micro mobile robot in water. In: Kawamoto, A., Arai, F., Matsuura, H. (Eds.), IEEE MEMS Micro Electro Mechanical Systems, pp. 300–305.

Index

Note: Page numbers followed by *f* indicate figures and *t* indicate tables.

position-based impedance control, 426–428
robot manipulator, 412–413
structure of, 415*f*
PHYSIOTHERABOT/w1, 396–398, 409–410, 428–435, 441
Denavit-Hartenberg parameters, 431, 431*t*
dynamic analysis, 430–432
electronics hardware, 429, 429*f*
human-machine interface, 428–429
hybrid impedance control, 432–435, 434*t*
kinematic analysis, 430–432
mechanical parameters of links, 433*t*
robot manipulator, 429–430, 431*f*
structure of, 430*f*
position-based, 395–396, 402–403
robotic system, 396–398
variable, 405–406
Yoshikawa' impedance model, 403–404, 404*f*
Impendence force control, 166
Independent noise, 190–191
Inducible nitric oxide synthase (iNOS), 459
Inertial measurement unit (IMU), 233
Inertia matrices, 164
Inherent state-space explosion problem, 146
Injection pipette, 165–167
IntelliArm, 396–398
Intelligent controller, 410–412
block diagram, 412*f*
functions of, 413–414*t*
PHYSIOTHERABOT©, 411–412, 412*f*, 413–414*t*
Intensity-based approach, 117
Interaction variables, 19
International Society of Electrophysiology and Kinesiology (ISEK), 261–262
Internet of everything (IoE), 386–387
Internet of nano things (IoNT), 386–387
Intracranial therapy, nanorobots, 456
Intracytoplasmic sperm injection (ISCI), 143
Intravascular environment, 202
Intrinsic mode functions (IMFs), 268–270, 270–271*f*, 276
Inverse kinematics, 43
In vitro fertilization (IVF), 143
In vivo cell transportation, 283–286
collision avoidance strategy, 307
controller, 311–312
cotton operator, 312–316, 313*f*, 315–316*f*
experiments, 316–322, 320*f*, 322–323*f*
vector methods, 307–311, 310*f*
control system, 311*f*
disturbance compensation controller, 297–298, 297*f*
enhanced disturbance compensation controller, 299, 299*f*
experiments, 299–306
fluorescently labeled microparticles identification, 289–290, 289*f*
nonfluorescently labeled microparticles identification, 291–293, 291*f*
position errors, 304–305, 306*f*, 324*f*
P-type controller, 295–297
robot-tweezer manipulation, 282
simulation result, 300–302, 301*f*
tracking microparticle, 293–294
trajectory, 300–302, 301*f*, 307*f*
Isokinetic exercise, 398–401
flow chart, 409*f*
impedance control, 408–409
intelligent control unit, 413–414*t*
Isometric exercise, 398–401
flow chart, 409*f*
impedance control, 408
intelligent control unit, 413–414*t*
Isotonic exercise, 398–401
flow chart, 409*f*
impedance control, 408
intelligent control unit, 413–414*t*

J

Jacobian matrix, 418–419
Joint torques, 423–424
Joystick-controlled prototypes, 231–232

K

Kalman filter, 40–41, 247–248, 270, 272, 272*f*, 276
Karakuri ningyo toys, 329–330
Kinematic analysis, 430–432
Kinetic exercise. *See* Isotonic exercise
Knee joint, PHYSIOTHERABOT, 416, 426

Printed in the United States
By Bookmasters